CW01262636

SIR HUGH PLAT

THE SEARCH FOR USEFUL KNOWLEDGE
in
EARLY MODERN LONDON

To
Jane
for her many years of
encouragement and support

SIR HUGH PLAT

*The Search for Useful Knowledge
in Early Modern London*

Malcolm Thick

PROSPECT BOOKS
2010

First published in Great Britain in 2010 by Prospect Books,
Allaleigh House, Blackawton, Totnes, Devon TQ9 7DL.
(http://www.prospectbooks.co.uk)

© 2010, Malcolm Thick.

The author, Malcolm Thick, asserts his right to be identified as author of this work in accordance with the Copyright, Designs & Patents Act 1988.

No part of this publication may be reproduced, stored in a retrieval system, or transmitted in any form or by any means, electronic, mechanical, photocopying, recording or otherwise, without the prior permission of the copyright holder.

BRITISH LIBRARY CATALOGUING IN PUBLICATION DATA:
A catalogue entry of this book is available from the British Library.

Typeset and designed by Tom Jaine.

ISBN 978-1-903018-65-1

Printed and bound in Malta by the Gutenberg Press.

Table of Contents

Introduction and Acknowledgements		7
1.	Biography	11
2.	Gardening	41
3.	Agriculture	80
4.	Military Food and Medicine	115
5.	The writing of *Delightes for Ladies* and *Sundrie new and Artificiall remedies against Famine*	147
6.	Alchemy	169
7.	Medicine and Medical Practice	206
8.	Scientific Thought and Technique	242
9.	Inventions, Technology and Practical Applications of Technology and Natural Science	285
10.	Moneymaking	323
11.	Conclusion	366

APPENDICES

I	Plat's manuscripts	379
II	Plat's alchemical poem	388
III	Transcription of the printed broadside on food and medicines for seamen, c. 1607	397
IV	A manuscript advertisement for medical and other supplies for seamen	401

List of Illustrations	404
Bibliography	405
Index	418

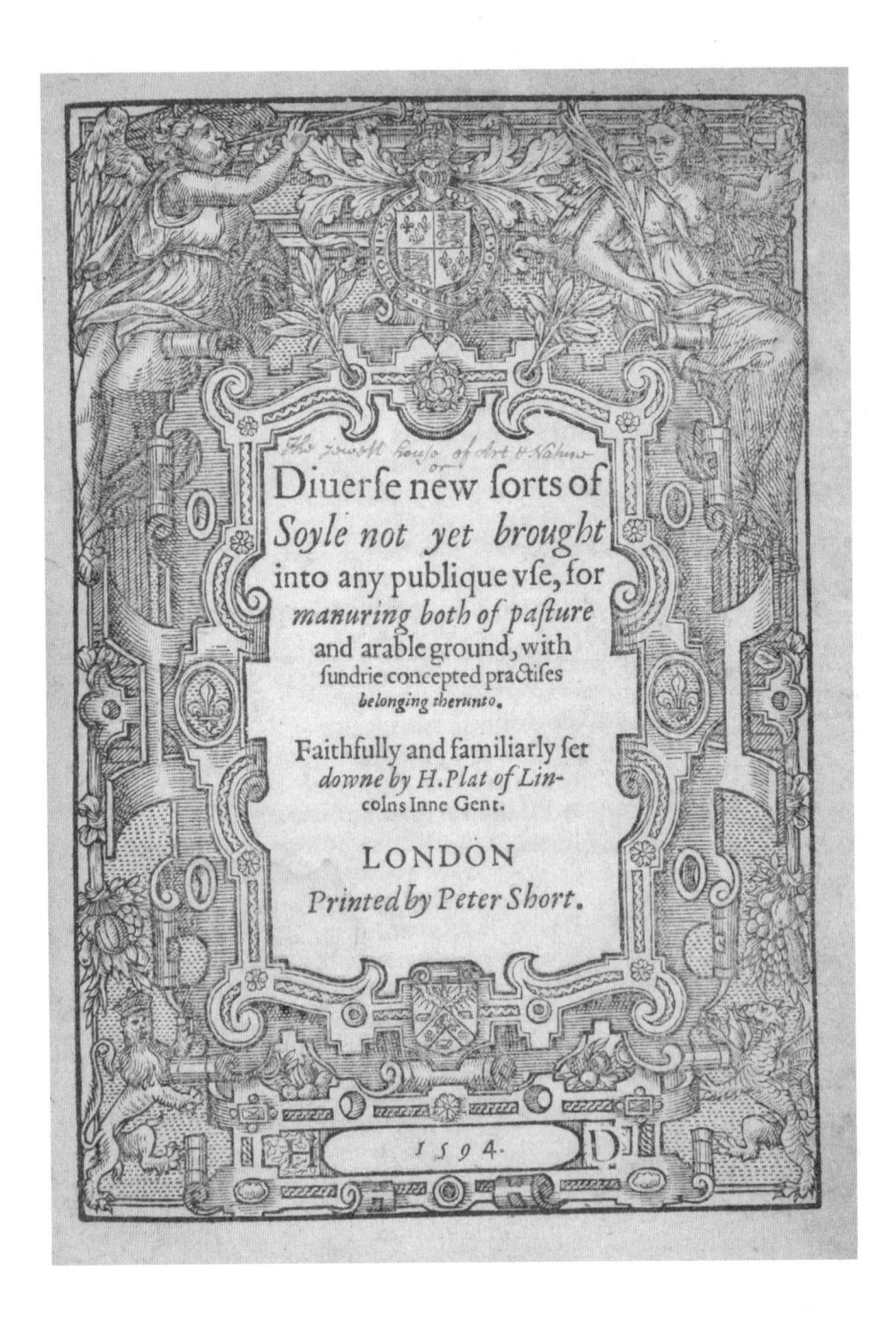

Diuerse new sorts of
Soyle not yet brought
into any publique vse, for
manuring both of pasture
and arable ground, with
sundrie concepted practises
belonging therunto.

Faithfully and familiarly set
*downe by H. Plat of Lin-
colns Inne Gent.*

LONDON
Printed by Peter Short.

1594.

Figure 1. Title-page of Diverse new sorts of Soyle, 1594.

Introduction and Acknowledgements

In Plat's time the introduction to a book was always written last, and the same is true today. The introduction contained the latest news about the book and its contents. Plat's introductory message to his readers in his last book was sobering, he did not know the length of his days but he was sure 'they are drawing to their period'. He died that year. I have no such news, *absit omen*, but there is relief that the book is finally finished and a feeling that a period of my life has come to an end. Another important purpose of the 'front matter' of late-sixteenth-century books was to advertise the work: browsing through the unbound pages of new works in stationers' shops, potential buyers read introductions in much the same way as we scan the dust-jackets of today's books. I, too, need to capture your interest (and reward the patience of close friends who have waited many years for this book to appear). Recent interest in early scientific research and the identification of a shifting group of Londoners in Plat's time who pursued an interest in many areas of natural science and technology has made my task easier – this book will, I hope, stimulate that interest. But, for all that, Plat is quite an obscure figure; why write a book about him and his work?

One obvious reason is that he left a body of written and printed information to be researched, and there is nothing an historian likes better than a pile of books and papers on a library desk to wade through. My initial intention was to concentrate on his printed works, writing perhaps an introduction to an edition of all, or some of his books and pamphlets. Then I discovered the extent of his surviving notes in the British Library and realized that to ignore these would distort any review of Plat's work. These papers include major topics of research little touched upon in his published works, most notably on alchemy and medicine. The manuscripts are also invaluable because they reveal how some of the published works were created – what was included, what discarded or altered, where some of the information in the books was obtained. They reveal Plat's editorial process. (I had not, at the outset, appreciated how rare an insight into the creation of published works in the period this was. What would a Shakespeare scholar give to see the first draft of a play, or even of a sonnet?) Although dates in the manuscripts are few, it is also possible to discern how Plat's interests and research developed over time, as well as gaining some idea of how his encounters with fellow-Londoners shaped his thinking.

When reading Plat's published works and manuscripts one is struck most forcibly by the variety of material he amassed. He collected information throughout his adult life and spent much time reviewing his notes and speculating upon what he had learnt. He also carried on what we would today call experiments in such fields as gardening, agriculture, alchemy, medicine, food preservation, winemaking and cosmetics as well as collecting much practical information on how many trades in London were carried on. Furthermore, he was a practising physician.

One result of the variety of his interests is that he has been frequently cited by historians, and writers wishing to add some historical background to their subject, as the source of very early mentions of a bewildering number of topics. I first encountered Plat in early books on the history of gardening : such authors as Alicia Amherst in 1895 and Eleanour Sinclair Rohde in 1924 quote him with approval (Rohde calls him 'the most famous of Elizabethan gardening authorities'). Early agricultural historians were impressed by Plat's published work on fertilizers and soils and historians of cookery praised his book on that subject.

I, like many others before me, have dipped into Plat in order to cite his interest in one topic or another. And such citations are commonplace, if one is to judge from the variety of hits his name scores in an Internet search. One site describes his way of etching designs onto eggshells; another recites his recipe for sugar plate; a U.S. firm supplying steam heating cites Plat as an early exponent of this technology; and one devoted to early cryptography notes that he wrote instructing 'How to write a letter secretlie that cannot easilie be discovered, or suspected'.

All this piecemeal use of Plat has led to 'serial stereotyping'. Each discipline tends to regard him as their own. Most recently he has fallen amongst historians of science and whilst they have carefully examined his written and published works they have, in some cases, interpreted almost all that he wrote as a quest for scientific knowledge, in the same way that the gardening writers thought him primarily a gardener or the cookery writers treated his cookery book as his most important work. I hope, by devoting a whole book to his multifarious interests, that I can show him as it were in the round, as a gentleman of varied interests, a Londoner trying to make his way in the world. I also hope to show Plat as a man of his time and place. My chapter on military inventions, for instance, reveals Plat as an inventor who talked to military commanders and bent his mind to their most pressing military needs. His work on famine relief was an immediate response to a run of bad harvests which threatened the food supply of by far the largest city in the country. His medical practice and the medicines he developed aimed to cure the diseases most feared by his friends and neighbours. Even something as seemingly frivolous as his work on cosmetics was of great value to those at Court, where appearance might dictate fortune. It may be argued that two important aspects of his research, alchemy and enquiries about the current technology of various trades, were not so immediately dictated by the needs of the time. Whilst his alchemical writings are the most esoteric and complex of his surviving manuscripts, much of his work in this discipline had a practical end in view, to develop powerful and effective medicines. His work on the technology of trades was by no means disinterested. In more than one instance he developed better ways of carrying out industrial processes than was then practised and tried, by patents or other means, to make money thereby.

The book that follows is by no means definitive. Every time I dip into his manuscripts I am amazed by some piece of information he has gathered or an experiment he has carried out – and conscious that it is deserving of a mention and I have not done so. Simply by dividing my book into chapters on specific topics, I have excluded matters on which he wrote which do not come under those heads. So, consider this book a glimpse into Plat's 'Jewell House', a mere scanning of the open shelves of his cabinet of curiosities. Believe me, hidden away in his books and notes are many more gems.

I must disappoint those looking for a book-length biography of Sir Hugh. The first chapter outlines what is known of his life. I make some small speculations to fill the many gaps in our knowledge of him as a person but these are restrained – many a time I have bought a biography of a man or a woman of Plat's time only to find long runs of sentences beginning, 'He/She *would have* done this or that…', the book being little more than a few facts strung together by many conjectures. As is the case with most gentlemen of his time, almost all his personal papers have disappeared, bar a couple of letters and some personal asides in his research notes. Those who eagerly sought his notes and preserved them for posterity were not interested in his shopping lists or laundry bills. They believed his notes, in particular those on alchemy, contained valuable secrets. Only in death are some personal facts revealed. A few court papers relating to a legal dispute over debts charged to his estate tell us more about his personal life than any notes he himself penned. (If I might speculate, perhaps Plat would have approved of such anonymity, regarding his funeral monument and his books as sufficient memorial.)

The publication of this book has been assisted by a grant from The Scouloudi Foundation in association with the Institute of Historical Research. I am most grateful for the generosity of the Trustees in support of my work.

I owe a debt of gratitude to many people for help and encouragement in writing this book: rather than try and name them all (and risk offending anyone omitted) may I just thank all librarians, archivists, booksellers, teachers, historians, friends and relatives who have provided information and services, proffered advice, given encouragement, and shown forbearance whilst I proceeded slowly towards completion of my manuscript?

Harwell, Oxfordshire
April 2010

CHAPTER ONE

Biography

Sir Hugh Plat was indebted to his father Richard for his status as a gentleman. Hugh's father passed much of his substantial wealth to Hugh, his eldest surviving son, and educated him suitably for his place in society.[1] Richard Plat, born in 1528, was the son of a Hertfordshire yeoman. He was probably a younger son, sent to London as an apprentice.[2] He became free of the Brewers' Company in 1550, paid a fine in lieu of the Stewardship in 1562 and the following year was elected Warden: at the age of 34 he was making his mark amongst his fellow-brewers. In 1566 he leased the Old Swan brewery in Thames Street for 60 years from the Company for £27 per annum, an act which probably laid the foundation of his fortune. This brewery was bequeathed to Sir Hugh and on *his* death in 1609 was sublet at £127 per annum, a clear profit of £100 each year. Richard's success may have been helped by business acumen but the phenomenal growth of London, in terms of population and spending power of its inhabitants, over his lifetime made it hard not to make money as a brewer, supplying a staple food to one of the fastest growing concentrations of population in Europe, most of whom did not have the resources to produce their own beer and ale.[3]

1   As Richard Grassby has shown, in providing a university and legal education and the income to live as a gentlemen, as well as probably introducing Hugh to both his brides, Richard Plat was following the practice of many of his fellow London merchants, *Kinship and Capitalism*, Cambridge, 2001, pp. 343, 362–6.
2   Richard's brother, Henry Plat of Aldenham, who made a will when seriously ill in 1575, was probably the eldest sibling, had inherited the family land, and remained a modestly-off yeoman.
3   John Stow, *A Survay of London* [1598], ed. Henry Morley, p. 239; The National Archives (TNA), E134/8 Jas 1/East 2.

In the sixteenth century, wealth not used as capital in commerce or agriculture usually had to be invested in land or buildings to provide an income from rents. The growth of London's population and its predominance as a centre of commerce, government and society meant that those who could invest in London property were likely to become steadily richer as rents increased. By the time he died Richard Plat had accumulated a considerable amount of property in London and elsewhere. Wealth brought social advancement and his rising status is reflected in the public offices he held – City of London Alderman and sometime Sheriff of London, as well as a regular member of the Brewers' Court. A grant of arms confirmed his rise into the gentry and, in the last year of his life, his portrait was painted. Now lost, the picture once hung on the south wall of the Brewers' Hall and depicted the 76 year-old 'in a black furred gown, holding a prayer-book in his hands crossed before him. The ruff is diminutive, and thrust above the ears, a little black cap on his head, a square red beard, and feminine features'. There is no likeness of Sir Hugh, so that of his father must serve to give at least an idea of his appearance.[4]

Richard Plat died in his 70s, only eight years before Hugh, who was appointed to administer his father's will. Both men spent all their adult lives in or near London and one would expect that such proximity meant Richard had a strong influence on his eldest son. (Elizabeth Foyster has found that there was often a close relationship between parents and grown-up children of the middling ranks of society at this time, especially when they lived close to each other.[5]) What specific traits or outlooks on life Richard passed to Hugh is, however, almost impossible to say. The father was a successful businessman who amassed a fortune. Hugh, as a gentleman, should not have been too closely associated with trade but, as we will discuss, he had a restless spirit of enterprise throughout his adult life. Was this a legacy of his father? We will discuss anon Hugh's religious beliefs but here we may again speculate that he was influenced

[4] James Peller Malcolm, *Londinium Redivivum*, vol. III, 1807, p. 141. For examples of Richard's property holdings see: Guildhall Library, MS 5485; London Metropolitan Archives, M92/III; A29341.

[5] Elizabeth Foyster, 'Parenting was for life, not just for childhood', *History*, vol. 86, no. 283, July 2001, pp. 313–327.

by his father. Richard, like many of his contemporary business colleagues, was a Puritan. In the preamble to his will he justifies the need for the document by explaining that he is sick and wants to settle his affairs, 'that I may hereafter apply myself to Godly and Spiritual Meditations'. He commends his soul in turn to God, Jesus, and the Holy Ghost 'who sanctifieth me and all the elect people of God hoping and assuredly believing' (that he will have full remission of sins) through the 'glorious Death Passion resurrection and ascension' of Christ and enjoy 'Life and joy everlasting in the Kingdom of Heaven'. Like many rich and godly citizens, he balanced the austerity of his faith with the obligations of his class. Numerous charitable bequests were made to relieve poverty and encourage education in his own parish and elsewhere. He wished to be buried in the parish church of St James Garlickhythe next to his wife and son Richard (who predeceased him), 'in decent comely and Christian manner without any vain or Superfluous pomp'. He did, however, wish Hugh to erect a monument upon his grave, and he bequeathed gowns to twenty 'poor men to attend on my Corps to burial'.[6]

A major bequest in Richard's will was the endowment of a grammar school and six alms houses in his birthplace, Aldenham in Hertfordshire. The school, for poor pupils, reflects a godly desire to educate the poor that they might learn more about Christianity. The bequest also emphasizes a difference between Richard and Hugh. Richard was an immigrant Londoner who still had strong ties to his old country home. Many relatives still lived there, including his brother Henry Plat, who died in Aldenham in 1575. His will, 'trusting by the meritts of the death and passion of his dearly beloved sonne our savior Jesus christ to have remission of all my sines', was also that of a Puritan. A Hertfordshire yeoman, Henry may have been the elder brother, inheriting family lands whilst his younger sibling was apprenticed in London. Richard left a bequest to Alice, his brother's daughter.[7]

Hugh also had lands in Hertfordshire but he was a Londoner and proud so to be. He was born in the city, baptized in the church where

[6] Guildhall Library, MS 5485.
[7] British Library, Sloane Mss, 2177, f. 6. Hugh's forebears may not have been deeply rooted in Hertfordshire. There is some evidence that they may have moved to

his father was later buried, St James Garlickhythe, on 3 May 1552. Apart from three years as an undergraduate at Cambridge, he spent all his life in London or the suburban village of Bethnal Green. He regarded London as 'the most renowmed citty of Englande' and was content to be regarded as a cockney. He went to school in London but we know not where.[8] He may have attended one of the prominent grammar schools, St Paul's, or Merchant Taylors'. Alternatively he may have been sent to one of the new parish grammar schools set up with the aid of city businessmen and Puritan clerics: 'Civic pride and protestant piety spurred educational foundations'. Most likely he went to St Anthony's, after being taught to write by 'Master Conradus, that teacheth over against saint Anthonies schoole'. This grammar school was, by 1598, 'decayed, and come to nothing' because its property income was removed by the dissolution of the monastic house of which it had once been part.[9]

Another proud Londoner, the antiquarian and contemporary of Plat, John Stow, chronicled the progress of the city's schools and often came across the local schoolboys in his perambulations of the city. He witnessed disputations between schoolboys 'on the eve of St. Bartholomew the Apostle … upon a bank boarded about under a tree' in the churchyard of St Bartholomew's, Smithfield, over principles of Latin grammar: 'in the end the best opposers and answerers had rewards'. He also noted rivalry between schools, including horseplay in which Plat may have indulged: the boys of St Anthony's were called Pigs by the St Paul's boys, they in turn were called Pigeons. After arguing in the street on points of grammar 'they usually fell from words to blows with their satchels full of books, many times in great heaps'. The growth of schooling in London improved general literacy in the city and in Plat's case this early education equipped him for

---

the county from Hassop in Derbyshire in the early sixteenth century (personal communication from Peter J. Platts, who has researched the Plat family and discovered several pieces of circumstantial information to support this view).

8   Sir Hugh Plat, *Diverse new sorts of Soyle not yet brought into any publique use*, 1594, p. 50; BL, SL 2209, f. 2; SL 2216, ff. 50v–52.
9   Roy Porter, *London, a Social History*, 1994, p. 60; Sir Hugh Plat, *The Jewell House of Art and Nature*, 1594, pt. 1, p. 41; Stow, p. 196.

university, giving him an ease with Latin which he displayed throughout his later writings and which was essential for any form of advanced study in this period.[10]

At sixteen Hugh was sent up to St John's College, Cambridge as a pensioner, from whence he graduated in 1571–2. As a pensioner (in effect an 'ordinary' student), he would not have been on intimate terms with aristocratic boys, who ate with the college fellows and were not encouraged to mix outside their station. Nevertheless, university, and especially one's college, was a place where firm and lasting friendships were made which might, at a time when personal connections were vital in government, commerce, and society as a whole, be of use in adult life. St John's, a fairly recent foundation (1511), provided a relatively advanced education in comparison with some Cambridge colleges and puritan ideas already instilled by his family were no doubt reinforced by this 'breeding ground of early Protestantism'.[11] To gain his BA Plat followed a four-year course of rhetoric, logic and philosophy, taught to an extent by voluntary attendance at University lectures but mostly under supervision of a college tutor: one-to-one tuition and some disputations with other students.

Soon after returning from Cambridge in 1572 his first book was printed, *The Floures of Philosophie, with the Pleasures of Poetrie annexed to them, aswel plesant to be read as profitable to be followed of all men.* This book, a collection of poems preceded by over 800 short, moralizing proverbs in the manner of Seneca, is unlike any of Plat's later works. It is a book by a precocious young gentleman. The sentiments of the proverbs are not original, and the poems follow current literary trends.' "The Pleasures of Poetrie" seem almost uncannily representative of themes and types preferred in the middle years of Elizabeth's reign.'[12] The poems reflect the education Plat obtained at St John's, revealing a young man proud of

---

10　Stow, pp. 98–102. 'Pigs' because St Anthony is associated with them, 'pigeons' because then (as now) there were a lot of them around St Paul's.

11　G.R. Elton, *England under the Tudors*, 1971, p. 432; Rosemary O'Day, *Education and Society, 1500–1800*, 1982, pp. 90–91.

12　Sir Hugh Plat, *The Floures of Philosophie*, 1572, intro. Richard J. Panofsky, New York, 1982, p. xiv.

and satisfied with his attainments. 'The "Pleasures of Poetrie" represents the *juvenilia* of a talented university student, and many of the poems derive from common school-room exercises. Here are laments for social disorders and injustices and celebrations of constancy and marriage; here are set-pieces of student life, rusticity and travel; here friendship opposes flattery, virtue opposes vice, moderation opposes greed, and consolation is offered for life's emergencies.'[13] The poems employ contrast and paradox in thought and language, many are literary exercises and Plat 'exemplifies an Elizabethan poet's delight in the arts of rhetoric and versification employed in ingenious verbal recreations.'[14]

The philosophy in the work is unadventurous, with no sign of the attack on Aristotelian metaphysics or of the logic known as Ramism (after its originator Pierre de la Ramée). Ramism, anti-hierarchical and based on the idea the knowledge was inherently useful and accessible to all men, appealed to extreme Protestants, and was taking hold in some Cambridge colleges, St John's among them.[15] Despite the conventional treatment of conventional themes we may find something of the young Plat in these pages. In the introduction and in some of the poems he displays a knowledge of gardens – their layout, flowers, and garden pests. In the final poem he erects a thorn hedge around his garden of poetry against his foes: 'caterpillars' and 'wormes'; and the first poem in the book, 'Agaynst those which will do nothing themselves, and yet envy at other mens dooings' also betrays a sense of insecurity and sensitivity to criticism which is echoed in later publications. A paradise garden is included in 'The conquering lover', a poem of passion and requited love which may have been a celebration of betrothal to his first wife. At this early age Plat was concerned by his lack of wealth. He explains in his introduction that he had 'not such welth at will' to bestow 'some goodly glorious golden gift' on his patroness and his introductory poem to his

13 Plat, *The Floures of Philosophie*, p. xiv
14 Plat, *The Floures of Philosophie*, p. xvi.
15 O'Day, pp. 127–9. Plat does make mention, in 1593, of Ramism at Cambridge. He wonders if he pesters unresponsive London brewers too hard with the benefits of a new brewing vessel, whether 'they would be ready to hisse me out of the schooles for a Ramist'. Sir Hugh Plat, *A Briefe Apologie of Certaine New Inventions*, 1593.

readers hopes they will 'Grudge not to spende a little coine, in buying' the book. Moneymaking, we shall see, will be a constant theme in his later writings.[16]

The moral tone, although leavened here and there with poems of love and youthful exuberance, reveals a young man content to be thought one of the 'godly' but by no means dour and narrow-minded. Dedicated to Anne Countess of Warwick, wife of Ambrose Dudley, brother of Robert Dudley, Earl of Leicester, the work puts Plat within the ambit of this Protestant peer. It would be over twenty years before Plat next appeared in print and he provides little clue within this book as to why it was published. Was it youthful bravado? Probably not. There is pride in the work but also many references to his inexperience and humility in putting it forth. Ambition may have played a part – a recent graduate, son of newly-rich father, he may have hoped to come to the notice of the Dudleys and thereby secure some administrative post. If so, he was disappointed, as he was in many of his later attempts to obtain income from public service.[17]

By the time the *Floures of Philosophie* was published, Plat had entered Lincoln's Inn and commenced the second phase of his higher education.[18] Lincoln's Inn was one of four Inns of Court which, together with the lesser Inns of Chancery, provided legal education. From beginnings in the twelfth century they had evolved by Plat's time into an important part of the educational establishment of London: according to Stow, 'a whole university, as it were, of students, practisers and pleaders, and judges of the laws of this realm'. Perhaps a thousand students resided at the Inns in term-time: in 1574 Lincoln's Inn had 169 members, amongst them Hugh Plat.[19] The prime purpose of these institutions was to produce lawyers. Internal organization was typically hierarchical: Lincoln's Inn had a master, senior members called benchers, barristers who had been judged competent to practise law, and clerks (students) – young men sent to the

16   Plat, *The Floures of Philosophie*, A4–5, pp. 77, 150, 171.
17   Plat, *The Floures of Philosophie*, p. xx.
18   At the end of the 'Epistle Dedicatorie' of *The Floures of Philosophie* he signs himself: 'Hugh Plat of Lincolnes Inne'.
19   Stow, p. 103; Roy Porter, *London A Social History*, 1994, p. 61; W.R. Prest, *The Inns of Court 1590–1640*, 1972, p. 6.

Inn to study. The hierarchy was made visible in the relative positions of these groups when dining in Hall and consuming dinners was (and is) a condition of membership. Stow estimated that serious law students took seven years to be called to the bar, having undertaken disputations (known as moots) and attended lectures (readings), given by benchers, as well as spending some years dining in their Inns.[20]

The demand for a legal education was high in the second half of the sixteenth century – rising population, increasing prosperity of the gentry, a thriving land market and increased commercial activity all created jobs for professional lawyers, while many laymen also wanted to know the rudiments of civil law. The Inns of Court recruited students predominantly from the universities, taking young men at an age when today's students begin university courses. Some found the transition between university and the law difficult – legal studies were 'harsh and barbarous', not helped by the absence of a tutorial system and emphasis on formal lectures and exercises. Richard Brathwaite, who came down from Cambridge to Gray's Inn in 1609, found 'these thorny plashes and places of the law' entirely repugnant after 'the fresh fragrant flowers of devine poesie and morall philosophy.'[21] One can see the 'fresh fragrant flowers' of Brathwaite's Cambridge in Plat's first book but there is nothing on how he took to his studies at Lincoln's Inn. Students intent on the law as a profession were however in the minority, for during 'the late sixteenth and early seventeenth centuries a stay at the inns was part of the conventional gentlemanly education and it seems to have been generally accepted that young men, whether or not destined for the bar, should acquire some knowledge of the law while they were there.'[22]

Rosemary O'Day discerns changes in teaching at the Inns caused by the increasing numbers of graduates who came on from university to attend them, 'this use of the universities as preparatory schools for lawyers had an incalculable influence upon the educational system of the Inns themselves. These same students came to the Inns heavily imbued with what they had learned in college and university, and accustomed to given

---

20 Stow, pp. 104–5.
21 Quoted in Prest, pp. 141–3.
22 Prest, p. 23.

study methods which differed radically from those traditionally practised at the Inns. The student was already used to private study – to relying on books, reading guides, private discussions with tutors and fellow students, commonplacing, cataloguing and listening to informal lectures rather than upon lectures and learning exercises.'[23]

The Inns were finishing schools where, as well as law, young gentlemen might learn how to behave as adults and acquire the necessary skills required by their station in society – 'The serious law students; the sons of the gentry … the place hunters, fortune-seekers, and social climbers – all mingled together at the schools of dancing, fencing and music scattered around the inns, to acquire the non-academic accomplishments which befitted the role of gentleman'. As at university, but more so because the young gentlemen were older and closer to deciding on a career, friendships made at the Inns helped in later years. Acquaintance with those higher in the social scale could lead to later advancement as could acquiring manners of a higher social class: 'just as a nineteenth-century industrialist might send his son to public school for processing into the figure of a gentleman, so merchants and yeomen of Elizabethan and early Stuart England achieved the same object by entering their sons at one of the inns'. Midway between the city and the Court, the Inns were an ideal place for spirited young men. Rules put some restraint on youthful exuberance but the students were known for their love of pleasure – playgoing, dancing, gambling, drinking, and dining in town. It has been estimated that in Plat's time a parent would have to find at least £40 a year to keep a young man at an Inn. Some of that money went on clothes, Sir Thomas Overbury thought 'a pair of silk stockings and a beaver hat' most obviously distinguished 'An Inn of Court Man' from a university student. Businesses near the Inns profited from these free-spending youths, especially gullible newcomers who might be tempted by 'these leaden daggers with their golden sheathes' in 'the Cutlers shops … [which] make a goodlie shew, if they were hung up in a Michaelmas Terme.'[24]

Given Plat's condemnation in his poetry of dancing, fencing, out-

23 O'Day, p. 159.
24 Prest, pp. 23–24; Sir Thomas Overbury's Characters, in *Character Writings of the Seventeenth Century*, ed. Henry Morley, 1891, p. 61; Plat, *Jewell House*, Preface.

landish food, fashionable dress, gambling, drunkenness, lewd talk and sexual licence, one might expect him not to fit into life at the Inns of Court but there is no evidence that he was greatly discomfited. In fact, he indulged in some of the activities he had condemned, as did many other youths from godly families, if they could afford to.[25]

We have no knowledge of his apparel at this time but his later researches into colouring cloth and shining footwear indicate an appreciation of fine clothes. Overbury's 'Inn of Court Man' was 'ashamed to be seen in any man's company that wears not his clothes well', and, in later life, Plat dressed fashionably on occasion with 'a certen Chayne of gould wch he did sometyme weare about his neck' and 'a brooche or Jewell wch he usuallie did weare in his hatte'.[26] Play-going was popularly supposed to be an Inn of Court student's favourite pastime: London's first theatre was erected at Shoreditch in 1576 and there were eleven by the end of Elizabeth's reign, many of them in the unregulated suburbs on the south bank where they competed with pleasure gardens offering beer, bear- and bull-baiting, as well as with bawdy houses and large inns. Plat was *au fait* with such entertainments. He conversed amicably with dramatists: 'I have been recently informed by Mr Lodge the play maker that i Ton of Mallasses mixed wth 7 Ton of fayre water maketh very good beverage for the Sea,' he notes. In 1603 he joked that the forthcoming details of a secret would be much more worthwhile than 'the second part of M. Venners Tragedie, lately acted at the Swanne on the bank side, with better profite to himselfe then pleasure to the beholders'. (Venner had announced, and taken money for, an entertainment which did not, in fact, take place.) At the end of the preface to the *Jewell House* in 1594 he parodied a theatrical speech, a memory, no doubt, of plays he had attended:

> But now it is high time for the prologue to give place because the Actors are at hand, who are readie to present such choice and varietie of matter, as that not withstanding they may happily faile in gesture or action, yet I doubt not but they wil either procure a friendly & thankfull plaudite,

[25] Prest, p. 154.
[26] *Character Writings of the Seventeenth Century*, p. 61; TNA, E134/8 Jas 1/East 2; Diane Scarisbrick, *Tudor and Jacobean Jewellery*, 1995, chap. 4.

which is the most that I can desire, or a most free & liberal pardon, which is the least that I can deserve.[27]

Neither was he a stranger to the pleasure gardens south of the river. In 1594 he looked back a dozen years to his student days, describing in great detail a recipe for coating hands so that they could be plunged into molten lead, as performed by 'a Dutchman called Hounce, a pretty nimble Chimist'. This, Plat explained, 'hee did for a pot of the best Beere in a garden in Southwarke about ten or twelve yeeres sithence, in the presence of myself and divers others, at which time I writ the receit even as I did both see him make it, and use it my selfe, disbursing the charge both of the Beere, and the ingredients.'[28]

Plat knew enough of dancing to string dance-terms together in an elaborate pun: 'Meethinks I am now in the midst of a stop galliard, & were it not, that I should heere offend so great a concourse of people, as I have nowe gathered together, in mine owne conceit, I could finde in my hearte to commaunde the Violands to cease, and so to breake off in the midst of a rough Cinque passe.'[29]

He was cautious about card playing; reporting without comment a way to cheat taught him by two gentlemen involving reflections of opponents' cards in drops of water on the table or in highly polished rapier pommels. In the *Jewell House* he refined the trick, using a ring with a glass or crystal stone cut and set so that it:

> will give a livelie representation to the eye of him that weareth it, of all such Cardes as his companions which are nexte to him doe holde in their handes, especiallie if the owner thereof doe take the upper ende of the Table for his place, and leaning nowe and then on his elbowe, or stretching out his arme, doe applie his Ring aptlie for the purpose.

The secret of setting the stone was then known only in France. 'There be some English knights that can sufficiently testifie the truth hereof by

27 Prest, p. 159; Porter, p. 62; BL, SL 2189, f. 135v; Sir Hugh Plat, *A new, cheape and delicate Fire of cole-balles*, 1603; Plat, *Jewell House*, Preface.
28 Plat, *Jewell House*, 1594, pt. 1, p. 30.
29 Plat, *Jewell House*, p. 39.

that which they have seen amongst French gamesters.' Plat provided these details as a warning to the young,

> I have discovered this secret rather to discorage yong Novesses from Card-play, who by one experiment may easily ghesse, how manie sleights & cousonages, are dayly practised in our dicing and gaming houses, not doubting but that the general publication there-of wil make the same so familiar with al men, as that I shall not iustly be charged of anie to have taught old knaves newe schoole pointes.[30]

He did not approve of drunkenness but conceded that, on occasion, a gentleman had to indulge his companions and drink to excess as a social duty. Before a party one could fortify oneself with a draught of olive oil or new milk before drinking wine,

> But howe sicke you shall bee with this prevention, I will not heere determine, neither woulde I have set downe this experiment, but onely for the helpe of such modest drinkers as sometimes in companie are drawne, or rather forced to pledge in full bolles such quaffing companions as they would be loth to offend, and wil require reason at their hands as they terme it.[31]

Overall, Plat's godliness does not appear to have prevented him enjoying the social activities pursued by most young men at Lincoln's Inn. Indeed, as Master of the Revels in 1574 he was in overall charge of entertainments over the Christmas and New Year holidays. In choosing this institution for his son's legal education however, Richard Plat ensured that his religious and moral education was looked after. Although Lincoln's Inn had no rules curtailing students' playgoing, gaming, or other possible vices outside its walls, 'The benchers of Lincoln's Inn consistently manifested far more positive concern for the spiritual well-being of members than

28 Plat, *Jewell House*, 1594, pt. 1, p. 30.
29 Plat, *Jewell House*, p. 39.
30 BL, SL 2216f. 42a; Plat, *Jewell House*, pt. 1, pp. 6–7.
31 Plat, *Jewell House*, pt. 1, p. 62.

their colleagues in the other houses.' This Inn 'maintained the longest succession of puritan preachers and the leading reputation as a hotbed of militant Puritanism.' So here, as at Cambridge, Plat came into contact, via sermons, lectures, and senior members, with a godly interpretation of Protestantism.[32]

Summing up Sir Hugh Plat's formal education, we may observe that, although more extensive than many young gentlemen of his day, it was in line with a move towards a better-educated (male) gentry. 'Probably the majority and quite possibly a large majority, of the Elizabethan and early Stuart gentlemen had received at least a grammar school education' and increasing numbers followed Plat in attending a full undergraduate course at University, followed by some time at an Inn of Court. This trend, evident from the first half of the sixteenth century, began as the gentry realized that continued access to the controls of power depended less on birth or military might and more on administrative skills and a rounded, cultured mind, for which education was essential. English society tends to accord gentry status to those with the wealth to maintain it and Plat was one of a growing number of young men whose fathers' wealth from trade had elevated into the rank of gentleman.[33]

Any inclination to youthful frivolity as a bachelor student lodging at an Inn was curtailed by marriage, setting up a household with wife, servants, and, in due course, children, coupled with the need, prior to his inheritance, to acquire the money to pay for these responsibilities. The years between 1574 and 1592 were spent in the city of London, probably in a house in Candlewick Street Ward of the city (in Stow's time known for its drapers), and form a distinct period in his life.[34] He married twice in these years. His first wife Margaret, daughter of a member of the Grocers' Company, bore him at least three children. The marriage may have been a love-match as well as an alliance between two city families. It has been suggested that the love poem 'The Conquering Lover' in Plat's first book celebrates his betrothal. Repeated magical symbols incorporating the

---

32 Prest, pp. 38, 92, 205.
33 F T Levy, 'How Information spread amongst the gentry, 1550–1640', *Journal of British Studies*, XXI, 1982, pp. 11–12.
34 *Harleian Society Publications*, vols. 109/10, 1963, pp. 152, 163.

words 'margaret I love' or variations on 'margaret I lyff' in one of his manuscripts may confirm affection for Margaret but some uncertainty as to her feelings. Was Plat attempting to conjure her affection for him?[35]

Margaret's father, Richard Young, brought Plat into contact with political and religious controversy and may have been instrumental in his introduction to the Queen's Court. Recent historians have not portrayed Young sympathetically. A minor player in the spy networks of Cecil and Walsingham, he has been described as: 'one of London's leading priest-hunters'; 'the notorious J.P. for Middlesex whose busy ferreting-out of religious suspects had made him odious to Catholic and Puritan alike'; and 'one of the most virulently anti-Catholic judges of the age.'[36] Young was even more odious to those he sought out – the captured priest John Gerard described him as an abusive interrogator who took bribes. Gerard also called him 'the devil's confessor' and ascribed his death to his anti-Catholic zeal: 'Day and night he toiled to bring more and more pressure on Catholics, drawing up lists of names, giving instructions, listening to reports. Then one rainy night, at two or three o'clock, he got up to make a search of some catholic houses. The effort left him exhausted; he became ill, contracted consumption and died.'[37]

From mid-century Young held the post of 'Packer' in the Port of London, which brought him a steady income of £300 to £400 a year. Soon after Plat married his daughter Margaret in February 1573, Young gained the more lucrative post of Customer, and his diligence was said to have gained the Crown additional customs duty of £100,000 over the succeeding eighteen years. He extended the range of Customs activities, at his own cost, to 'divers outports' to increase yields. He was, whilst Plat's first wife was alive, prosperous, marrying a widow said to be worth at least £3000. 'He gained greatly in trade, got together, in lands and leases, above 200*l*. a year, and maintained himself well; yet living within 400*l*. a year, so

35 Ibid., pp. 100, 152; *The Floures of Philosophie*, p. xvi; BL, SL 2203, f. 88; *ODNB*, Sir Hugh Plat.
36 Alan Haynes, *Invisible Power*, Stroud, 1992, p. 43; *Sir John Harington's A New Discourse of a Stale Subject called the Metamorphosis of Ajax*, ed. Elizabeth Story Donno, 1962, p. 167n; *The Diaries of John Dee*, ed. Edward Fenton, 1998, p. 343.
37 *John Gerard*, trans. Philip Caraman, 1951, pp. 77–8, 92–3.

that he might leave a yearly surplus.' He was also appointed as a Justice of the Peace for Middlesex, a key post because of the populous London suburbs within the county. As a JP and Customs official, he became, in effect, the head of an immigration service, trying to intercept Catholic priests and their supporters on entry to England or arrest those who had made it to the London area. He was also involved in detecting spies and potential assassins, such as the Queen's physician Lopez.[38] On a more mundane level, he was asked by the Privy Council to try and stem the growth of illegal building in the suburbs, part of a long-running (and futile) attempt by the Crown to limit the growth of London.[39] His heavy work-load led to neglect of his financial affairs and this, together with high personal expenditure on Crown business, greatly damaged his wealth. In 1588 he was granted the monopoly of the manufacture and import of starch but he had difficulty in enforcing this unpopular measure despite a Privy Council proclamation against illegal starch imports in 1592 so that 'instead of gain he reaped a loss, until a little before his death'.[40] Shortly before he died he petitioned the Crown to cancel his debts to the Queen and he came to some arrangement with her in return for his surrender of the residual right to the starch monopoly.[41]

Sir John Harington, in *The Metamorphosis of Ajax*, published in 1596, makes slighting allusions to Young's zealous pursuit of papists and his starch monopoly, remarks which draw a threat from the anonymous author of a rejoinder published in the same year, *Ulysses on Ajax*: 'a *Young* that will be old (saith thus) in behalf of old *Young*; that except you presently put on a habit of more conformity, if some his enemies may promote you, you shall be the next dog that shall be sacrificed in the Lupercalia, and therefore provide yourself for it (except you get a better tongue in your head, or a modester pen in your hand).' In other words, Harington's attack on Young might be repaid by harassment for religious nonconformity.[42]

38   Sir John Harington, ed. Donno, pp. 244–5.
39   *Acts of the Privy Council (APC)*, 1590, pp. 324, 326.
40   *APC*, 1592, p. 45; *Calendar of State Papers Domestic (CSPD)*, 1595, pp. 103–5.
41   *CSPD*, 1595, pp. 103–5; Sir John Harington, ed. Donno, p. 167.
42   The author implies he is a younger relative of Richard Young, although a recent article has made a case for the author to have been Plat himself. Sir John Harington,

One can readily see how a witty and intelligent courtier like Harington might despise a single-minded official such as Young, who had the approval of the government to use strong-arm methods against those thought to be a threat (often, if petitions to the Crown are to be believed, imprisoning and interrogating on little evidence – a practice not uncommon in our own times when governments are uncertain about the strength or intentions of their enemies). But Young was good at his various jobs. As Customer he refused bribes and 'discovered great concealments of Customs'. He also administered the collection of subsidies (Crown direct taxes) in London. For his efficiency in collecting both direct and indirect taxes he 'got much hatred, and sustained many bitter curses'. His efforts to stop illegal building in the suburbs would not have endeared him to London landlords or the poor in search of housing.[43] Plat makes a puzzling allusion in print to his first father-in-law in 1593, apparently implying that Young's unpopularity might affect him, but all his manuscript mentions of Richard Young are favourable. He described a tank for keeping small freshwater fish in a living-room for pleasure which he had seen 'in my ffather youngs howse' and recounts that Young was philanthropic during the plague visitation of 1593, procuring pills against the disease from Plat and distributing 'abroad many hundreths of them'.[44]

Young shared Plat's interest in medicine and alchemy and they both went to see John Dee, the most famous English alchemist of his day, in 1582. Dee had many amicable dealings with Young, describing him as his 'brother' which might imply either kinship or close association. Young came on another occasion, with others, to Dee to discuss a foreign voyage, Dee regularly wrote to him when he went abroad and, when he returned to England in November 1593, he apparently stayed with Young for at least three weeks. Later, Young supported Dee at a meeting with the Archbishop of Canterbury to discuss benefices. In sum, the picture of

---

    ed. Donno, pp. 17, 167, 244; *Ulysses on Ajax. Written by Misodiaboles* [1596], 1814, p. 32; 'Sir John Harington, Hugh Plat, and Ulysses upon Ajax', R. Bowers & P.S. Smith, *Notes & Queries*, vol. 54, no. 3, 2007, pp. 255–259.

43 CSPD, 1595, pp. 103–5; 1596, pp. 184, 282. The author has personal experience of hostility towards those who administer taxes.

44 Plat, *A Briefe Apologie*, 1593; BL, SL 2209, f 33; SL 2210, f. 80.

BIOGRAPHY

Young as a narrow-minded 'secret policeman' is the one-sided view of his many adversaries and ideological opponents, men who had cause to rue the efficiency with which he carried out his government offices.[45]

Plat's son John by Young's daughter Margaret was born in 1581 and Plat is described as Young's son-in-law by Dee in September 1582, so Margaret died between the latter date and June 1584 when he married his second wife Judith. When she became his bride he had three children, probably aged between four and eleven, so a second marriage was necessary if for no other reason than to provide a mother for these young sons. Judith was 22 when she married Hugh (he was 32) and their marriage produced a further two sons and three daughters. She survived him by many years, dying in 1635/6. (On Plat's death in 1608 five sons and one daughter were alive, a favourable survival rate for London at this time.)[46]

This marriage was another alliance with a city family albeit not one given to controversy. Plat's new father-in-law, William Albany, was a member of the Merchant Tailors' Company, rising to the Livery in 1550–1, elected to the Court of Assistants in 1562, and taking a turn as Master in 1568. He remained a member of the Court for many years, dying sometime after 1589. Stow records a memorial to him in Allhallows, Bread Street – he was probably buried there in the choir next to his wife. With the new marriage came a new set of kin – one of his brothers-in-law, Robert Albany, spent time in Lincoln's Inn and, if it was another love-match, may have introduced Plat to his new wife.[47] Plat himself provides a glimpse of Judith. She invented a salad which he notes with approval and she also perfected a way of making cheese which Plat set great store by. He dropped hints about it in 1594 in the *Jewell House*, finally giving full details a few years later in *Delightes for Ladies*. Plat implies that his wife kept the secret for as long as she could against his desire to publish it, an indication, perhaps, that she was a strong character.[48]

45 *The Diaries of John Dee*, ed. Fenton, pp. 47, 53, 231, 233, 237, 240, 246, 248.
46 TNA, E134/8 Jas I/East 2; Harleian Society, vol 93, 1913, pp. 6, 100; *ODNB*, Sir Hugh Plat.
47 Guildhall Library, M'film MF 324 (records of the Merchant Tailors' Company); Stow, pp. 324–5; TNA, E210/10401 20 Eliz; *The Diaries of John Dee*, ed. Fenton, p. 47.
48 See Chapter 6.

The surviving papers concerning a law-suit against Judith for recovery of an alleged debt to the Crown from Sir Hugh's estate also give the impression that she was a strong woman, capable of complicated financial deals in an effort to defend the estate and also used to fashionable attire – she was wont to wear several small jewels in her hair. Two questions put to witnesses by the Attorney General asking what marriage portion Plat received with Judith and what further portion came to him on the death of his father-in-law hint that Plat gained financially from his second marriage.[49]

Whatever the extent of these windfalls, before he received the residue of his own father's estate in 1600 Plat may have needed an income over and above any allowance from his father. This the law may have provided. Plat's sojourn at Lincoln's Inn was not the two years or so of social polishing with a smattering of legal education experienced by most young gentlemen. He trained properly as a lawyer. He was called to the Bar in June 1581, henceforth he could practise as a barrister.[50] His legal education took a couple of years longer than Stow thought usual (marriage and children may have slowed him down) but after nine years of frequenting 'readings, meetings, boltings, and other learned exercises, whereby growing ripe in the knowledge of the laws' he joined those qualified 'to be common counsellors, and practice the law, both in their chambers and at the bars'. He listed standard legal texts amongst his books and, like other legally trained writers of his day, he peppered his non-legal writings with legal puns and terminology. Moreover, he took his turn at teaching the law. He was Steward of the Readers' Dinner in Lincoln's Inn in 1586, and paid a fine to avoid the duty again in 1589. A Reader lectured on aspects of the law and supervised moots. The survival of a few pages amongst Plat's papers, apparently summaries of cases reported from the reigns of Henry VIII and Edward VI, may be part of his lecture notes.[51]

Although little on the law is to be found in his surviving manuscripts, there are stray references to now-lost papers. He apparently wrote books

---

49 TNA, E134/8 Jas 1/East 2.
50 *The Records of the Honorable Society of Lincoln's Inn, The Black Books*, vol. I, 1897, pp. 422, 440; vol. II, 1898, p. 12.
51 L.W. Abbot, *Law Reporting in England, 1485–1585*, 1973, pp. 32–4; BL, SL 2210, f. 1; Stow, pp. 104–5; BL, SL 2209, ff. 10–14a.

on the law (he mentions one manuscript, 'My title of Misceiffes', remarking 'this may bee gaynfull in releiffinge thextremities of Law') and he hoped to make money from their publication, selling them to fellow lawyers. One of his many lists contains types of legal case he thinks he might make money from.[52] Plat clearly kept in contact with the law, visiting law courts and the Inns of Court throughout his life, noting in his commonplace books recipes and suchlike information gleaned from fellow lawyers. He hoped to use his extensive legal contacts to help sell his inventions and ideas.[53]

Learning the law must have occupied Plat more or less full time but, after his call to the Bar in 1581 he was increasingly involved with research, both practical and theoretical, one of the fruits of which would be his books and pamphlets published from 1593 onwards.[54] There was a long gap between his first, youthful excursion into print and his next publication, the small pamphlet *English Wants* of 1593. Thereafter, pamphlets and books appear steadily almost until his death in 1608. Moreover, these works are, with the exception of a poem in Latin in 1594, all on the diverse 'practical' subjects which have attracted the notice of historians. He intended to put forward his ideas earlier – the 1593 pamphlet explains that he has been prevented by 'infirmitie' from publishing 'diverse necessarie & pleasing experiments both of Art and Nature' (i.e. *The Jewell House of Art and Nature*, published in 1594).[55]

Legal work and the domestic and emotional upheaval of the death of Margaret and marriage to Judith must have played a part in delaying his publications. For the rest, he was simply doing long hours of research. Like a present-day PhD student, he accumulated a good deal of material from many sources before he was prepared to launch into print. We may regard the *Jewell House of Art and Nature* as the equivalent of his doctoral thesis. Later, as his ability as a researcher and writer grew, he produced other substantial works, such as *Delightes for Ladies* (1600) and *Floraes Paradise* (1608) interspersed with several pamphlets. As outlined in more detail in the appendix on his manuscripts, we can trace through his papers how his

52   BL, SL 2197, f. 21v; SL 2223, f. 60v; SL 2247, f. 2.
53   BL, SL 2216, f. 153; SL 2210, f. 74v, 93; SL 2171, f. 13.
54   See Appendix 1.
55   Plat, *A Briefe Apologie*, 1593.

research developed and how he adopted the note-taking techniques taught at grammar schools and, in more depth, at Cambridge in his day.[56]

Plat enjoyed his early years of private study: despite reference in his dedication of the *Jewell House* to 'the painefull and chargeable travells of my youth' he confesses to 'having spent som of my sweetest hours in reading & many of them in conference, and more in practise, but most in contemplation'. He clearly recognized this work as the first fruits of his labours. Despite being 'a young Novess in the schools of Philosophie, a slender Husbandman in the rights and culture of the ground, a man of civell education, that hath drawn the longest line of his life within the lists and limits of London', he has decided to, 'adventure as boldlie as the rest, to commend the flowers of my youth, to the courteous view of all well disposed Readers'.[57]

Plat moved out of London in 1593 or 1594, renting Bishop's Hall in the suburban village of Bethnal Green. The residence of the Bishops of London until 1553, the house was granted by Edward VI to the Wentworth family. Plat may have moved later to another substantial property in the same village,[58] but he remained in close contact with the city and Westminster despite the move to Bethnal Green (describing himself as 'of London' in a contract of 1599) and he also had a residence near Garlick Hill in London in 1600.[59] Plat was what Margaret Pelling described as a 'skirter' – a Londoner who moved frequently between the suburbs and the metropolis, retaining the status of a Londoner but seeking to avoid some of the risk of disease that London presented.[60]

Bishop's Hall was a substantial house and Plat created a garden there: he described in 1600 how, using horn shavings as manure, 'I obtained a most flourishing garden at Bishops Hall, in a most barren and unfruitful plot of ground, which none of my Predecessors could ever grace or beautifie either with knots or flowers.' It is likely that Plat used this garden

56 See Appendix 1.
57 Plat, *Jewell House*, Introduction and Dedication.
58 George F. Vale, *Old Bethnal Green*, 1934, pp. 82–3; Henry G.C. Allgood, *History of Bethnall Green*, 1905, p. 96.
59 BL, SL 2172, ff. 12, 29v.
60 Margaret Pelling, 'Skirting the City', in *Londonopolis*, ed P. Griffiths & M.S.R. Jenner, 2000, pp. 154–175.

as a proving-ground for his horticultural and agricultural theories, and that much of the information in *Floraes Paradise* and in his manuscripts on gardening not gleaned from others is linked to his own efforts. We may further speculate that his writings on indoor gardening refer to Bishop's Hall, where he may have adorned a 'faire gallery, great chamber or other lodging ... that openeth fully upon the East or West sun ... [with] sweet hearbs and flowers, yea & fruit if it were possible' and trained vines and fruit trees through windows into the house. His ideal summer banqueting house may have graced the garden, along with the fanciful planting he suggests in his books.[61]

A suburban retreat rather than a country house, Plat nevertheless kept a couple of cows there in the summer. He probably exasperated his wife with the strange smells, smoke and noises from his alchemical laboratory, and the sudden requisitioning of chambers to create a *camera obscura* or set up a mechanical toy. Such practical experiments had been carried on in his earlier London residence, resulting in an exhibition of new inventions and practical demonstrations of his new 'cole balles'.[62]

Both houses were also, no doubt, used for consultations with gentry patients who came to seek advice on cures when Plat set himself up as an unlicensed physician.[63] He tells us on several occasions that he 'did entertain divers of my friends' at home as one would expect a gentleman to do, as well as entertaining his family and in-laws and, on occasion, putting forward his best food and wine for those he wished to impress. Being a Londoner, however, striving to get on in the world both financially and socially, many of his social interactions took place outside his house.

London gentlemen, a social category becoming increasingly notable in this period, lived not from rural rents, squires in their own villages accorded deference in their part of the country, but from renting urban property, often commercial or industrial buildings such as Plat's father's brewery, and sometimes from professional fees, as lawyers, doctors, or civil servants. In the cosmopolitan world of London, a gentleman was

61 Sir Hugh Plat, *The new and admirable Arte of setting of Corne*, 1600, f. D; Sir Hugh Plat, *Floraes Paradise*, 1608, p. 31.
62 Plat, *Jewell House*, pt. 3, p. 12, and see Chapter 9, below.
63 See Chapter 7, below.

likely to mix with all sorts of people on a daily basis, albeit that careful maintenance of status was achieved by dress, mode of transport (Plat had a coach), attendance of servants and general speech and carriage of one's person. This mixing was more pronounced in Plat's case because he clearly relished talking to people, picking their brains for information and enjoying the unique opportunity which London, as the national centre of commerce and the arts, presented for social intercourse.

He had many linked circles of acquaintances. The most important, the Inns of Court and law courts, the royal Court, his medical practice, his alchemical experiments, and his commercial collaborators we have, or will look at more fully, but the profusion of names and topics of interest associated with them in his books and manuscripts illustrate his eagerness to talk to people. We are as likely to encounter him talking to an itinerant woad-grower about medicine, an aged gardener about plants, a Dutch entertainer about tricks with molten metal, an Italian woman about preserving nuts, as to hear that he has conversed with fellow gentlemen at the Bar or at Court. He quizzes London tradesmen about their secrets, extracting information from, amongst others, saltpeter men, metalworkers, sailors, vintners, goldsmiths, apothecaries, bakers, dyers, builders, comfit makers and cooks. His neighbours provide him with ideas and lend him books. His father's close connection with brewing, reinforced by his inheritance of the long lease of a valuable brewery from him, gains him expertise in brewing. He is used to the hurly-burly of London streets, 'the lying and forswearing Huswives' who sell all manner of sub-standard produce in the streets, perjuring themselves on the Sabbath, 'maintaining their sales, with such bold countenances, and cutting speeches, with such knavish practices and such forlorn Consciences, as that they have both driven away many honest Matrons from their stalls, and corrupted a number of young maiden Servants with their bold and lewd lying'.[64]

Plat's position as a gentlemen, however, dictates that he spend much of his time in better company than that of the streets. As an active member of Lincoln's Inn, he mixed with professional lawyers and young gentlemen

64 Sir Hugh Plat, *The second part of the garden of Eden*, 1675, p. 130.

finishing their education. Gentlemen from all over the country came to the law courts themselves to litigate and Plat undoubtedly obtained much information about other parts of the country, especially agricultural practices, from these visitors.

As a resident London gentleman, son of an Alderman and sometime Sheriff, he mixed with prominent citizens. He invited 'a Court of Aldermen' to his London house to see his 'Coleballs' and demonstrated his bolting hutch to some of the 'choycest' members of the Bakers' Company together 'with diverse other Citizens of good worship and account'. He presented, one Christmastime, 'unto Sir John Allet L. Maior of the Citie of London 8. greene and fresh Artichokes upon the twelfe day, with a score of fresh oranges, which I had kept from Whitsuntide'. New Year's gifts were an occasion for ostentation and Plat was no doubt pleased to demonstrate his horticultural prowess in this way.[65]

His desire to maintain and advance his position, drew him, like many of his peers, to the royal Court, an institution succinctly described as 'a natural goal for any man of ambition'. The royal palace at Whitehall was the seat of government, where the Queen made decisions, the Council met and issued orders, where patronage was dispensed, and where political intrigue occurred. These functions merged with the social activities of the Court, as the home of the Queen, her favourites, personal servants and Councillors; a flamboyant display-case of the arts and fashion; a visible manifestation of the social order of the gentry and nobility; a place for pleasure: dancing, music, feasting, plays and poetry; and for advancing up the social and economic ladder by lobbying the Council and civil servants for contracts or office, and attracting the notice of the Queen or those close to her for honours, sinecures, or other social advantages.[66]

Dressed and clearly acting like a gentleman, Plat would have found the physical area of the Court relatively easy to enter. Here, much time was spent in idleness, waiting for an opportunity to impress the Queen or those close to her, ever ready to bend a knee or press a suit should she or her entourage appear in the more public areas to which he would have had access. Time was passed in games, such as card playing, music-making and,

65   Plat, *A Briefe Apologie*; Plat, *Jewell House*, pt. 1, p. 5.
66   David Starkey et al., *The English Court*, 1987, pp. 1–2.

most importantly, conversation. Such talk might take the form of idle chat or flirtation but it might be more productive. Clearly Plat had many useful talks with courtiers and servants who had interesting information to pass on: royal cooks, rat-catchers, and Court painters divulged their secrets, Drake and Raleigh talked of the problems of long sea-voyages; gentlemen from Ireland vouchsafed miscellaneous information; foreign diplomats brought knowledge from distant lands, and English gentlemen talked about their ailments and recommended cures, their gardens and their local agriculture, and the peculiarities of their local counties. Conversation could be destructive as well as constructive: a gathering of rich, idle, status-seeking gentry, often well educated, and well versed in the art of argument, was a hot-house for bickering and personal animosity. In the 1590s, when Plat seems to have begun attendance at Court, there was the added spice of factional politics occasioned by the ill-conceived attempts of the Earl of Essex to extend his political power.[67]

Plat clearly put a good deal of effort into life as a courtier, and the experience was at times wearing on the emotions. He had to learn the fickle arts of a courtier: in the acid words of Sir Thomas Overbury, 'He follows nothing but inconstancy, admires nothing but beauty, honours nothing but fortune: Loves nothing'. He was unlucky, or unwise, in his attempts in the latter years of Elizabeth's reign to make any impact at Court. He tried acquaintance with Raleigh and his circle – visiting his house in the Strand in 1591, only to see him in disgrace a year later. He had conversations with Drake about naval victualling, but Drake died at sea in 1596. A more serious setback was Plat's decision to ally himself with the Essex faction, fulsomely dedicating *The Jewell House of Art and Nature* to the Earl in 1594, hoping to prosper under 'so honorable a Patron' (whom he likened to Achilles) and in the following year presenting him with home-made wine at Christmastime. Any hopes here were dashed with the rebellion and execution of Essex in 1601. Foreign contacts might also prove fruitless. Plat talked intimately with the Spanish Ambassador Mendoza on more than one occasion, prior to his expulsion in 1584. Mendoza showed Plat a lamp with a long-lasting wick and a mixture of wheat straw with grain as flour 'was commended by Mendozza himselfe, wherewith he assured

---

[67] Simon Adams, *Leicester and the Court*, Manchester, 2002, pp. 46–67.

me upon his honor that he had relieved a Spanishe towne, in an extreame dearth, and scarcity of victual, and therewithal shewed mee a loafe of that composition, which was of wheate straw, chopt into short peeces, and grounde with som proportion of wheate into meale.'[68]

We learn of Plat's efforts in Court circles from his pamphlet advertising his inventions, the *Briefe Apologie* of 1593. He has exhibited inventions to some 'of her Maiesties privie Councell, and diverse other Gentlemen' and hopes his artificial coal will 'finde many furthering friends at Court'. He attended a major Court function, a gathering of 'two hundredth Gentlemen at least' including 'that most learned senator the Atlas of England' (possibly William Cecil, Lord Burghley). But this pamphlet is full of his frustration at those, especially within the Court, who have derided his efforts. He protests against 'the sundrie and sharpe calumnations wherewith these few inventions have bin of late dayes most wrongfully and vehemently charged, by the ignorant and yet malitious enemies of all ingenious devises'. Allusions to university teaching may indicate his detractors were known to him at Cambridge, and he further identifies them (to his contemporary readers but not, alas, to us) as 'senslesse blockes and penifathers of our time, (whose pictures & portraitures have bene so lively drawne of late, by two good perspective Painters, in theire several and sharpe invectures both of puisne date)'. His artificial coal 'must needes finde many furthering friends at Court' and he is discontented that other of his inventions 'to my great griefe, loss, and discredit (I know not by what froward fates or misfortune) hath beene staied, crossed, or coldly commended in Court'.[69]

In the book adumbrated by this pamphlet, *The Jewell House* of 1594, he returns to the attack on Court detractors, singling out those who wasted time as students, being deficient in Latin, the law and university-taught philosophy despite being 'brought up even in the same schooles, yea and under the same Tutors' as Plat. These critics were not, Plat sarcastically (and long-windedly) claims, willing to 'put their money where their mouths were': 'Yet herein I must of necessitie allowe of their grave wisedoms, who

---

68   Morley, *Character Writings of the Seventeenth Century*, p. 31; Sir Hugh Plat, *The Jewell House*, 1594, A2; Plat, *Sundrie new and Artificiall remedies against Famine*, 1596, B2; BL, SL 2210, ff. 91, 154v.

69   Plat, *A Briefe Apologie*, 1593; 'puisne' is a legal term meaning later or subsequent.

being not able to rule their rash and riotous tongues, have always shewed so great a government over their enchanting purses, as that at al times when the question came to be tried by any round wager (which hath beene oftentimes offered them in my behalfe, and shal stil bee maintained in despight of their ignorance) they have presentlie shifted themselves by an Apostrophe to a new matter, as if they had spoken unawares they knew not what.'[70]

The identity of at least one of Plat's Court detractors becomes clear with the publication in 1596 of Sir John Harington's *A New Discourse of a Stale Subject called the Metamorphosis of Ajax*. This celebrated work, a manual on the flush-lavatory contained within a dense mass of literary and classical allusions, full of mocking jibes at friends and enemies alike, is a satire which displays the clever, playful mind of a courtier, much richer and more assured of his position than Plat, who could afford to play at Court life safe in the knowledge that any indiscretions would be eventually forgiven by his godmother, the Queen. 'The Boy Jack', as the Queen affectionately called him, claimed 'no great acquaintance' with Plat, although they were both Cambridge men who went on to Lincoln's Inn. Harington playfully suggests that the Inn is now famous not only for legal minds but 'two such rare engineers' as himself and Plat. Having studied Plat's publications to date, Harington concludes that both Plat's secrets, to produce better barley and artificial coal, depend on urine and dung, and so both inventors are working with similar materials. Harington attacks Plat for wanting a monopoly to protect his inventions, alluding to the unpopular monopoly of starch held by Plat's first father-in-law (Harington makes another slighting remark about Young later in the work). Alluding possibly to Plat's exhibition of inventions to courtiers, Harington reminds Plat that it is difficult to beguile the new Lord Keeper, Sir Thomas Egerton, 'with goodly shews' because he is both incorruptible and no lover of monopolies. Plat is advised to 'impart his rare devises gratis, as I do this'. The reference to Egerton's incorruptibility may also be a swipe at the corruption of Court life, a corruption Harington implies that Plat is a part of, in contrast to his own detachment.[71]

70 Plat, *Jewell House*, 1594, A2–B1a.
71 Sir John Harington, ed. Donno, 1962, pp. 165–172; J. Scott-Warren, *Sir John Harington and the Book as Gift*, 2001, pp. 74–5.

BIOGRAPHY

Harington's jibes, coming on top of verbal abuse at Court, stung Plat into quick retaliation in print. He makes a series of puns on dung and urine, reminding Harington of the royal censure *Ajax* has provoked and mocking Harington's flowery writing style, 'his glib paper & gliding pens'. Harington's implication that dung binds his cole-balls is dismissed,

> not regarding the censures of those ignorant, or malicious spirits of our age, who presuming to know the simples of my fire, may happtly range into base and offensive matter, and thereby labor to discredit that secret, whose composition they could never yet reach unto, nor, if they had the particulars, were they able to combine & knit them with their lefthanded workmanship.

Plat reminds his readers that his inventions 'do as yet attend some courtly favours'. He continues,

> And for the better satisfaction of my welwishing friends, & the confutation of mine underserved foes, I would have them to understand, that seeing the premised secrets, have not onlie bin seen, and allowed, but at this present are also countenanced by those which are right Honorable in their places: that from hencefoorth they will scorne the malice both of viperous tongues, as also of slanderous pens, if any man should happen to bee so extreamlie, or desperatelie mad, as to take upon him to argue upon that proiect, whereof he can neither finde a medium, nor communes terminus, and therefore impossible to conclude Sillogistice sinon in Bocardo against it.[72]

An anonymous pamphleteer, probably a relative of Richard Young (or possibly Plat himself), supported Plat in *Ulysses on Ajax* which also appeared in 1596. With crude allusions and a threatening style, this pamphlet defended both Plat and Young (who died in that year). Harington was here threatened with arrest for heresy, with implications that he had Catholic sympathies, and the pamphleteer uses a string of alchemical terms in his defence of Plat.[73]

---

72 Plat, *Sundrie new and Artificiall remedies against Famine*, 1596, D4–E1. 'Bocardo' was the name given by medieval logicians to a type of categorical syllogism.
73 *Ulysses on Ajax. Written by Misodiaboles* [1596], 1814, pp. 32, 57.

The heated exchanges cease after 1596, although in an advertisement for 'Cole-balls' in 1603 Plat again condemns 'rash pens' who have written that the secret of their composition is cowdung. He claims that his wooden boiling vessel was exhibited to 'the right honourable the Lord high Treasurer of England, my good Lord and master, who vouchsafed me his honourable presence at my house.... Whose good approbation is able to counterpoise and overway the shallow and light conceits of all my adversaries.' The overall impression is that he was struggling under Elizabeth to make an impression at Court. Plat, an earnest and serious gentleman who hoped to gain from Court contacts, found his efforts thwarted both by rich dilettantes like Harington, who ridiculed his ideas, and the vagaries of politics and death which robbed him of a succession of possible patrons.[74]

The absence of bitter comments in print after 1604, combined with his success in gaining a knighthood in 1605, indicate that the Court of James was more congenial to Plat. To judge by the names he casually drops into the advertisements at the end of *Floraes Paradise*, he enjoyed good company here. They appreciated his home-made English wines,

> I am content to submit them to the censure of the best mouthes.... I could bring-in the French Embassador, who (now almost two yeeres since, coming to my house of a purpose to tast these wines) gave this sentence upon them; that he never drank any better new Wine in France. And Sir *Francis Vere*, that maritiall Mirrour of our times, who is seldom without a cup of excellent wine at his table, assured me that he never dranke the like unto mine, but once, and that in France. So that now mee thinks I begin to growe somewhat strong in my supporters; & therefore I make some doubt, whether I shall need to bring in that renowned Lady Arabella, the Countesse of Cumberland, the lady Anne Clifford, the Lady Hastings, the Lady Candish, & most of the Maides of Honour, with divers Lordes, Knights, and Gentlemen of good worth, that have generally applauded the same; or leave it here to worke out his owne credit in his due time, because it is rich, and of a strong boiling nature.[75]

74 Plat, *A new, cheape and delicate Fire of Cole-balles*, 1603.
75 Plat, *Floraes Paradise*, 1608, ff. O7–P3.

By this time Plat was not only a knight, he was much richer following the death of his father in 1600. The ease provided by his father's legacy probably accounts for the production of two further books, *Delightes for Ladies* in 1600, and *Floraes Paradise* in 1608 as well as two pamphlets. Other works, principally on alchemy, were in preparation at his death and may date from this period.

One more strand of activity which brought Plat into contact with a wide range of people and also augmented his income was medicine. A practice-book lists patients from 1593 until 1606 and to judge from the names listed, his medical work was quite extensive. With the usual irony we may note that his own health was not good at times, he twice claimed illness as a reason for delays in publication and he foresaw his death in 1608 when excusing shortcomings in his last book which appeared in the year he died. He may have had a long-running illness which he knew would inevitably kill him, although a question about the cost of medicines bought from apothecaries taken 'in his sickness', posed in the dispute over his estate, may suggest an illness for which there was hope of cure. When he made his will shortly before he died his mind was clear although he was 'sick in bodie'.[76]

It is also ironic that the legal wrangle over his estate after his death provides many of the few scraps of intimate information we have about him and his family: his gold chain worn on important occasions, the jewel he wore in his hat, and his 'seal ring'. The papers reveal the age of his surviving sons and we glimpse his second wife 'dame Judith', a woman capable enough to handle the complexities of a large estate and astute enough to try and hide some assets from potential creditors. A stylish woman who wore small jewels in her hair on occasion, Sir Hugh refers to her as 'my loveinge Wife' in his will.[77]

Sir Hugh Plat died late in 1608, a godly member of the Church of England believing himself bound for heaven, his soul bequeathed 'to Almightie God and trustinge to be saved by the only merritt and passion of Jhesus Christe'. To judge by his calm and businesslike will, he was at

[76] Plat, *A Briefe Apologie*, 1593; Plat, *Floraes Paradise*, 1608, A3; TNA, PROB 11/112/114; E134/8 Jas I/East2.
[77] TNA, E134/8 Jas I/East2.

ease with his approaching end. If he had been asked to judge the success of his life and his likely place in history his answer would most certainly have been blank incomprehension. In the chapters which follow, I hope some light will be shed on his work but I too am wary of so large a question.

CHAPTER TWO

Gardening

An important part of Plat's posthumous reputation rests on his two published books on gardening: *Floraes Paradise* which appeared as he was dying in late 1608, and a further collection of his gardening notes, *The second part of the garden of Eden*, produced by a relative in 1660. This later work appeared at a time when there was renewed interest in Plat's publications. Much of this centred on his agricultural works but John Evelyn cites Plat several times in connection with growing forest and fruit trees and others also noticed his garden writings.[1] Some years after Evelyn, in 1727, Stephen Switzer recommended gardeners to read Plat, but thereafter his horticultural legacy has been judged by garden historians. His stock was high in the nineteenth and early twentieth centuries. Amelia Amherst in 1895 commended him for collecting material from the best contemporary gardeners, and she perpetuated the myth, started by Charles Bellingham in his introduction to the second edition of Plat's gardening book, of his selfless acknowledgement of material: 'He was intimate with all the chief gardeners of his day, and is most conscientious in giving the credit of any piece of information to the friend from whom he learnt it.' In 1908 Donald McDonald developed this theme, claiming that as an agricultural and gardening writer, Plat 'was looked upon in his day as a learned and keen observer and clear-headed writer, and a valuable pioneer in the trenches of knowledge, yet so great was his modesty that most of his works seem to be posthumous. He held a general correspondence with all the most enlightened exponents on agriculture in his day, and such was the justice of his methods that, in direct contradiction to many other writers of his time, he always named the originator of every discovery communicated

1   John Evelyn, *Sylva*, 1670, pp. 32, 72–3, 179, 189.

to him.' McDonald's text is almost all invention: Plat makes no mention of corresponding with anyone about gardening and, as for modesty, Plat's *Floraes Paradise* was posthumous (or almost so) simply because he died of an illness before he could complete his planned publications on gardening. The numerous attributions at the ends of paragraphs are in the book because Plat had no time to edit his text and remove them. (McDonald goes on to cite two to 'correspondents' in *Floraes Paradise* who are not in fact mentioned in the book!)[2]

In the 1920s Eleanour Sinclair Rohde mentions Plat favourably, but in more recent times he has been little noticed as a gardening writer. Miles Hadfield made one passing reference to him in his *History of British Gardening* in 1960; he is not mentioned at all in *The English Garden* by L. Fleming and A. Gore of 1979. He is absent from the essays on the English gardens of his time and on English gardening books in the catalogue to the major exhibition *The English Garden* held at the Victoria & Albert Museum in 1979. He fares slightly better in *London's Pride*, the catalogue of the 1990 exhibition at the Museum of London on London gardening, but, apart from one lengthy passage (copied from a paper by me), his gardening works are not mentioned. Martin Hoyles takes little notice of him in his books published in 1994 and 1995 but Susan Campbell has made use of his work in her recent book on kitchen gardening. His star, perhaps, is on the rise again.[3]

The circumstances of the appearance of *Floraes Paradise* in 1608 are explained by Plat in his introduction:

> Having out of mine owne particular experience, as also by long conference with diverse gentlemen of good skill and practice, in the altering, multi-plying, enlarging, planting, and transplanting of sundry sorts of fruites &

---

2   Martin Hoyles, *Gardeners Delight*, 1994, p. 37; Alicia Amherst, *A History of Gardening in England*, 1895, p. 172; Donald McDonald, *Agricultural Writers*, 1908, pp. 55, 60.

3   Eleanour Sinclair Rohde, *Old English Gardening Books*, 1924, pp. 31–2; Miles Hadfield, *A History of British Gardening*, 1985, p. 143; L. Fleming & A. Gore, *The English Garden*, 1979; *The Garden*, ed. John Harris, 1979; *London's Pride*, ed. Mireille Galinou, 1990, pp. 58, 61, 88; Susan Campbell, *Charleston Kedding*, 1996, pp. 56–7, 64, 69, 139; Hoyles, *Gardeners Delight*, pp. 4, 13, 37; Martin Hoyles, *Bread and Roses*, 1995, p. 77.

flowers, at length obtained a pretty volume of experimentall observations in this kinde: And not knowing the length of my dayes, nay assuredly knowing that they are drawing to their periode, I am willing to unfolde my Napkin, and to deliver my poore talent abroad.

He described the contents of the book as 'these two hundred experiments (wherof some are but mine owne Conceits and Quaeres, and some, the reports of other mens practices)'. The 'experiments', mostly in short paragraphs, are divided into two sections. The first is untitled but is mainly on flowers and vegetables (apart from a long first section on a philosophical garden), whereas the second, headed 'Secrets in the ordering of Trees and Plants', deals mostly with trees. Apart from this division, there is little order in the book. The notes on gardening which Plat acquired from an earlier writer (those by 'T.T.', whose identity is discussed in a later chapter), are split between the two sections and the material for the book as a whole is drawn from manuscript SL 2189 now in the British Library. As I have mentioned, the hurried nature of the composition probably derives from Plat's assessment of his ill health and anticipation of approaching death although, curiously, he announced in the *Jewell House* in 1594 that 'my conceyted booke of gardening' was 'readie for the Presse' and was held up only because of the poor response to his publications so far.[4]

Perhaps the answer is that he had planned a more comprehensive and organized work which he had to hastily abandon as death approached. The 'General Rules' on gardening and the diagrammatic layouts of advice on planting and grafting in BL, SL 2210, a notebook completed before 1594, may be part of this aborted work. The same manuscript also contains much of the information in the posthumous publication, *The second part of the garden of Eden* of 1660. This book is also loosely organized, although some of the source material has been edited – similar topics are often grouped together. Plat may have done this himself at some time (for the style remains his after the editing), producing a manuscript from which this posthumous work was printed and which is now lost. The editor of

4   Plat, *Floraes Paradise*, 1608, 'To the studious and well-affected reader'; *Jewell House*, 1594, pt. I, pp. 5–6.

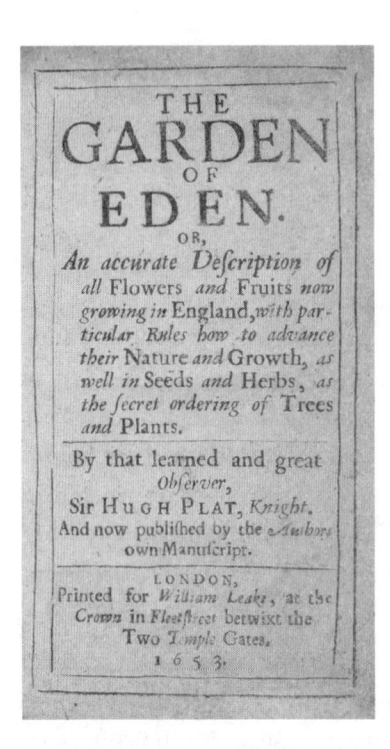

Figure 2 (above). The title-page of The Garden of Eden, *the second edition of* Floraes Paradise.

Figure 3 (below). Pages from Floraes Paradise *showing secrets taken from T.T.'s manuscript.*

The Second Part, a relative of Plat's, Charles Bellingham, offered doubting purchasers sight of 'the Original Manuscript under the Authors own hand, which is too well known to undergo the suspition of a counterfeit.' Bellingham had previously re-issued *Floraes Paradise* as *The Garden of Eden* in 1653, complaining that he had done so to protect his late kinsman's reputation against those 'who were ready to violate so usefull a Work'. Both the books were popular in the later seventeenth century, being regularly republished up to 1675.[5]

Even after the manuscripts were mined for both books, much remained in them which has not appeared in print and I will draw on these notes in the course of this chapter. Although one, or maybe both, of Plat's gardening books appeared posthumously he had been interested in gardening throughout his life. Many of the poems in his first book, *The Floures of Philosophie* of 1572, display a knowledge of gardening and there are dated entries on gardening in his notes from the 1580s and 1590s.[6] Although himself a keen gardener (he describes several gardening experiments in his own garden), Plat derived most of his information from others. Some were professionals, either working for gentlemen or institutions, or market gardeners trading on their own account. More and more people in the London suburbs were entering the relatively new avocation of market gardener in Plat's time, contributing to the supply of food to a metropolis which at least doubled in population during his lifetime and providing plants, trees and seeds for private pleasure gardens. Amongst them were Protestant refugees from the Low Countries, attracted to London by the demand for all types of garden produce: Plat talked to Jeames, a 'duch gardener' about layering vines and preserving fruit and heard further advice, second hand, from a French gardener.[7]

5 BL, SL 2210, ff. 23, 171; Plat, *The Garden of Eden*, 1653, Introduction; Plat, *The second part of the garden of Eden*, 1660, Introduction; Blanche Henrey, *British Horticultural and Botanical Literature*, 1975, vol. I, pp. 264–5. The existence of a now-lost manuscript possessed by Bellingham would explain how Bellingham produced *The second part of the garden of Eden* whereas circumstantial evidence indicates the original source, document SL 2210, was not in his possession. See Conclusion, below.
6 BL, SL 2210, ff. 35v, 41, 106; Plat, *Floraes Paradise*, p. 91.
7 Malcolm Thick, *The Neat House Gardens, early market gardening around London*, Totnes, 1998, pp. 39–61; BL, SL 2210, f. 23, 36; SL 2189, f. 189.

Plat would have seen commercial gardens on his journeys from Bethnal Green to the city and viewed their produce for sale in London markets: his neighbouring village of Hackney was famed for garden turnips in the 1590s.[8] Although the closest he comes to mentioning market gardens is to put forward a way of producing peas two weeks earlier than they do at Fulham (another early area of market gardening to the west of London), in a speculative passage in his notes headed 'profitable practizes in fruites & flowers' he sets out how he believes gardening for the market could produce a profit:

> yf you wold make an extraordinary profitt owt of gardein and orchard and other grounds for the better mainteyning of your charge then place and sow these fruits and flowers following. plante abricotts against your walles or pales, place stoare of Artichokes and Skirretts wth all ordinary rootes for foode. also sow earlie hastings, especially the Italian or suger pease which being sowed wth the forwardest pease that wee have will ripen 14 dayes before them. & they are to bee eaten wth theyre shales after the manner of french beanes. Mr Dacus of dorset shire hath of this kind growing they must be carefully defended from byrds. plant and graffe all earlie fruits make a dwarf orchard. multiplie carnations by theyre seedes. graffe carnations that you may have of all kinds uppon one roote. plant the provence Rose, doble stock gelliflowers, doble honny sucker also practize to keepe orenges, lemons, quinces, berberries, grapes, plommes, cherries &c. all the yeere by arte, for such rare fruits and flowers will prove very saleable both in cheapside & in the courte. Mr flower of bednall greene when hee soweth his earlie pease hee ploweth his ground at the least 5 or 6 tymes before hee sow. I think a lieght sandy ground is best for them.[9]

This informative passage reveals Plat looking at market gardening from a gentleman's perspective. 'Ordinary roots' are quickly passed over for the more sought-after early peas, especially mange-tout 'to bee eaten wth theyre shales'. Wall fruit, as well as a range of exotic fruit, would sell well, both in the open market and to the gentry at Court, as would roses,

8   John Gerard, *Herball*, 1633, p. 232.
9   BL, SL 2189, ff. 158a, 185.

gillyflowers and honeysuckle. As early as the 1590s there was clearly a demand for out-of-season and exotic fruits, vegetables and flowers over and above that provided by the gardens of resident gentry. Mr Flower and his colleagues were meeting this demand.

Nothing further is known about many of the gardeners named by Plat. Judging by the information they gave, they were probably gardeners to the gentry: the ever-helpful Smith, who appears frequently (he is identified only as 'S' in *Floraes Paradise*), gave advice on such topics as planting and grafting fruit trees, over-wintering artichokes, hot-beds, layering roses, grafting carnations, and growing aniseed in England, not the concerns of an average market gardener. Goodman talked to Plat of fruit trees, keeping grapes, planting strawberries and carnations, another list of topics more likely to concern a private gardener.[10]

We know more of some who helped Plat. He sought advice from 'old Spencer' who tended the garden at the Merchant Taylors' Hall, and he also talked to 'Jefferies the Gardiner at Sheene'. Jefferies had planted a 'growing Diall', described by Plat as 'a Circle [divided] into pieces by thadvise of somme good Horologian, & then planted… with apte herbes' (clock-dials made of herbs are still planted today). Could Jefferies have been the Royal Gardener at Sheen Palace? The gardener of Mr Nicholson in Hackney advised Plat, as did the 'fine planter' Throgmorton, who was 'Sr Thos Cornwallis man'. Lord Wentworth's gardener gave advice on planting peas. Wentworth lived in Stepney close to Plat's suburban home. Wentworth and Cornwallis were cousins and Plat may have met one through the other.[11]

The occupation of nurseryman had, from the sixteenth century, a status equal to that of a gardener to the gentry. The skills required were similar and, by the eighteenth century, there was a considerable movement between the two spheres. Nurserymen, especially those who dealt in trees and exotic plants, had a large amount of capital invested in their stock and many were prosperous tradesmen (some achieved gentry status in later life). Plat consulted a tree-nurseryman from Twickenham, a village

10 Plat, *Floraes Paradise*, p. 27, numerous paragraphs are credited to 'S' – e. g. paras 53–74.
11 BL, SL 2210, ff. 12v, 24v, 25, 82, 147v, 160v–161, 187. See *ODNB* entries for Thomas, first Baron Wentworth and Sir Thomas Cornwallis.

in Middlesex up-river from London, where several nurseries were sited by the 1630s. Thus 'Maister Pointer' gave advice, predictably, on tree-cultivation. Vincent Pointer's garden housed, according to John Gerard in 1597, 'the greatest varietie of those rare Plums', and he was 'a most cunning and curious graffer and planter of all manner of rare fruites'. He died a gentleman of some wealth in 1619 and his son became a bishop. Kirwin of Newington Butts, who advised 'To have your Nursery full of stocks to graft on, sow the stampings of Crabtres which are commonly full of Kernells', may also have been a nurseryman.[12]

Many of Plat's acquaintance among the London gentry, professional men and artisans proffered gardening tips. One who has already been mentioned, John Gerard, is a figure of some importance in the history of gardening and botany (although it has recently been argued his publications have given him a prominence which has unjustly put contemporary botanists in the shade). Born in Cheshire into a gentry family, he was apprenticed to a London barber-surgeon in 1562 but became known principally for his deep knowledge of medicinal herbs and general botany. He grew what he studied, gardening on his own account as well as supervising the gardens of William Cecil, Lord Burghley at his town residence in the Strand and country mansion, Theobalds. Gerard also acted as curator for the first physic garden of the College of Physicians. His lasting legacy is *The herball or Generall historie of plantes*, first published in 1597, which, for all its faults, is a monumental work.[13]

Plat makes many references to Gerard in his notes, sometimes calling him Jarrett.[14] Plat saw in Gerard's garden at Holborn some of the rarities he grew there: potatoes, and 'a fayre young tree that grew of a pomegranate seede'. Gerard advised Plat in July 1589 on grafting trees in the bud ('yff

12 John Harvey, *Early Nurserymen*, 1974, pp. 41–2; BL, SL 2189, ff. 12, 173v, 183v; Plat, *Floraes Paradise*, p. 111.
13 Henrey, *British Botanical and Horticultural Literature before 1800*, 1975, vol. I, pp. 36–38; R.H. Jeffers, *The friends of John Gerard*, Falls Village, CT, 1967, pp. 9–10. Deborah Harkness, *The Jewel House*, 2007, pp. 50–56.
14 Plat's 'Jarrett' is John Gerard. At one point Plat writes 'Garr', crosses it through and writes Jarrett. He visited Jarrett's garden seeing pomegranate trees and potatoes growing there, both specifically mentioned as grown by Gerard in his *Herball*.

you graff a budd of an Almon tree upon a peach stock, the same will prosper exceedingly') and 'How too make dwarfe trees, whereby to have your orchard beare his fruite in the first yeere'. He also recommended a mould (compost) to grow 'whotte simples' (exotic medicinal plants), told how to graft carnations of various colours on the same stock, and how to increase carnations from slip cuttings.[15] His garden at Holborn grew 'all the rare simples which by any meanes he could attaine unto' including 'all manner of strange trees, herbes, rootes, plants, flowers, and other such rare things, that it would make a man wonder, how one of his degree, not having the purse of a number, could ever accomplish the same'. Perhaps the answer to this question was that Gerard was an enthusiastic gardener and, as Plat observed, grew much from seeds or cuttings. The extent of his collection of plants is evident from the catalogue he published in 1596, the first ever for an English garden.[16]

Gerard may have introduced Plat to Clement King (described by Plat as Burghley's gardener), from whom Plat obtained a list of basic rules for gardening. Gerard and Plat had several other acquaintances in common, including Richard Rich of Lee in Essex, James Garret the apothecary of Lime Street who criticized Gerard's *Herball*, Vincent Pointer, mentioned above, William Gooderons or Goodrus, Serjeant-Surgeon to the Queen, and Thomas Goodman, gardener in St Botolph's. If Plat and Gerard were not close enough to be called friends, they at least moved in the same social circles.[17]

As well as consulting the gardeners of some leading gentlemen, Plat had entrée to their gardens. We have mentioned the 'floral clock' he saw

---

Finally, a surgeon colleague of Gerard's, William Clowes, in 1591 refers to him as Maister Iaret. Both Plat and Clowes must have heard Gerard, in his Cheshire accent, pronounce his name as they chose to write it. Gerard, *Herball*, pp. 927, 1450; Jeffers, *The friends of John Gerard*, p. 38; BL, SL 2210, ff. 163v, 169v.

15 BL, SL 2210, ff. 23, 163, 169v; Plat, *The second part of the garden of Eden*, 1675, pp. 136–7.

16 Gerard, *Herball*, Introduction by George Baker; John Gerard, *Catalogus arborum, fruticum ac plantarum*, 1596.

17 Jeffers, *The friends of John Gerard*, pp. 26, 30–32, 65; BL, SL 2209, ff. 5, 9v; SL 2210, ff. 23v, 118v. For more on the scientific community in London of which these men were a part see Deborah E. Harkness, *The Jewel House*, 2007.

at Sheen. He also visited the garden of Sir Francis Walsingham and his wife at Barnes Elms, admiring the dwarf apricots and peaches grown there as wall-fruit, and the vines grown against walls heated by fires.[18] Plat reports how a near neighbour, Lord Zouche at Hackney, in 1597 'removed diverse apple trees, damson trees, &c, being of thirty or forty yeeres growth' successfully transplanting them so that, a year later 'al put foorth leaves at Michaelmas after'.[19]

Much advice was gleaned from gentlemen, merchants and legal and medical professionals, seemingly in casual conversation. Gardens, both as a source of fruit and vegetables and as a mark of social status, were obviously important to many of Plat's gentle acquaintances. Medical men, Sawte the physician and Watson, a surgeon who had an old book from which Plat took information, talked of fruit-keeping and hot-beds; clerics, Parson Simpson and Doctor Fletcher, gave gardening tips (Richard Fletcher, ecclesiastic and courtier who became successively Bishop of Bristol, Worcester and London, was probably a Court acquaintance). A miscellany of traders proffered advice, such as Garret the miller, Taylor the surveyor, and Master Jacob of the glasshouse (Giacomo Verzilini, a Venetian immigrant), a keen grower of indoor plants.[20]

Of the gentlemen, we may note names associated with the Court and government. William Hill, an Auditor of the Exchequer was a keen gardener with many secrets to give Plat. One informant, identified only as 'the musk mellon gentleman', was an expert in the difficult art of melon-growing – he may have been 'Mr Fowle' Gentleman-Keeper of the Queen's House at St James, described by Gerard as 'a diligent and curious' grower who produced many musk melons for the royal table. The courtier Sir Thomas Chaloner, a keen naturalist who knew Italy well, gave Plat information about a garden pump. He, like many gentlemen who had gardening knowledge to impart, were not only courtiers but also travellers with knowledge of foreign gardening. Señor Romero, from Naples, and one of Plat's most prolific informants on many topics, gave Plat advice on keeping grapes fresh. An idea on quince growing was provided by Lord

18   BL SL 2210, ff. 23, 79v; Plat, *The second part of the garden of Eden*, 1675, pp. 1–2.
19   Plat, *Floraes Paradise*, pp. 147–8.
20   BL, SL 2210, ff. 24, 28, 78, 140v, SL 2189, f. 49v; Harkness, 2007, p. 222.

Darcy, 'brought owte of Italy', probably during a tour of classical sights. A German called English advocated hot-beds for raising seeds and Mr Dakeham, a Dorset gentleman, promised Plat some 'white italian or suger pease' which he most probably obtained while abroad.[21]

The course of political events in the British Isles brought Plat into contact with gentlemen from Ireland in the 1590s and, following King James's accession, Scotland. A Mr Andrews, 'The greate saltmaker of Ireland', told how caterpillars might be killed and spring onions raised the year-round in pots. Sir Edward Denny, adventurer at sea, soldier in Ireland and MP, told Plat that he had, in Ireland, raised liquorice in 'such grownde as by Nature is stony or rocky underneath the earth'. Denny was enthusiastic in recommending almond trees, he 'assured mee of the same uppon his owne triall, that becawse they are deinty and shadie trees, are excellent good to make walcks of for Noble Men.' Plat incorporated new advice on orchards from 'Nepper the Skott' into one of his projected publications.[22]

Although, as Plat would no doubt have acknowledged, books on gardening were badly needed in the England of his day, he had recourse to the printed page for some of the knowledge he gathered. He copied notes from the works of major Continental natural philosophers, Giambattista della Porta the Neapolitan; Girolamo Cardano from northern Italy; and Bernard Palissy, a French expert on soils. He also took horticultural notes from a collection of secrets written by Thomas Lupton in 1579 which drew on many Continental authors. Two translations attracted his attention: Barnaby Googe's of a husbandry treatise by the German Conrad Heresbach and Leonard Mascall's of a French work on fruit and nut trees. As for English authors, he looked at the herbals of William

21 Hill: BL, SL 2189, ff. 11, 143, Plat, *Floraes Paradise*, paras 21 et seq., TNA, LR1/41-45; Fowle: BL, SL 2210, ff. 161, 168, Gerard, *Herball*, 1633 p. 918; Chaloner: BL, SL 2210, f. 177, see also *ODNB*; Romero: BL, SL 2210, f. 83, he was a Spaniard but from Naples, part of Italy then Spanish-controlled; Darcy: BL, SL 2189, f. 13; Dakeham: BL, SL 2216, ff. 157-8, SL 2189, f. 185. Joan Thirsk points out the increasing interest amongst the English gentry in foreign fruit and vegetable gardening from the time of Henry VIII, *Alternative Agriculture*, 1997, pp. 31–33.
22 BL, SL 2210, ff. 28v, 38, 137v, 139v; SL 2189, f. 167. 'Nepper' may have been the Scots mathematician and inventor of logarithms, John Napier; See *ODNB*.

Turner produced earlier in the sixteenth century, and two 'modern' works, Richard Gardiner's little book on kitchen gardening of 1599, by far the most practical work in English at the time, and a small book by John Taverner, *Certaine experiments concerning fish and fruit*, of 1601.[23]

Plat was himself a gardener and from his manuscripts we may observe his familiarity with flowers, plants for borders and hedges, earths and moulds for potting, plant propagation and garden design. He knew, by experience, about the management of fruit trees, although he was eager to obtain more knowledge from his acquaintances, and he was a keen grower of indoor plants.[24]

One manuscript, BL, SL 2189, contains many gardening paragraphs written in a hand other than Plat's, many signed 'T.T.' (one of the paragraphs reproduced in *Floraes Paradise* is attributed to 'T.T. a Parson'). They appear to be the work of a Catholic priest writing in the 1550s. Whoever T.T. was, his gardening notes impressed Plat sufficiently for him to draw on them extensively.[25] We may agree with Plat's own summary of his sources on gardening that he wrote 'out of mine own particular experience, as also by long conference with diverse gentlemen of good skill and practice.'[26]

Plat's most enduring interest in gardening was that which has always preoccupied gardeners – controlling nature and defying the seasons. The grail is to grow what one likes, when one likes, where one likes. Plat's manuscripts and published works constantly pursue this end: experiments to retard flowering and fruiting or to bring crops forward. Such mastery over the vegetable world strikes a chord with the aims of alchemists – the control of nature. Alchemists tried to harness nature in enclosed glass vessels, gardeners tried to do the same in an enclosed garden or in a clay pot. Both knew that their aims were likely to be unobtainable, but they hoped with diligence, care, close study and experiment to build on the experience of others and move closer to their ultimate goals. The inclusion

23   BL, SL 2210, ff. 161, 78, 139v, 39, 69; Plat, *Floraes Paradise*, p. 109; Plat, *The second part of the garden of Eden*, pp. 47, 86, 110, 140, 148–52.
24   See below for fruit trees and indoor plants.
25   See Chapter 5 for full discussion of 'T. T.'
26   Plat, *Floraes Paradise*, 'To the Studious and well affected reader'.

of 'Paradise' and 'Eden' in the titles of his gardening books emphasizes both the enclosed experimental space where it was hoped garden advances could take place and the attempt, in gardening as in alchemy, to recover the secrets of nature believed to have been lost in the Fall of Man. Little wonder, then, that Plat was both a gardener and an alchemist.

The alchemical/gardening links are made explicit in the beginning of the first edition of *Floraes Paradise* where 16 pages are devoted to the perfection by alchemical means of a growing medium. Plat describes 'A philosophicall garden,'[27] a small brick-lined vessel containing earth carefully prepared over a period of years by an alchemical process and enriched with astral influences which would, he hoped, grow 'any Indian plant, and make all vegetables to prosper in the highest degree,& beare their fruites in England, as naturally as they doe in Spaine, Italie, or elsewhere'. He opines that miraculous plants such as the Glastonbury thorn were planted in such earth 'or with a graine or Two of the great *Elixir* applied to the roote' and he reminds his readers that the fifteenth-century English alchemist Ripley made a pear tree fruit in winter.

> Nay if the earth it selfe, after it hath thus conceived from the clowdes, were then left to bring forth her own fruits & floures in her own time, & no seeds or plants placed therin by the hand of Man, it is held very probable (unlesse for the sin of our Parents, begun in them, & so mightily increased in us, the great God of Nature, even Natura naturans, should recall, or suspend those fructifying blessings, which at the first he conferred upon his celestiall Creatures) that this heavenly earth, so manured with the starres, would bring forth such strange and glorious plants, fruits & flowers, as none of all the Herbarists that ever wrote till this day, nor any other, unlesse Adam himself were alive againe, coulde either knowe or give true & proper names unto these most admirable simples.

Failing this miraculous pre-lapsarian earth he recommends a 'Philosophers aqua vitae' which will induce fertility in barren ground. But this liquid

---

27   Significantly, an early draft of this passage in his notes is headed 'a second paradise' and Plat thinks the philosophically prepared earth 'for richness will have somme resemblance wth the first paradize', BL, SL 2189, f. 49.

must be made with a special fire 'by Nature generally offred unto all, & yet none but the children of Art have power to apprehend it: for, being celestiall, it is not easily understoode of an elementall braine'. As a true alchemist, Plat does not disclose plainly the nature of his fire, concluding that he will 'proceede to write in plaine tearmes, of such a Garden & Orchard as will better serve for common use' – launching straight into general gardening advice.[28]

These philosophical passages reveal a deep interest in the secret of soil fertility; I will develop this further in the chapter on agriculture. As far as gardening was concerned, he noted a good many 'earths', 'moulds', steeps and composts to aid growth. Some were of general application, such as the 'notable rich and cheape earth for gardens & orchards' composed of rotted or burnt fern, but most were for particular situations or plants. He wondered if steeping peas in brine, seawater, or brine made from ashes would hasten their maturity in sandy soils. 'Sweete confections' was another suggested steep.[29]

Various potting composts were suggested. Gerard advised him 'one parte of tames sande, & one parte of good earth, is fitt molde for whotte simples to grow in'. The slips of carnations and roses should be potted into one part chimney soot, one part cow dung and two parts earth, and Plat wondered if 'deinty and costly flowers' would appreciate 'a good proportion of woad well rotted amongst common earth'.[30] Of the many prescriptions for particular crops, sheep's dung soaked in water made 'Artichokes and Strawberries to grow very greate', and pigeons' dung infused in water benefited pumpkins. Various preparations were advocated to promote the growth of trees; dead cats and dogs dug in near the roots; rainwater strained through fallow earth; rotted horse and cow dung mixed with fine earth and wine lees to benefit apricots. Old vines were revived with blood laid to their roots. Some of these concoctions were passed to Plat by friendly gardeners but others were the result of his own experiments.[31]

28   Plat, *Floraes Paradise*, pp. 1–16.
29   BL, SL 2210 ff. 29, 183; SL 2244, f. 39v; SL 2189, f. 158v.
30   BL, SL 2210, ff. 23, 175; SL 2189, f. 174v.
31   Plat, *Floraes Paradise*, pp. 24, 153; BL, SL 2210, f. 78; SL 2189, f. 174; SL 2245, f. 77v.

Plat was interested in the complex problem of the use of heat to promote vegetables. He discussed the hot-bed – soil heated by organic decomposition beneath – which was entering the gardens of English gentlemen in Plat's time. The earliest description of a such a bed in an English gardening book is found in Thomas Hill's *The gardener's labyrinth* of 1577:

> The dung in a Garden plot, for the planting of young sets ought not to be couched or laid next to the roots of the plants: but in such order the dung must be used, that a thin bed of earth be first made, for the setting of the young herbs, next laid to this a handsome bed of dung, as neither too thick or thin spread on that earth: above that let another course of earth be raked over of a reasonable thick-nesse: workmanly handled and done, see that your plants be set handsomely into the ground, and in a chosen time. For the earth and beds (on such wise prepared) help that the plants bestowed shal not at all be burned: neither the heat of the dung, hastily breath forth to them.[32]

Hill drew heavily on classical authors. Diligent gentlemen reading these sources in the many cheap editions which appeared in print early in the sixteenth century may have noticed the use of hot dung described in Columella. The Emperor Tiberius was so fond of cucumbers that he had them grown in baskets of dung sheltered by sheets of transparent mica. Such gentlemen may also, like Plat, have read French editions of Estienne & Liébault's *Countrey Farm* which were available from the 1570s and found their description of melons grown with the aid of hot horse-dung and sheltered by straw mats or boards against overnight frosts.[33]

As the Roman cucumber-growers realized, covering the hot-beds with a transparent medium greatly enhanced their effectiveness. Glass was the medium used in the sixteenth century and it was the price of glass – in the form of bells or sheets to put in frames, which determined the success or failure of dung as a source of usable heat. Glass was also used in garden

---

32   Thomas Hill, *Gardener's Labyrinth*, [1652 ed.], Oxford, 1987, p. 41.
33   C. Estienne & J. Liébault, *L'Agriculture et maison rustique*, Paris, 1586; C. Estienne, *Maison rustique*, London, 1616, p. 192; David Stuart, *The kitchen garden*, 1984, p. 39.

buildings to shelter and keep plants warm. There are some records of the use of heated garden houses on the Continent in the fifteenth century, and Lord Burghley had an orangery by 1561, but glass – to let in light while retaining heat – remained a prohibitive expense despite the efforts of glass-makers to increase production of the cheaper green glass. So, Plat was writing at a time when hot-beds and glass were being cautiously introduced into the vegetable gardens of the rich.[34]

Plat was conversant with the use of hot-beds from his own experience and the advice of his gardening friends. A long manuscript passage, later reproduced in *Floraes Paradise*, describes, 'the whole manner of planting and ordering the muske mellon' beginning with a hot-bed:

> gett a loade or two of fresh horsedonge, such as is not above 7 or 10 dayes old or not exceeding 14, lay it on a heape till it have gotten a greate heate and then make a bedd therof of an ell long and half a yard broad & 18 Inches high, in somme sonny place, treading every lay downe very hard as you lay it, then lay theron 3 inches thicke of fine blacke sifted mold, prick in at every 3 or 4 inches a muske mellon seede which hath first ben steeped 24 howers in milcke, pricke the Topp of your bedd full of litle forcks of wood appearing some 4 or 5 ynches above ground, upon these lay sticks and uppon them so much straw of thicknes as may both keepe owte a reasonable shower of rayne and also the Son and defend cold (somme streyne canvas slopewise only over theyre bedds) let your seedes so rest untill they appear above ground which will bee comonly in 6 or 7 dayes.

Gourds and cucumbers were to be grown in the same way while hot dung was also used to forward peas and spring onions.[35]

He described hot-bed experiments with substances other than dung, most notably unslaked lime, and he made a passing reference to the waste tan-bark which Dutch gardeners made so much use of. He wondered if

34 Stuart, *The kitchen garden*, p. 39. By the time of Plat's death (or soon after), melons were being grown commercially, probably under glass, at the Neat House gardens in Westminster, a group of market gardeners in the fore-front of kitchen gardening in the seventeenth century. See Thick, *The Neat House Gardens*.
35 BL, SL 2189, ff. 186–187v; SL 2245, f. 92.

GARDENING

soap ashes and unslaked lime strewn on early peas and other early crops would forward them while his German acquaintance 'English' told the secret of fast-growth: 'How to make parsley, beanes, & diverse other herbes to grow the length of ones finger in one day':

> make a lay, of powdered lyme & ashes, then a lay of earth & dounge, & then lay of lyme, & uppon yt a lay of earth & dounge, then sow your seedes wch must firste bee stiped in white wyne, & water them presently, & the heate of the lyme & dounge, will make a wonderfull springe in a few howres per English per experto, Aprill 1590.[36]

The expense of glass meant that Plat's friends were instead using straw or reed mats – 'cappes or covers … of matts like Beehives' – to cover delicate plants on hot-beds in cold weather and at night. For additional protection, 'cast dounge all over your covers so as no ear [air] may enter'. Plat himself used glass in the rather curious form of 'collars' around individual plants,

> Sett glasses over your plantes, whose bottoms are broken owte, & let them have a breathing place. these glasses defend of the cold aer, encrease the heate of the sonn, & keepe the plantes moiste, becawse the water, as yt ascendeth so yt descendeth presently agayne by reson of ye smalnes of the mowthes of ye glasses.[37]

The most prevalent source of heat was, of course the sun and Plat's love of ingenuity led him to propose many ways of enhancing its efficacy, such as tiles or tin plates placed behind growing fruit to reflect the rays upon it or gravel scattered under dwarf fruit-trees to reflect them back against the trees. We have noted above an early reference to glass covers: Plat was also taken with the idea of glasses over individual tree-fruits or bunches of grapes to maximize the heat of the sun.[38]

36   BL, SL, 2210, ff. 106, 184, Plat, *The second part of the garden of Eden*, 1675, pp. 152–3.
37   BL, SL, 2210, ff. 29, 141v.
38   BL, SL 2210, ff. 79v, 81.

His gardener friends taught him the most obvious way in England to capture the maximum sun – plant trees and other garden plants against a south-facing wall – and he frequently advocated this: 'Quinces growing against a waall lying open to the Son & defended from cold windes eate most delicately this secret my L. Darcey brought owte of Italie'. Plat visited, and was impressed by, the garden of Walsingham at Barnes Elms, where he observed 'yt is good also to binde upp your dwarf trees ... close to the wall, as you doo your vines, for so I didd see both peaches and apricocks grow exceedingly well at Mr Secretaries at barnes elmes.' Observing wall fruit at Barnes Elms led Plat to consider an elaborate 'reflecting wall' for fruit:

> Sir Frauncis Walsingham useth to plante his fruit trees againste a sowth-wall that the reflexion of the [sun] may better digest the fruit. I thincke if everie tree were placed in a severall tabernacle in a brick wall wherin diverse glasses might bee sett, or els they might bee lined wth leade, that the same wold marvelously help.

On a smaller scale, he speculated that reflectors might be used for pot plants, 'Quere, if it be not good in the Summer and Spring time to place concave backs of Iron or Tin plates in every pot wherein you have planted either Dwarf-trees or Flowers, and so to remove your pots from time to time as they may best receive the reflection of the Sun, whereby to ripen them the sooner'. Such elaborate walls were built by gentlemen sometime after Plat. In the nineteenth century, 'The citrus trees in the Orange Walk at ... Combe Royal [near Kingsbridge in Devon], safeguarded in alcoves along a wall which faces south-westwards, were given added protection from the wind and frost by wooden shutters.'[39]

Pondering further the way heat might help the growth of plants, he explored extensively the use of fires to heat them. Here too, Walsingham's garden may have provided inspiration, for Plat saw what may have been the earliest heated walls in English gardens: 'my Lady Walsinghams vines

39 BL, SL 2210, ff. 23v, 79v; SL 2189, f. 13v; Plat, *The second part of the garden of Eden*, 1675, p. 48; Todd Gray, 'Walled gardens and the Cultivation of Orchard Fruit in the South-West of England', in *The Country House Kitchen Garden 1600–1950*, ed. C. Anne Wilson, 1998, p. 120.

at barnes elms be planted against the back of chimneys whose fiers doo greatly helpe the ripening of the grapes, so likewise of the Apricock wch are bounde upp to the wall wth the vines.'[40]

Plat was fortunate in having a friend, Master Jacob, who ran one of the London glasshouses, probably making the green glass which would eventually be produced cheaply enough for use in gardens. This man was an enthusiastic flower grower: 'I have knowen Maister *Jacob* of the Glass-house to have Carnations all the winter by benefit of a roome that was neere his glass house fire.'[41] Discussions with Jacob, his own knowledge of brewing on a large scale, and visits to the sites of other London industries which interested him, led Plat to the thought that there could be a symbiosis between industrial heat and gardening, involving 'all such as are forced in respect of there trade to keep any great or continual fires, as Brewers, Diers, Soap-boilers, Refiners of Sugar, and the owners of Glass-houses, and such like, who may easily convey the heat or steam of their fires (which is now utterly lost) into some private room adjoining, wherein they may bestow theire Fruit-Trees to their greater pleasure and contentment.'[42]

Plat further speculated that heat could be transferred from domestic fires to rooms set aside for garden plants and trees. In a delightful passage which also includes a novel way to heat bathwater, he writes,

> yf you hange a pott of water upon the fier wth a clos cover and a place to let a pipe of leade therin this pipe may bee conveyed thorough an other roome into the sides of a bathing tubb wth a false bottom full of holes wherin you may sitt and bath your self, you may put sweet herbs in the potte to sweeten your bath. Qre if such a pipe wold not serve to keepe a litle roome warme wherin you might place your gelliflower potts being prevented of theyre first bearing, and so likewise rose rootes being cutt

40 BL, SL 2210, ff. 79v.
41 BL, SL 2210, f. 24; SL 2189, f. 184; Plat, *Floraes Paradise*, pp. 38–9.
42 Plat, *The second part of the garden of Eden*, p. 10. In manuscript notes Plat expanded the use of waste industrial heat to include drying and keeping parsnips, 'pompions' and artichokes, drying meat for long voyages, and growing carnations, BL, SL 2210, f. 59v.

presently after theyre bearing and so have roses & gelliflowers & perhappes some other flowers when there is none to be hadd in the latter end of the yeere. [43]

Alternatively, he wondered if direct heat might applied:

qre iff a great large fonnell wth a pipe having an elboe at yt, beinge hounge over a cleere burning fier after the smoke is past, wold not serve to convey a sufficient heate into somme litle roome harde by to make a stove therof to serve for diverse purposes, for this may bee donn in everie chimney in a Mans howse. [44]

In a long note headed: 'How too keepe all things yt will grow in potts, fresh all the winter, as Carnations, Lemmons & orenge trees, especially dwarfe trees, & how to have a greene garden wthin doores. & perhapps to have such fruits growing and ripening in England, as otherwise our climate will not suffer,' Plat brought together his ideas on fires and gardens. He produced a rambling passage, reproduced in full below, in which his imagination takes over and leads to more and more fanciful ideas. What starts as a greenhouse with stove attached seems to become an outside orchard and garden with concealed underground heating. Canvas covers are advised, which Plat then proceeds to heat with steam. Insertions add dung-water to the system, and queries at the end of the passage add various new crops to those which may benefit from artificial heat. They also repeat advice on reflective plates behind potted plants and finally wonder if aqua-vite would work as fertilizer! Whatever the practicalities of these ideas, we may admire Plat's ingenuity and note the way he added to his manuscript as a single original idea led to further thoughts and speculations.

C you must plante your trees either inn great stronge bucking toobbes well piched wthin and wthout, or els ingreate stone potts wch muste bee made in .3. or .4. severall peeces & afterwarde well wyred together, sett these trees in a proportionable distaunce in a vaulte of bricke, & lett them hange

43   BL, SL 2189, f. 147v.
44   BL, SL 2210, f. 59v.

GARDENING

so, as the stemme of the warme water may play rownde abowte them. they may bee placed so cunningly as that they shall appere to stande in firme grownde iff all the toppes and sides bee covered over wth earth. There muste bee kepte a continuall fier (duringe colde weather) in a convenient furnace, over the wch there muste hange a close greate pann full of warme water, [*insert from side margin* – this warme water linnea 13. iff yt were such wherin doung had ben infused for 2 or 3 monthes perhaps yt wold woorke somme straunge effects./] In the midst of whose side there muste passe, a great large leaden pipe, to convey the stemme of the water to the rootes of all the trees, wch will bee donn, iff you make a vent at the farther ende of the valt, iuste opposite to the mowth of the furnace. This pipe may either bee shorte, and so the venthole to draw all the stemme alonge, or els farr better as I thincke, iff the pipe bee made full of litle holes. this may bee donn wth litle charge where the kichin ioynes to the orchard for so you shall always have use of your waste fire. or els yt is good to place the samme in somme litle garden howse in a fornace of bricke for the saving of fier, and so likewise to have the venthole, to breath owte into somme odd corner owte of sight, for so No man shall perceyve wch way you keepe the trees warme in winter. you may make many trialls of heates, till you finde a trew temperate heate. qre of dry fiers in this practise. you may also iff you thinke good sett upp high posts and so streyne oyled canvas both over the trees, and also square abowte them to keepe them from all violence of froste and colde weather. And iff you thinke that the warminge of them only at the rootes will not sufficiently defende them from colde, you may also convey, by a crooked peece of a pipe wch you may make to take of and on, the stemme of the water, from the other pipe, to flie upp & downe wthin the canvas, and so to keepe them warme, both abowte the boddies and armes also. Thus you may have strawberries and all manner of flowers growinge owte of their usuall seasons, and perhapps they will grow wthin doores, iff you suffer them firste too take good roote abroade, and then place them wthin your howse in a gentle stove. perhapps also yt will not bee amisse to make choice of a roome that has a very high roofe, whose sides opening sowthward, you may take away at your pleasure, to have the plants sonn & aer now & then if neede bee. Quere iff yt bee not good somme tymes to infuse the water firste in dunge 24 howres, before you put it into this balneo therby to give the greater norishement to ye trees and flowers. I wold wish this border to bee doble in plants or els to make a square border of such trees, for that I thinke the stemme of one furnace in this manner cold extend yt self very farr in keeping of a temperate heate at the rootes of ye trees or flowers./ prove this in pompions, cowcumbers; mellons.&c. …. qre iff yt bee not good in the sommer tyme to place a rownde back of yron plate at every pot that hath either flower or dwarf tree in yt, wch may bee made to remove, as the sonn doth./ qre also iff every of these potts or tubbes did stand in a tubb or pott of water not tuching the water iff so, the heate of the yron plates, together wth the moysture, wch the sonn wold draw upp from the nether tubbes, wold not wonderfully encrease the fruits or flowers. qre also if somme aqua vite were mingled in the water./[45]

Plat's stove, vaulted in brick and heated from below, may have sprung from his fertile mind or it may have been based on a Continental model. Susan Campbell has drawn attention to 'an overwintering gallery in Liège, 30 metres long [which] was heated by coal-fired hypocausts in 1635'; and by 1685 the greenhouse of the Apothecaries' garden at Chelsea was heated by underground coal fires.[46]

Discounting any speculation on artificial heat, Plat was keen on indoor gardening purely for the pleasure of having flowers and plants inside the house. The details in some of his notes provoke the thought that he is describing parts of his own residence at Bethnal Green. He reckoned it 'a most delicate & pleasing thing to have a faire gallery, great chamber or other lodging, that openeth fully upon the East or West sun, to be garnished with sweet hearbs and flowers, yea & fruit if it were possible.' For herbs he recommended sweet marjoram, basil, rosemary, for flowers in pots, carnations, gillyflowers and roses, with bay, germander, and sweet briar in the 'shadie places of the roome'. He wondered if 'Orenge or Lemon trees will not grow wthin doores planted in apte large and removeable vessels of wood'. The flowers and herbs in pots would 'stand loosely uppon faire shelves'. Window boxes of boards or lead (apparently inside the room)

[45] BL, SL 2210, f. 39v–40.
[46] Susan Campbell, 'Glasshouses and Frames', in *The Country House Kitchen Garden 1600–1950*, ed. C. Anne Wilson, 1998, p. 102.

were to be planted with herbs or flowers, 'if hearbes, you may keepe them in the shape of greene borders, or other formes'.[47]

Plat suggested a number of devices to maintain these indoor plants, including a hoist and pulley to let down pots from a chamber window to the garden below for air and moisture, and lead guttering over windows leading to pipes draining into window boxes. His pots were 'to have large loose squares in the sides; and the bottoms so made, as they might bee taken out at ones pleasure, and fastened by little holes with wiers unto their pottes, therby to give fresh earth when neede is to the roots'. Generally, 'you must often sette open your casements, especially in the day time, which would be also many in number; because flowers delight and prosper best in the open ayre'. Although he does not make it plain, I believe Plat envisaged his light and airy chamber full of flowers and plants between spring and autumn, and 'in sommer time, your chimney may be trimmed with a fine banke of moss, which may be wrought in works beeing placed in earth, or with Orpin, or the white flower called Everlasting.'[48]

Although not strictly indoors, Plat thought rooms could be perfumed and beautified with trees and plants planted outside but trained into a room. Vines, apricot and plum trees might be 'let in at som quarrels'(i.e. narrow openings) to 'run about the sides of your windowes'. More fanciful was his idea for 'An arbour wthin doores' formed as follows:

> you may lett in at somme parte of your window yt standeth aptley for yt, somme of the branches of ye vine & so let them ronn all over the roof of your roome or els make heere and there a seate, arbour, or somme other fantasticall devises wth them./ as of sweetbrier, jesomyne, privit, bayes &c.

More practical was his idea of brick or wood containers, leaded within, built against the outside of a house – with pipes inserted to water the roots of plants, containing roses, carnations, dwarf-trees or vines. These were year-round features, with canvas covers against sun and frost – a permanent window box.[49]

47   Plat, *Floraes Paradise*, pp. 31–2; BL, SL 2189, f. 184v.
48   Plat, *Floraes Paradise*, p. 36; BL, SL 2210, f. 187; SL 2189, f. 184v. 'Orpin' is the common Eurasian stonecrop, *Hylotelephium telephium*.
49   Plat, *Floraes Paradise*, p. 36; BL, SL 2210, f. 187; SL 2189, f. 184v.

Half-way between the garden and the house, so to speak, was a garden banqueting house. Banqueting – eating preserved fruits and sugared sweetmeats usually as the last course of a meal – in a specially built room in a garden or on the roof of a house was very popular with the gentry in Plat's time. Essentially a summer occupation, banqueting rooms set the diners in the midst of a garden, the sight and smells wafting in while they ate. Plat wished to go further and turn his banqueting house into a greenhouse full of flowers, plants and dwarf trees to delight the feasters. To keep them at their best the house could be turned on a central spindle to face the sun, the wooden walls and windows could be opened to let in air, and a removable roof gave access to light and allowed rain to fall upon the plants. This was to be an all-weather house, with a stove for the winter. Here is Plat's succinct description:

> How to make a Banquetting howse to have all sorts of flowers growinge fresh therin all the yeere
>
> Make choice off a place well fensed by Nature from the North & north easte, iff the grownde will give you that advantage. make your banquetting howse either rownde or square but rownde I take to bee better both in the sides, and on the topp, a thirde parte at the leaste of this howse wolde bee all glasse windowes both in the sides & the topp, you may also have a false cover, or a false linninge for these windowes. the topp wold bee made to uncover wth a pully to resceyve rayne uppon the flowers or els you shall bee driven to sett your potts abroad uppon somme shelffes rownde abowte the howse, the howse being made so in ioynts as yt you may open yt of all sides. this howse wold bee made to torne uppon a pinn as the windmilles doo, that the glasse windowes mighte always bee opposite towards the Sonn, wch in a sonny day (iff neede bee) you may turne alwayes towards the sonn, & in sommer you may alwayes turne your windowes to resceyve fresh aer. In this banquetting howse you may place all sorts of dwarf trees in convenient large potts & all other delicate kinds of flowers, & I thincke yt wilbee very good to leade the flower therof, so as the roof being open the rayne yt

50   BL, SL 2210 f. 133.

falleth may passe away by gutters. heere a good witt, & a large purse hath scope enough./ you may place a stove in frostie weather in this room.⁵⁰

Plat had much to say on the subject of pots and potting – demonstrating that he was a 'hands-on' gardener and maybe that his own garden was relatively small. He suggested making clay pots 'well glaz'd within' made in two vertical halves, with a hole in the bottom covered with a pierced lead plate to let in moisture but not pests. Such pots could be planted in the ground when appropriate and the two halves gently eased from the root-ball. The divided pots might also be used as covers for delicate plants and larger versions could be used for exotic dwarf trees – oranges, lemons, pepper-trees and pomegranates. A suggested variation was to make the pots bottomless. Instead, 'strong & double oyled Paper' would form the bottoms, the oil made bitter to discourage pests. Such pots would house deep-rooting plants (the 'Paper bottoms may very well decay and rot during the time that each plant will require for his deep rooting') planted in the soil and periodically removed for display.

He gave much thought to the mixtures of mould and earth in which to pot plants, advising gardeners to renew the mixture on occasion by removing divided pots, paring away the old soil and adding new. Plants in the pots could be watered by placing them in shallow 'baths', the water infused with mixtures of dung and other fertilizers. To catalogue a nursery of plants he proposed an idea which was either his own or borrowed from a nurseryman (or perhaps from Gerard the expert plantsman):

> I could also wish that each of the aforesaid pots should have some small holes in the lip of every pot, especially if they want ears, that thereby thin plated Lead might be fastened by smal wiers; in which leads, having your Prints for that purpose, you may strike two or three such Letters of the ABC as you shall think good, which letters you may always refer to some paper-Book, wherein you may set down in particular the name of the Seed or Plant, the ordering, the season wherein you set them, and all other circumstances whatsoever, whereby you may learn either to iterate or avoid the like practice the next time.⁵¹

51   Plat, *The second part of the garden of Eden*, pp. 28–35

Plat used pots, both indoors and in the garden, for flowers, particularly carnations, of which he was very fond. Many passages in his notes are devoted to this flower, including methods to ensure double flowers and the conceit of 'How to graff upon one root of Carnations all manner of Carnations, Gilliflowers, Pinks, etc' which he learned from John Gerard. Roses were the other flower much favoured by Plat, he gave advice on growing them early or late, at Christmas, and grafted onto strange stocks. He notes that in one year 'I didd cutt my gelliflowers the 11th of may and roses the 6 of may and they bare abowte michaelmas'.[52]

Pots were prominent in Plat's ideas on garden design, exemplified by his ideas for 'Most delicat carnations or Geliflower potts to bewtifie a gardein wthall':

> Cawse potts of 3 qtrs of a yard high, & of a good breadth to bee made in what fashion you will wth 2 eares easte & west, & 2 pipes north & south at the wch you may water your flowers let the pipes bee full of litle holes at the entring into the potts, & let your potts bee made full of holes abowte the sides, one hole distant an ynch from an other and everie hole as bigg as your thumbe or therabowte, in these holes you may sett younge plants of tyme ysopp or Lavender, and as yt groweth keepe the samme even wth cutting or you may leave somme part of the herbes longer then others to make therof diamonds, ffretts, &c. in these potts you may place lillies, Carnations &c. then you may place them uppon square pillars, also you may make your potts like flowerdeluces, rownde balls, diamonds &c. these you may place on the toppes of high pillers to the great admiration of the beholders.[53]

Nowhere does Plat set out a design for an ideal garden or spend time discussing the principles of design, but we may gather much from scattered remarks on garden features (such as the carnation pots on pillars). We may also speculate that at least some of the features were present in his own gardens, either the city garden of his early married life or that surrounding his suburban home at Bethnal Green. We have noticed his detailed design for a garden banqueting house and, given the delight he took in pleasure

52 Plat, *The second part of the garden of Eden*, pp. 26, 51, 69, 136; BL, SL 2210, f. 40.
53 BL, SL 2210, f. 78v.

gardening, I have no doubt that he had such a structure at Bishop's Hall. When processing through the garden to the banqueting house his guests could admire borders of rosemary or box gilded by painting the leaves and branches with a mixture of egg and mastic as a fixative for gold leaf. To delight them even more Plat might tempt his guests to pick and eat candied flowers in the garden:

> Make gum water as strong as for Inke, but make it with Rose-water; then wet any growing flower there-with, about ten of the clocke in a hot Sommers day and when the Sun shineth bright, bending the flower so as you may dip it all over therein, & then shake the flower well; or els you may wet the flower with a soft callaver pensill, then strewe the fine searced powder of double refined suger upon it: doe this with a little box or searce, whose bottome consisteth of an open lawne, and having also a cover on the top; holding a paper under each flower, to receive the suger that falleth by: and in three howers it wil candie, or harden upon it; & so you may bid your friends after dinner to a growing banquet.[54]

Plat's ideal pleasure garden was surrounded by a whitethorn hedge and was probably fairly small, divided into quarters. The garden was shaped as much by carpentry, wirework, brick and stone, artfully shaped clay pottery and tubs on wheels as by flowers, trees and plants. The quarters, instead of being demarcated by privet hedges, might be surrounded by 'a fence made with lath or sticks, thinly placed, and after graced with dwarf apple, and plomme trees, spred abroad upon the stick'. Alternatively, a division 'to grow in one yeere to the higth and breadth of a faire privy hedge' could be achieved, with much use of carpentry, as follows:

> yff you boorde upp earth to the breadth and higth of a privy hedge yt is of .6. or .7. yeeres groath wth bord very thick and well seasoned, and boared thorough full of holes, [added – or rather being full of longe slittes] after the earth is well settled you may plante ye topp of the border & sides likewise wth ysopp, tyme, lavender &c. or els you may plante the sides wth somme contrary plants, to make the one to sett of the other. this way

---

54 Plat, *Floraes Paradise*, pp. 27–30. The 'soft callaver pensill' was a thin paintbrush.

you may make brave borders of Carnations yff you keepe the sides cutt in woorke or fretts, and plante the Carnations on the topp of the borders. your holes, muste bee bored slopinge downwards or els yff you cutt owte square holes like a checker boord, or faire Romaine letters in poses or emblemes, in the sides of the boords, and plante them wth tyme, ysopp &c. and keepe them accordinge to the woorks, coloring and oylinge all the other part of the boords & setting yt owte in woorks, I thincke the samme will make a very dayntie shew.[55]

Within the quarters Plat recommended all manner of artificial shapes covered with herbs and flowers,

make severall stages one above another wth boordes and plante them wth what plants you thinck good, and as the plants grow upp, cutt them so as everie stage, may hide the boords of the nexte stage wth the plants wch will comme to passe in a few yeeres. and so may you make mounts or pirams wch will seeme straunge to bee made of plants that doo usually grow low. heere a good witt will finde woorke. or iff you will have your stages to laist the better then make them of bricke.

More fancifully, he describes, 'An artificiall tree or Arbor that may bee forced to grow greene wthall sorts of herbs or flowers therin in a few monethes space. so likewise of Artificiall men, bease, fowle, fishes &c.':

In the springe tyme of the yeere, cawse somme hollow substance to bee made of wood wch may have diverse branches fastened unto yt either in the ffoorme of a Tree or Arbor, bore the same full of holes sloping them downward, fill the sayd device wth earth & being well settled, sett the holes wth hearbes or flowers & keepe the same in good order wth cuttings. To cover the secrete you may hide the armes and belly wth the barck of Trees or mosse./ Pyramedes of 6 or 8 square & of a greate higth wold seeme strange, beinge all growne over wth thime or ysopp./ This way may you have lions, doggs, beares, bulles men and fishes and fowles being made hollow wthin, & well piched wthin and wthout, & then filled with the beste fatt mold you can gett, & after the mold is well settled then plant

55 Plat, *Floraes Paradise*, p. 63; BL, SL 2210, f. 38v.

therin ut ante, such parts of the beaste or fowle as are too litle to bee made hollow, may bee made of wood, wier, &c. & paynted. they may also bee made to goo uppon wheeles to remove them wherever you please. there may bee also pipes of leade conveyed thorough the boddies of the beasts, wch muste bee stoppte at the ends and have diverse litle holes in them, wherby water may bee conveyed wth a fonnell into the pipe and by the pipe, unto every parte of the earth. A good witte hath heere a large feild to walk in.[56]

Note the recurrence of the recommendation of these ideas to 'a good witt'. Plat revelled in the ingenuity needed to construct these garden conceits. In other notes he describes artificial borders (with holes for growing plants) made from wood and lead, square tubs on wheels for dwarf trees, artificial green arbours in winter, knots and mazes of barberries and the 'growing Diall' he saw at Sheen. The foreign traveller Sir Thomas Chaloner brought Plat news of a trick fountain which he had probably seen in a Continental garden: 'make the pipe of your pumpe to rise upp iuste under the pumpe, & make a false pipe a prety way off to desceyve those wch comme into your gardein & so they shall wett themselves exceedingly.'[57]

Giving due allowance to Plat's love of 'conceits', his ideas on garden design were very much of his time. The artificial banks and borders, arbours, trellis and woodwork, beasts, pyramids and other artificial structures would be familiar to the illustrator of Hill's *The Gardener's Labyrinth* of 1586, or William Lawson's *New Orchard and Garden* of 1618. They also would have been familiar to his Queen: a Swiss visitor to England in 1599 saw at the royal palace at Nonsuch 'a maze or labyrinth surrounded by high shrubberies to prevent one from passing over or through them. In the pleasure gardens are charming terraces and all kinds of animals – dogs, hares, all overgrown with plants, most artfully set out, so that from a distance, one would take them for real ones.' At Hampton Court he saw similar sights, 'There were all manner of shapes, men and women, half men and half horse, sirens, serving-maids with baskets, French lilies and delicate crenulations all round made from dry twigs bound together

56   BL, SL 2210, ff. 38a, 77.
57   Plat, *Floraes Paradise*, pp. 68–70; BL, SL 2210, ff. 140, 161, 176.

and the aforesaid evergreen quick-set shrubs, or entirely of rosemary, all true to life and so cleverly and amusingly interwoven, mingled and grown together, trimmed and arranged picture-wise, that their equal would be difficult to find.'[58]

Plat's remarks on garden design are practical. He tells us how to make the banks, borders, trellises, knots, mazes, fountains, and arbours, as well as the artificial shapes – emblems, mounts, pyramids, fake trees, beasts and men to be placed in the garden. But he must have been aware, as a literate courtier, of the aesthetics which would control the disposition of these features, creating perspective or theatrical effects, and the symbolism in the choice of artificial shapes and creatures to adorn the garden. His visits to royal and noblemen's gardens in the London area would have provided him with first-hand experience of the importance of gardens as cultural statements.[59]

Much of Plat's garden writing was taken up with trees, predominantly fruit trees. *Floraes Paradise* is divided into two sections, the largest of which, entitled 'Secrets in the ordering of Trees and plants', contains advice on fruit trees, pruning and grafting, and grape vines. Fruit trees and vines occupy most of the posthumous *The second part of the garden of Eden* and, as one might expect, his manuscript notes (on which the books were based) have a great deal to say on the subject.

His interest in fruit trees reflects a trend amongst gentlemen which started at least a couple of generations earlier. In 1525 Fitzherbert 'spoke of propagation of apples, pears, cherries, damsons and plums, using crab seedlings as apple rootstocks, pears for pears.' A few years later, Henry VIII, an avid admirer of Continental tastes in food and gardens, set up an orchard in Kent growing many different types of apples, pears, and cherries. Leonard Mascall, a Sussex gentleman, is said to have imported the first apple Pippins from France in 1525. Introducing a book of

---

58 Thomas Platter, quoted in Laurence Fleming and Alan Gore, *The English Garden*, 1979, p. 34.
59 The whole of *Garden History*, vol. 27, I, 1999 is devoted to Tudor gardens and greatly amplifies my remarks here, but see particularly within it Elisabeth Woodhouse, 'The spirit of the garden', pp. 10–31 and 'Kenilworth, the earl of Leicester's pleasure grounds following Robert Laneham's letter', pp. 127–144.

instructions on tree culture translated from French and Dutch originals in 1572, he commended gentlemen who 'hath given themselves to Planting and Graffing', and praised other travellers who 'hath replenished this our realme with divers straunge Plants, herbes, and Trees'. He recognized that some parts of the country were now 'greatly commended and praysed' for their fruit growing.[60]

Commercial nurseries near London were starting to cater for the gentry's demand for fruit trees, although some gardeners were supplying fruit trees to royal and noble gardens in London as early as the thirteenth century. John Gerard in 1597 mentioned three suppliers of such trees near the capital. One, Henry Banbury, ran a business in Westminster founded by his father which had supplied trees since before 1560. Another was Mr Warner of Horsley Down in Bermondsey, an area of early commercial gardening, and the third was known also to Plat, Richard Pointer, from Twickenham, an area of specialized tree nurseries, particularly fruit trees, which expanded rapidly in the early seventeenth century. The London nurserymen soon learnt how to pack trees securely for long-distance travel, and supplied gentlemen in many parts of the country. Other customers relied on local gardeners. In the West Country, far from the influence of the pioneer London tree-nurserymen, the Court, and energetic gardeners like Sir Hugh, we find enthusiastic fruit growers. John Hooker of Exeter in 1600 wrote that many varieties of tree-fruits were grown in Devon, and Richard Carew said the same of Cornwall in 1602, writing that the gentry 'step not far behind those of other parts, many of them conceiving like delight to graft and plant'.[61]

Significantly, a textbook for children on French conversation from the 1570s includes, in a section on dining, the reply to a visitor enquiring whether a London gentleman has finished dinner that, 'Hee is not at the Fruit yet'. When this course was due, the gentleman commands, 'Now serve the fruite: laye here those rosted paires, and the scrapped cheese:

60 Joan Thirsk, *Alternative Agriculture*, Oxford, 1997, p. 31; F.A. Roach, *Cultivated Fruits of Britain*, 1985, pp. 34–35; Leonard Mascall, *A Booke of the Arte and Maner How to Plant and Graffe all Sorts of Trees*, 1572, The Epistle.
61 John Harvey, *Early Nurserymen*, 1974, pp. 40–42; Todd Gray, 'Walled gardens', p. 116.

set those appels lower: they bee pepins, as it seemeth unto me: did you ever see fearer pepins?'. A guest bets that they come from Normandy and is told 'No no, it is growen in England'. A rival French conversation book also has a dining scene where the lady of the house serves a guest. 'Come on, let me give you some of this Quince pye, of this Tarte of Almonds, of that of Cherrie, of Gooseberries, of Prunes.' Plat's own cookery book, *Delightes For Ladies*, has many recipes involving fruit and nuts including cherries, damsons, strawberries, apples, grapes, oranges, lemons, almonds, mulberries, quinces, pears, pomegranates, walnuts.[62]

The most important activities in growing fruit trees are planting and pruning. In most cases (apples, pears, and most stone fruit), planting involves grafting – fixing securely part of the stem (a scion) from an already established tree onto the truncated stock of a growing tree, usually of the same genus. The grafted scions benefit from a developed root system and the rootstock may confer beneficial characteristics, such as disease resistance, hardiness, reaction to soil conditions or dwarf stature to the resulting mature tree. Various methods are used to graft the scion to a stock, the object is to ensure that no disease enters the wounds made in the tree, and that the cambium cells between the bark and wood of stock and scion unite to secure the two parts of the graft. Some fruit trees can be propagated from seed, others from layering, or cuttings.[63]

Plat amassed a body of knowledge on grafting, and, whilst some of his ideas would not find favour with modern gardeners, much of his advice is sound. Here are his notes describing propagation by T-budding, headed,

> How to graffe in the budde all manner of fruites
>
> In the smooth parte of the stocke wheruppon you meane to graffe, you muste firste slitte the barke abowt half an ynch overthwartewise, then slitte the barke also downward from the midste of the overthwart slitte, somewhat more then an inche into the wch convey your budd wth the leaf

---

[62] M. St Clare Byrne, *The Elizabethan Home*, 1949, pp. 28, 30, 70; Hugh Plat, *Delightes for Ladies*, 1602, The Table.

[63] *Royal Horticultural Society Encyclopedia of Gardening*, 1992, ed. Christopher Brickell, pp. 372–4, 541–2.

at yt, the leaf beinge cropte of at the ende and sides, binde the barke of the stocke abowte the budd, wth such bands as they use to binde brawne wth [*side margin* – & close upp the ioynt wth tempered lomme & mosse./] everie moneth you muste loosen the bande, becawse the budd will swell, & then you muste binde the same againe more easely wth a new bande. Note that in the gatheringe of your budd, you muste bee carefull that you hurte not the budd in the inner side of the barcke. when you gather your budd, you must first make an overthwarte slitt above the leaf wch leaf muste have a faire budd by it, then slitt the barck on either side of the leaf, and make the side slitts too meete in foorme of a scucheon in the base pointe. [*crossed through* – when ye budde hath taken sufficiently, then you may cut of the budds of the tree, on the barke side of the budd, somewhat neere the budd,] slopinge the samme upwarde, unlesse you will have the self samme tree to beare severall fruits. yt is good a litle after midsommer too graf in this manner.[64]

He went into similar detail on pruning fruit trees:

In pruning of your trees or of any other shrubb or plante bearing fruit always have respect whether it beare his fruit uppon the first, second, or third yeeres sprowts, for you must never cut away all the bearing sprowts, for so you prevent the fruitfulnes of your trees. and in pippins the third yeere sprowt doth onlie beare fruit, and in somme other fruit trees only the second yeere sprowts, in gooseberries, the last yeere sprowtes beare most, and the second yeere do beare somme also, and therfore somme hold, that if you take away the last yeeres sprowts from them or the greatest parte of them, that so the former yeeres sprowtes will beare much bigger gooseberries. you may easelie know which yeeres sprowts doo beare in every kind of fruit, by observing them in their times of bearinge.[65]

Plat leant heavily on expert advice when compiling notes on fruit trees – notably from John Gerard. This was particularly valuable because Plat could often see for himself the results of the master's work: figs propagated

64 BL, SL 2210, f. 35v.
65 BL, SL 2189, ff. 11–11v.

from seed 'wherof somme hee shewed mee growing in his gardein in holborne'; an almond tree grafted in the bud 'will prove exceedinge well as I myself can witness by the one wch I saw in Jarretts gardein'; an almond grafted upon a peach stock 'I didd see proved in Jarretts gardein.' Plat's other friends such as Auditor Hill also took a keen interest in fruit trees and shared advice with him.[66]

Dwarf fruit trees were an especial favourite. Developed in medieval times by French monks, Susan Campbell credits Protestant refugee gardeners from the Low Countries with their introduction into England. Plat may have learned secrets from refugee gardeners of his acquaintance. Dwarfing was effected by grafting scions onto stocks which would promote low growth, although Plat also seemed to think it could be achieved by ring-barking, grafting onto willow stocks or layering. He saw the latter performed by John Gerard:

> you muste have store of potts these potts may also consist of 2 parts & bee slitt thorouge the midst wherof everie one muste have a hole in the bottome, thorough wch draw severall bowes of such trees as you meane to plante, then fill the potts wth earth, and so the bowes will roote in the potts wth earth, and so the bowes will roote in the potts in one yeere wch you may remove and plant in your orchard. bee sure to beare upp the potts, so as the weight of them doo not hurte your trees. this I didd see performed in Jarretts garden in July.1589. this way you may make infinite store of Roots of Trees in one yeere, wch iff you plant againste a sowth wall, bindinge them upp, as wee use too doo in vines they will ripen very speedely. thus you may have trees growinge in great potts or tubbes which you may defend from weather at your pleasure.[67]

For Plat one of the chief delights of dwarf fruit trees was that their size allowed them to be managed in sophisticated ways. He envisaged a dwarf orchard contained in a small brick-walled enclosure, well sheltered from the wind to encourage early fruit. Further encouragement would be given by paved or gravelled walks to reflect the sun. As the trees were

66 BL, SL 2210, ff. 23, 23v, 35v, 38v; SL 2189, f. 169v.
67 BL, SL 2210, ff. 23, 35, 142; Susan Campbell, *Charleston Kedding*, pp. 62–3.

small, at blossom time they could be covered with coarse canvas 'such as Painters use' if there was a danger of frost. The same canvas could be used to 'backward' apricots, 'after they be knit, if you wold have them to beare late when all other trees of that kinde have done bearing.' The origin of the 'backwarding' idea is explained in a frequently quoted passage from *Floraes Paradise*:

> a pretty conceit of that delicate knight, Sir Francis Carew; who, for the better accomplishment of his royall entertainment of our late Queene of happy memory, at his house in Beddington, led her Maiestie to a Cherrie tree, whose fruite hee had of purpose kept backe from ripening, at the least one month after all Cherries had taken their farewell of England. This secret he performed, by straining a Tent or cover of canvas over the whole tree, and wetting the same now & then with a scoope or horne, as the heat of the weather required.[68]

Dwarf trees were even more manageable if they were planted in great pots of stone or vessels of wood, for then 'you may remove them if you see cause, and so preserve them from all injury of the weather'. Placing the containers on pavements or gravel maximized the reflected sun they might receive. Recognizing that they might get parched, Plat recommended an automatic watering system 'by way of filtration out of apt vessels placed for the purpose'.

> But if to have early fruit, we do neither regard labour nor charge, then let us build a square and close room, having many degrees of shelves, one above another, in which we may aptly place so many of these Dwarf-trees as we shall think good; in time of cold weather, we may keep the same warm in the nature of a Stove, with a small fire being made in such Furnaces, and in such manner as I will at all times be willing to shew to such as are willing to make any use therof; and if the weather be fair and open, and that room be made full of windows or open sides, we may for such time use the benefit of the Sun-shine, or carry them abroad at our pleasure.

68 BL, SL 2189, f. 185; Plat, *Floraes Paradise*, pp. 40–42, 73–5.

Plat implies that his own garden contained such a green house. Dwarfing was particularly useful for exotics – oranges, lemons, and pomegranates which needed to be potted and moved into stoves in winter-time.[69]

Grapes, both for the table and for wine, demanded careful culture in the variable English weather. Plat provides a good deal of practical advice on viticulture: shallow-ploughing a vineyard to keep down grass; recovering old vines with a fertilizer of blood; applying turpentine to cuts to stop vines bleeding; nipping off new sprigs after 'the grapes are knit' to improve quality; planting layered vine cuttings to have a harvest of grapes in the first year of a vineyard. Much of this was based on Continental experience which Plat tried to adapt to English conditions:

> qre if vines planted in England and kepte after the French manner low, wth their leaffes and sprouts nipte away wherby the sun shall shine as whott in England upon the grapes as yt can possiblie in ffrance upon them being shadowed wth theyre leaffes, iff so wee may not make very good wyne of them in England.

Earthing-up vines to protect against frost was a tip passed on by Plat's Spanish/Neapolitan acquaintance, 'this of Sr Romero for a French practize'.[70]

Table grapes were much prized in Plat's time and he has several suggestions for keeping dessert grapes fresh through the winter – he was especially keen to have them fresh for the Christmas table. To keep grapes fresh on the vine he advocated layering a branch with grapes growing on it into a pot, earthing up vines in winter up to the bunches and covering these with straw, or a more complex method which he found in the works of Cardano: 'Put in the bunches after they are knitt into greate & apte glasses having two mowthes or holes, thone opposite to thother, wherby both the water, & the parching sonn may have issue. And when you feare frosts, you may stopp upp the ends clos.' To preserve grapes off the vine he suggested gently drying them in bunches in a warm room or carefully

69 Plat, *The second part of the garden of Eden*, pp. 9–10; BL, SL 2210, f. 188.
70 Plat, *Floraes Paradise*, pp. 144–5, 149, Plat, *The second part of the garden of Eden*, pp. 15–16, 142–3; BL, SL 2210, ff. 141v, 142.

packing them in sand. He was advised to hang a vine-cutting with a bunch of grapes attached in a cool place with an apple stuck on both ends of the vine, changing the apples every three days.[71]

English gentlemen such as Plat consumed most of their grapes in the form of wine, almost all of it imported. Plat sought to encourage English winemaking, on grounds of import substitution and patriotism, 'that we shall not need either to be beholding to the *French-men* our doubtful friends, or to the *Spaniards* our assured enemies'. Drawing on Barnaby Googe's translation of the *Four Books of Husbandry*, Plat pointed out the flourishing vineyards in Germany on the same latitudes as England, and from Holinshed he noted that vineyards had been common in England in the past. The present lack of vineyards he put down to ignorance, 'the extreme negligence and blockish ignorance of our people, who do most unjustly lay their wrongful accusations upon the soil, whereas the greatest, if not the whole fault justly may be removed upon themselves.' He accused English gardeners of not realizing that vines had to be carefully tended, especially in our climate. He suggested careful reading of Googe on the subject, supplemented by the advice of 'an experienced French Gardner' plus that of an English gardener on special techniques needed to cope with the English climate. Significantly, when estimating the yield from an acre of vineyard, Plat had to rely on Cato, Varo, Columella and Seneca because he had no English data to follow. Finally, Plat drew on his own winemaking experience to provide recipes for bolstering grape juice from unripe English grapes by adding raisins or currants before wine-making, or fortifying the wine with boiled and concentrated grape-juice.[72]

I conclude this chapter conscious that Plat wrote much on gardening which has been but lightly covered. He was fond of strawberries and passed on several hints on their growth. His gardener friends gave him some sound advice on growing kitchen-garden stuff – for instance, grow alternate rows of beans and peas and let the peas climb up the bean stalks, or sow spinach and radish every month on the wane of the moon before midsummer to have them in succession, then sow on the midsummer wane for the winter. He was interested in the potential of a new source of food,

71 Plat, *Floraes Paradise*, pp. 87, 93; BL, SL 2210, ff. 81, 83, 161, 185v.
72 Plat, *The second part of the garden of Eden*, pp. 56–68.

potatoes: 'I have seene of theise rootes grow in [England – *crossed through*] holborne as Mr Jarrett assured mee in his owne garden, qre iff the same will grow of seedes for that weere a proffitable cooarse.' He experimented in his own garden with exotic herbs and spices: 'I sowed Anniseeds and Fenigreek the 26 of March, 1594 and they prospered exceedingly well … I do find by experience that Anniseeds and Fenigreek delight in ground that is enriched with Sope-ashes.' 'In March 1595 I sowed English Wormseed… in ground enriched with horn, and it grew very ranck, and full of blossoms'. Plat is here describing his own efforts, not those of servants he directed. He would have sown the aniseed and fenugreek earlier, he tells us, 'but that the beginning of March was so showring, that I could not garden any sooner'.[73]

In pursuing his gardening, Plat had, on occasion, to buy seeds commercially in London markets and his disappointment when only one in a hundred of artichokes grown from bought seed came up drew from him a fine stream of abuse against London seed-sellers and petty traders generally:

> [it] is the ordinary practice in these days, with all such as follow that way, either to deliver the seeds they sell mingled with such as are old and withered, or else without any mingling at all to sell such as are stark naught. I would there were some fit punishment devised for these petit cozeners, by whose means many poor men in England, do oftentimes lose, not only the charge of their seed, but the whole use and benefit of their ground, after they have bestowed the best part of their wealth upon it. Cheapside is as full of these lying and forswearing Huswives as the Shambles and Gracechurch street are of that shameless crew of Poulterers wives, who both daily & most damnably; yea upon the Sabbath day itself, run headlong into willful perjury, almost in every bargain which they make … maintaining their sales, with such bold countenances, and cutting speeches, with such knavish practices, and such forlorn Consciences, as they have both driven honest Matrons from their stalls, and so corrupted a number of young Maiden Servants with their bold and lewd lying, with

[73] BL, SL 2210, f. 169v; *Floraes Paradise*, p. 85; *The second part of the garden of Eden*, pp. 133–5.

GARDENING

their desperate swearing and forswearing, that they have made all plain and modest speech, yea all kind of Christianity to seem base and rustical unto them. I would inveigh more bitterly against this sin, if my text would bear it; but now I will leave it unto the several Preachers of the Parishes where they dwel, who can present this matter more sharply, and with less offence than I may; I pray God that either by them, or by the Magistrates, or by one means or other, this great dishonour of God and of Religion may be speedily removed amongst us.[74]

74   *The second part of the garden of Eden*, pp. 130–133.

CHAPTER THREE

Agriculture

Plat's manuscript notes and publications on agriculture reflect the importance of food production in the England of his day. He, like almost all gentry, had an interest in farming as a landowner and he was at some time directly engaged in husbandry on land which he had in hand. He met many gentlemen from all over the country who came to London on business or pleasure and talked to him about farming in their native regions. Although a city-dweller, he was therefore knowledgeable about agriculture and well aware of the constantly growing demand for all types of agricultural produce for the ever-expanding capital. He also appreciated the success or failure of farmers in meeting this demand – as reflected in prices in the London markets. He observed the agricultural scene in the early 1580s, when grain supplies were more plentiful, prices fell, and the government was alarmed by the quantities of starch made from grain and the enthusiastic adoption by some farmers of woad-growing to increase their profits. In the middle of that decade he probably shared the concern of many about the return of poor harvests and high grain prices. Shortly after he published the *Jewell House of Art and Nature*, which contained a section on agriculture (mainly a discussion of soil fertility), England was faced with traumatic years of severe shortage between 1594 and 1598, when harvests failed year after year. His response was a detailed, if hastily produced, pamphlet on ways to alleviate hunger. Another pamphlet, in 1600, dealt with setting corn by hand, and digging as opposed to ploughing the land, as well as with manures and fertilizers. Plat's notes on agriculture reflect, as well as his own enthusiasm for the subject, a generally increasing interest in agricultural innovation by gentlemen willing to debate, research written sources, and use their home farms for practical experiments. The results of these endeavours were seen in a growing diversity in English agriculture – new crops and new types and breeds of livestock were introduced and new methods of working the land tried out.[1]

Plat's later reputation as a writer on agricultural matters rests in large measure on his publications on soil fertility. He covers, in published works and manuscripts, the theory of soil fertility, current practices in manuring and fertilizing – as gathered from acquaintances who told him what husbandmen did in their own areas of England – and new fertilizers he has discovered or invented. Neither John Fitzherbert nor Thomas Tusser, the only English agricultural writers of note who preceded him, dealt with these matters in such depth. Fitzherbert relied heavily on classical sources, acknowledging his debt to Virgil. The most important Continental works to be translated into English in the sixteenth century, *Maison Rustique* of Estienne and Liébault and the *Foure Books of Husbandry* of Conrad Heresbach, paraphrase the Roman writer Varro on the merits of various dungs. Varro, Virgil and Columella pervade these books, which the English historian George Fussell described as bringing 'the classical tradition to the English literature of farming'.[2]

On the strength of his *Diverse new sorts of Soyle not yet brought into any publique use* (1594), Fussell credits Plat with 'some sort of breakaway from the classical tradition, and the small beginning of independent and scientific, or pseudo-scientific thinking.'[3] Plat had an interest, through his alchemy and, more specifically, from his reading of the works of Paracelsus, in chemistry and the possibility that chemistry might contribute to the progress of industry and agriculture. Apart from some material in *Floraes Paradise* and a few paragraphs in other publications, what he published on soil fertility is contained in *Diverse new sorts of Soyle*, a 60-page section of

---

1   Joan Thirsk, *Alternative Agriculture*, 1997, pp. 23–4, 27–29.
2   John Fitzherbert, *Booke of Husbandry*, [1598 ed.], pp. 28–9; Marcus Terentuis Varro, *On Agriculture*, Loeb Classical Library, 1967, pp. 263–5; C. Estienne, *Maison Rustique*, 1616, pp. 534–7; Gervase Markham, *The Whole Art of Husbandry*, 1631, pp. 32–3.
3   George Fussell, *The Classical Tradition in Western European Farming*, 1972, p. 112.

The Jewell House which was re-issued as a separate pamphlet in 1594 (see the illustrated title-page, figure 1, above). This was surely a sign that it had generated some interest among the book-buying gentry.[4]

Almost all the theory contained in the work is derived from Continental writers: some material is quoted from Franciscus Valetius' *de sacra Philosophia*, but Plat relies most heavily on the works of Bernard Palissy, a French potter. At the outset, Plat mentions Palissy's *Discours admirables de la nature des eaux et fontaines* and proceeds to translate almost the whole of his *des sels diverses* followed by more selective extracts from *de la marn*. Plat quotes with acknowledgement (and apparently with approval), telling us when he does not agree with the original author. As Alan Debus points out, much of Palissy's theory is in turn underpinned by the theories of Paracelsus. Plat's object in translating Palissy is to enlighten gentlemen who are 'true infantes of Art' that they may receive 'a full light into nature, which dooth heere present herself in all hir royaltie, with her Cornucopia in her hande'. Only as an afterthought, he adds that 'the ignoraunt Farmers may also gleane with them, a fewe lose and scattered eares, to make so much breade of, as may relieve their hungrie bellyes'.[5]

Palissy believed there was an essential element which gave fertility, life and structure to animate and inanimate objects. This element he called a salt. He wrote that there were vast numbers of salts each of which animated different substances. Without its salt, the animal, vegetable, stone, etc., would disintegrate. He gives examples of salts which can be isolated and used by man – the salt in tanbark tans leather, that in ashes whitens cloth. When these salts have been extracted by the tanning and whitening processes, the bark and ashes are spent and useless, having been deprived of their essential ingredients. Turning specifically to fertility, Palissy thought that there was a 'salt of the earth' which embodied fertility. Without dunging, crops exhausted the soil. Land must then be dunged or lie fallow 'to the ende that it may gather a newe saltnesse from the cloudes and the raine that falleth upon it. But I speake not here of common salt,

---

4   Plat, *Diverse new sorts of Soyle*, 1594; My comments on soil fertility lean heavily on A.G. Debus, 'Palissy, Plat, and English Agricultural Chemistry in the 16th and 17th Centuries', *Archives Internationales d'Histoire des Sciences*, 21, 1968.

5   Plat, *Diverse new sorts of Soyle*, p. 9; Debus, 'Palissy, Plat', pp. 69, 72, 74, 76.

but of vegetative salt.' He explained 'it is apparaunt, that no dungue, which is layde uppon barraine groundes, coulde anie way enrich the same, if it were not for the salt which the strawe and hay left behinde them by their putrifaction'. Thus, where dung heaps are open to the rain, the salt leaches out and the best grass is seen to grow on spots where the heaps were piled.[6]

From the treatise on marl, Plat extracted Palissy's idea that the secret of this type of earth was a fifth element, 'which is a generative water, cleer, subtile, mingled inseparably with other waters, which water being also brought among common waters, doth indurate and congeale itself with such thinges as doo happen to bee mixed with it.' The fifth element enters what was originally simple earth, congeals there after ordinary water has gone, and turns this earth into marl. With the aid of ordinary water this element is conveyed from marl which has been spread on sown land to the seeds and enables them to grow. Palissy wrote that this fifth element was an essential element in men, beasts, trees and plants, indicating that it was, if I am not mistaken, the same as his 'salt'.[7]

So far, we have emphasized the derivative nature of Plat's writing on fertilizer. It is when he leaves theory for practice that he shows originality. He describes a great many types of fertilizer, providing details on local practices in parts of England, exhorting gentlemen to experiment with fertilizers, describing his experiments on the essential element governing fertility, and advertising his own fertilizer produced from a secret formula. He tried to build on Palissy's idea of an essential salt, using alchemical means to produce 'an excellent way to enrich ground':

> C gather great store of Sal terrae owt of the aptest groundes wth aqua celestis, dissolve it in aqua coelestis & water therwth [*marginal note* – or imbibe seedes therin] any plant or flower, but this is most excellent to water your plants that are planted first in a philosophicall Earth. but if this salt be imbibed in great quantity wth aqua celestis for 6 months & so be the first earth you work in, then you shall have a strange Earth for fructification.[8]

6   Plat, *Diverse new sorts of Soyle*, pp. 11–16.
7   Plat, *Diverse new sorts of Soyle*, p. 25; Debus, 'Palissy, Plat', p. 78.
8   BL, SL 2245, f. 69v. The capital 'C' at the beginning of this paragraph is a device frequently employed by Plat. It is a *pilcrow*, defined by the OED as: 'A symbol marking the start of a paragraph'.

To capture the congelative water of Palissy he suggested: 'evaporate away good store of thexhalative part of rayne water & wth the congelative part that remayneth water your plants.... or els make a litle square pond of bricke in your garden to restreyne the rayne water that falleth in the wch infuse all the herbes wch you cutt downe & water therwth.' He added a practical question: 'qre iff this water will not stincke exceedingly iff yt stand longe.'[9]

A more ambitious attempt to collect this water was by a gentle distillation, utilizing the earth's heat, a practice Plat used for alchemical purposes.

> C make a hole as in somme gardein or secret place as depe as a pitfaall, and so large onlie as a greate pewter still head may well cover set the same in the evening and take such water as you shall finde in the morning to have fallen into the resciever before the son rise (qre of doing this in a sonny day.) doo this in the springe time. and you shall have greate store of water. qre of the strength of this water for fructification, & whether the earth spend his strength this way. even as a man by opening of a vayne may bleede to death.[10]

The only experiment Plat published is the mysterious first section of *Floraes Paradise* headed 'A Philosophicall Garden':

> First, pave a square plot with bricke (and if it be covered with plaister of paris, it is so much the better) making up sides of bricke also plaistered likewise: let this bee of a convenient depth, fill it with the best vegetable [*Saturne*] which you can get, that hath stoode two yeeres, or one at the least, quiet within his own Sphear: make contrition of the same, and be sure to avoide all obstructions, imbibe it with Aqua coelistis in a true proportion, grinde it once a day till it bee dry: being dry, let it stand two or 3. daies without any imbibition, that it may the better attract from all the heavenlie influences, continuing then also a philosophicall contrition everie day (this grinding must also bee used in the vegetable worke where

9   BL, SL 2210, f. 182v.
10 BL, SL 2189, f. 10.

the [*Mercury*] of hearbes is used insteade of aqua coelistis) during all the time of preparation: then plant what rare flowers, fruites, or seedes you please therein. And (if my Theorie of Nature decieve me not) this [*Saturne*] so enriched from the heavens, without the helpe of any manner of soyle, marle, or compost (after one yeres revolution) will make the same to flourish and fructifie in a strange and admirable manner.[11]

This experiment seeks to capture the essence of fertility directly from the heavens, without the need for any medium such as rainwater, the grinding and pounding of especially prepared soil presumably laying it open as much as possible to these invisible rays. Plat would have come into contact with theories of astral influences and astral magic from his reading of Cornelius Agrippa, Paracelsus and Bruno, as well as contacts with such alchemists as Dee and Harriot. The experiment was of interest to John Evelyn later in the seventeenth century: he described a similar one in *Terra*, noting,

And to this belongs Sir Hugh Platt's contrition, or philosophical grinding of Earth, which upon this exposure alone, without manure of soil, after the like revolution of time, will, he affirms, be able to receive an exotic plant from the farthest Indies, and cause all vegetables to prosper in the most exalted degree: and, to speak magnificently with that industrious man, to bear their fruit as kindly with us as they do in their natural climates and this Dr Munting pretends to have done in Holland.[12]

Both Valetius and Palissy had spent some time discussing whether common salt had a fertilizing or devastating effect on soil but Plat chided Palissy for not concluding clearly that common salt *was* beneficial and complained that Valetius had merely opined that the exact quantity per acre was important – too little would be ineffective and too much would poison the ground. Plat uses a common-sense approach to the problem, theorizing that the four elements (air, earth, fire and water) are merely differing states of one element,

11   Plat, *Floraes Paradise*, pp. 1–3.
12   See Chapter 6, below; John Evelyn, *Terra*, 1787, p. 28.

> which being admitted is a manifeste proofe that there is a greate, and neere affinity betweene the lande, and the sea … And yet further wee see that of the earth, and water together are made one globe, so as a smalle matter will make them friendes being so neerely united together.

So, it is likely that salt water poured on land will do good, it being essentially the same element. He backs his contention with several examples: a 'Sillie swaine' who dropped a bag of grain in the sea, retrieved it and sowed the seed, harvesting a rich crop; 'him that of his owne mother witte sowed a bushell of salte long since upon a smal patche of barren grounde at Clapham which to this daye remayneth more fresh and greene, and full of swarth then all the reste of the fielde about it'; farmers in the West Country who fertilize their barren grounds with salty sand from local beaches; husbandmen at Nantwich in Cheshire who pump brackish waste water from salt mines onto nearby land and find that 'no soile or dunge is comparable unto it, for the manurance of their groundes'; and the fertility of land at Erith that was found to be the result of inundation when the sea-dykes burst.

> But to sette downe one experience that may serve for a thousand because it consisteth of nothing els but salte…. Before you sow your ground, do but only mingle two bushels of bay salte, amongest foure bushels of winter graine, and so disperse them together uppon the ground, and you shall finde a good encrease of corne and the land it selfe muche bettered, and cleered of weedes, as I have beene very credibly enformed.

(In his notes Plat recorded slightly different instructions: 'uppon the first breaking upp of your ground vz at alhallowtide, sow half a bushell or 3 pecks of salt uppon every acre let it rest and about mich [i.e. Michaelmas], plowgh again and again in april & then sow barlie & you shall find good encrease.' These directions were provided by 'Mr Nepper the Skott' who 'hath this secrete privileged in Skoteland'. Plat concludes that common salt spread on land, especially as brine or seawater, putrifies, turning to the 'vegetative salt' essential to fertility. Careful experimentation is advised to determine the benefits of salt, perhaps by a controlled inundation of land with seawater. Plat urges farmers to be cautious, first reading his treatise 'over againe and againe', then they should 'begin with small practises, and first upon arable groundes, before they proceed to pasture', using barren ground for initial trials and not trying to salt land already sown or about to be sown. Despite this advice, Plat records that he had talked to a Mr Robenson who watered his ground after it was sown with a brine of 18 or 20 water to salt strength: 'this is performed in a hogshedd drawne upon somme sledd or litle cart having one of the hedds full off small holes.'[13]

Plat followed up Palissy's remarks on marl by exhorting English farmers and gentlemen to follow Palissy's plain directions on finding and spreading marl 'seeing wee may have such wealth for the digging, let us not spare the Shovell and Mattocke, till wee have founde out some Marle-pittes in our owne demesnes. For the veynes of Marle are more in number, much longer and broader, and deeper than wee thinke ... It is a small adventure to hazarde a shilling to gaine a pounde'. He advises digging any likely veins, making trials to find the best marl, experimenting with quantities, seasons of sowing, and types of ground on which to spread it.

Sensibly, he urges husbandmen 'untill you have attayned unto the verie pricke of proportion, learne first all the experience which you can drawe from other men'. And there were 'other men' to be consulted. Back in 1523 Fitzherbert had noted:

> in Cheshire, Lankishire, and other Countreys, thay use for manure a kinde of blewe Marble-like earth, which they call Marle. This is for those Countries an excellent manure and though it be exceeding chargeable, yet through good neighbour-hood it quiteth the cost: for if you manure your grounds once in seaven or twelve yeares, it is sufficient, and looke how many yeares he beareth Corne, so many yeares he will beare grasse and that plenty.

Eric Kerridge has noted an upsurge in the use of marl in many parts of England in the sixteenth and early seventeenth centuries – Plat and other writers must have stimulated interest in marl and at the same time

---

13   Plat, *Diverse new sorts of Soyle*, pp. 38–43, 45–47; BL, SL 2189, ff. 174–174v.

reflected its growing use. Plat ends his comments on marl in *Diverse new sorts of Soyle* with another exhortation: he was aware of the high initial cost of marling and these remarks seem to be aimed at convincing doubting farmers of the long-term gains to be made,

> And therefore you must never rest till you have made a full tryall of all the inward veynes of the earth, in all the seasons of the yeare, in all degrees of proportion, in all kindes of graine, upon all sortes of grounde, with all such like necessarie circumstances, and so in the ende you shall finde out those differences, and make such observations to your selfe, as the sluggish and idle loyterers of our time (though they have the same matter to worke upon) shall never bee able to reach unto, or imitate.[14]

On the use of manures and fertilizers other than marl and salt, Plat again writes both as a reporter of practices he has observed or been told about, and as an innovator, tendering his own ideas. Joan Thirsk has speculated that the 'remarkable interest shown in the sixteenth century in using a multitude of unusual fertilizers on the land' may well have been stimulated by a sentence in Columella on the variety of vegetable matter which could be composted by farmers. Whilst Columella may have caught the eye of many gentlemen, they did not have to search hard to find contemporary examples of the diverse substances spread on the land by husbandmen to increase fertility. Even if farmers in their own locality were constrained to use only animal dung from their own farms, they would have met others with interesting tales to tell.[15]

Plat gathered ideas from gentlemen passing through London: 'Mr Rotheram of Bedfordshire comendeth malteduste wonderfully, for barley grownd but there is no store of this to bee hadd but in maltinge Townes'. A Mr Kimberley told him about burnbaking in the Midlands: 'In warwick shire they cut upp sworde of the earth into greate thin turfes, & pile the same hollow in greate heapes & when they are drie they burne them, &

---

14   Plat, *Diverse new sorts of Soyle*, pp. 29–31; John Fitzherbert, *Husbandry*, 1523, pp. 28–9; Eric Kerridge, *The Agricultural Revolution*, 1967, pp. 241–48.
15   Joan Thirsk, 'Making a fresh start', p. 22 in *Culture and Cultivation in Early Modern England*, ed. M. Leslie & T. Raylor, 1992.

wth thashes they enrich theyre barren grownd and yt lasteth a dozen yeeres fertil'.[16] Mr Wortley, a Yorkshire gentleman, described to Plat in detail the use of waste slag from ironworks as a fertilizer for wet meadows and Plat speculated that 'there is plenty of this stuff to bee hadd in Sussex, Essex, Wales &c. where there hath ben any longe continuaunce of an yron works, and those hilles that consist of this matter are worth the breaking upp though they bee auncient and have lien long uncovered.'[17]

Mr Harsley, a neighbour of Plat's in Bethnal Green, told him, 'The River of Trent in Lincolnshire is suffered once in seven years to overflow a great Marsh, whereby it carrieth as much Swarth as can stand upon the ground'. One Webb, Mrs Campion's brother, told of the use of waste woad as fertilizer, Watson the surgeon recommended spent tanners' bark, and Plat's cousin Duncombe commended steeping wheat in urine for a day prior to sowing it in sandy ground. Mr Bostock from Oxfordshire shared the results of his experiments in steeping seed-grain in liquid dung, and the same man 'hath also proved that 12 bushells of malt dust bestowed uppon an acre of ground is as good as one folding of sheepe for corne, & uppon pasture ground it lasteth 4 croppes.' (This was probably Lyonell Bostocke, an Abingdon maltster active in the 1590s.) Other gentlemen told him that farmers in the West Country brought salty sand up to five miles inland to improve their ground, while others spread waste pilchards, from which oil had been extracted, on their land.[18]

No English writer prior to Plat had brought together so many regional variations in soil fertilization. Many in the seventeenth century followed his lead. Not long after Plat's death John Norden mentions diverse practices in Sussex, Surrey, Hertfordshire, the West Country, Shropshire, Denbigh, Hampshire and near London. Others followed suit. Donald Woodward has drawn attention to the main limitation on the use of most local fertilizers – overland carriage was very expensive and useful substances were often only economic close to their sources. Plat was aware of such limitations. Introducing a variety of fertilizers in

16 BL, SL 2210, f. 55.
17 BL, SL 2189, f. 174v.
18 BL, SL 2189, f. 175; Plat, *The second part of the garden of Eden*, 1675, para 118; Plat, *Diverse new sorts of Soyle*, pp. 42, 59.

Diverse new sorts of Soyle, he cautions that although they 'be excellent in their kinde, but most of them [are] appropriate only to particular places, and some of them not to be had in any great quantity'. Later on he wrote, 'For howe cheap soever our other soile [fertilizer] bee, yet the transposing therof from place to place (if the land lie at any distance) doth make it so chargeable that the poorer sorte of farmors in many places of this realme wil scarcely aford the cariage therof to their grounds, though they might have the same freely given them'. However, publicity, albeit only at gentry level, may have spread new ideas to areas with conditions similar to the original innovators'.[19]

Plat wished every possible substance to be tried as fertilizer. His recommendations often are little more than lists of possibilities. In *Diverse new sortes of Soyle* he makes brief mentions of street soil, river and pond ooze, ferns, hair, ashes, malt dust, soap-ashes, and earth found in willow tree trunks. Chapter 6 of *The new and admirable Arte of setting of Corne* ends with short paragraphs on a variety of other fertilizers not yet touched upon: sea kelp, dregs of beer, soot, waste from saltpetre making, shavings of horn, hulls of oats shed in making oatmeal, rotted ferns, dregs from woad vats. Elsewhere 'wollen shreds' are mentioned. This comprehensive approach was adopted by later writers. Robert Child writing in *Hartlib's Legacie* in 1655 and Adolphus Speed in 1659 both introduce lists of fertilizers in a similar way, and the fervent advocate of agricultural improvement Walter Blithe devotes a chapter to many different types of fertilizer.[20]

Many of the substances Plat (and others) recommended were waste products from trades and industries based in towns and cities. As a Londoner, he appreciated that waste-disposal was a problem in large centres of population. Hair combed from hides by tanners, soot scraped from brewery chimneys, waste shaved from horn, woollen rags, spent

19 John Norden, *The Surveyor's Dialogue*, 1607; Samuel Hartlib, *Hartlib's Legacie*, 1655; Donald Woodward, 'An Essay on manures' p. 277, in *English Rural Society, 1500–1800*, ed. J. Chartres & D. Hey, Cambridge, 1990; Plat, *Diverse new sorts of Soyle*, pp. 49, 51.

20 Plat, *Diverse new sorts of Soyle*, pp. 34–5, 47–59; BL, SL 2210, f. 182v; *Hartlib's Legacie*, pp. 33–7; Adolphus Speed, *Adam in Eden*, 1659, p. 117 et seq.; Walter Blithe, *English Improver Improved*, 1655, p. 133.

woad from dyers' vats, blood from butchers, all were noxious or at least inconvenient by-products which Londoners would be well rid of, especially if farmers could be persuaded to collect them for use as fertilizers. His connections with brewing alerted him to the potential of malt dust. As well as the Mr Bostock of Abingdon noted above, his cousin Duncombe recommended '24 bushells of malt duste uppon an acre of grownde and then sow your wheat, and turne them both together within the furrow', and 'Mr Rotheram of Bedfordshire comendeth malteduste wonderfully, for barley grownd,' but cautioned, 'there is no store of this to bee hadd but in maltinge Townes'. Significantly, both Abingdon and Bedford were malting towns in Plat's day.[21]

Plat had high hopes of the fertilizing powers of another industrial by-product, soap-ashes. He writes as if this waste was not then used by farmers near London. Eric Kerridge noted, 'From about 1590 soap-ashes came into increasing use in the West Country; and not long after a certain Mr. Broughton introduced into the Vales of Hereford the practice of buying them from the Bristol soap-boilers.' Plat most probably heard of this provincial use of ashes and he was anxious to convince farmers near London of these 'further & sweeter helps for her barren groundes, then shee hath bin hitherto acquainted withal'. He was saddened, 'that I daily do see a most rich commodity trampled under foot,' introducing his readers to, 'sope ashes which our sope boilers for the most part, wil give for the cariage, and some of them, also doo pay for the cariage when they are conveyed from their houses'. Plat acknowledges, 'Flemmings to be our first teachers, in the use of them,' explaining that, until they were forced to desist by the wars in the Low Countries, Flemings took London soap-ashes away for fertilizer, paying '3.s or ten grotes a lode, besides ye carriage of them into their own countrey'. They used them as follows:

> For their manner about Bridges [Bruges] was after they had sowed the same with greine, to strowe these ashes thereon with their handes till the grounde did seem to have gathered a whitish garment upon it, and that was sufficient for that yeare, and by this practise they might sowe the

21 Plat, *Diverse new sorts of Soyle*, p. 59; Plat, *The new and admirable Arte of setting of Corne*, 1600, Chap 6; BL, SL 2210, ff. 2v, 55, 164v; SL 2189, f. 174.

ground yearly without leaving it fallow at any time; yea their ground being helped in this manner would yeald them a most rich crop of flax, whose seed of al other doth most burne, & pill the ground.

Plat claims that soap-ashes kill the rushes, weeds and worms that infest barren grounds and speculates that they might also destroy fern and broom. They fertilize and cleanse pasture, especially moist and cold grassland and he wonders if they might also be used on ground prepared for both woad and hops. Plat found the ashes better for winter than for summer corn. As an example of the efficacy of soap-ashes, *Diverse new sorts of Soyle* was illustrated by a life-sized picture of a large ear of summer barley 'drawne truely and sharpely' which, together with many others, Plat grew in barren ground at his house in Bethnal Green 'by the helpe and means of those sope ashes … to the great admiration of the beholders'. Plat also successfully experimented with this fertilizer to grow 'outlandish' seeds: aniseed, fenugreek and cumin. Another 'well experienced Citizen of London' had frequent success in using soap-ashes to fertilize artichokes. A possible problem might be a build-up of a barren crust of soap-ashes after several years, but Plat concludes that frequent rainfall will prevent this. Soap-ashes were a strong fertilizer, two loads would dress an acre and so the cost of carriage was reasonable.[22]

The secret of soap-ashes, Plat reasoned, was the salt which remained in them, even after the soap-boilers had extracted most of it for their use. This same salt made them useful in whitening linen and other types of cloth. The ashes could be utilized to scour trenchers and other wooden surfaces, and clean windows. They might also be used to surface streets and many hundred loads were used on London bowling alleys, with many more used as the basis for mortar in London houses. He was at pains to make a virtue out of this inconvenient by-product of the soap industry.[23]

Plat considered at some length the efficacy of steeping seeds – soaking them in liquid prior to sowing in order to both improve germination and impart fertilizer to the seed at the very point of germination. In 1594

22 Kerridge, p. 243; Plat, *Diverse new sorts of Soyle*, pp. 50–58; Plat, *The second part of the garden of Eden*, para. 95.
23 Plat, *Diverse new sorts of Soyle*, pp. 55–58.

AGRICULTURE

he comments that 'studious practisers' have 'confidently affirmed' that corn steeped in water and cow dung for from 12 to 36 hours, once or twice, and then sown in barren ground will prosper well. He is sceptical, explaining,

> But this kinde of practise I have heard both maintained, and impugned aswel by reason as by experience, and that by men of good judgement on both sides, although I woulde sett downe mine own experience herein, I must needes confesse I could never yet attaine to any truth in this secret, or make any apparent difference betweene the corne that was husbanded in this manner, & that which grew of itselfe without any such helpe, yet will I not (for the credit of the Reporters) altogether discredit the invention, for that peradventure I mighte faile in the nature of the graine, or in the time of imbibition.

Despite his reservations, Plat carefully sets out an experiment using a dung-water steep which produced a good crop of wheat in 'an hartlesse peece of ground, which lacked also one tilth and which no man durst adventure to sow with any grain'. Plat may have been impressed by the experiment because it was conducted by a Bishop, 'a spiritual Lord that died of late'.[24]

Like many of his fellow gentry, Plat was conversant with the work of 'that learned and poeticall Husbandman' Virgil. In the sixth chapter of *The new and admirable Arte of setting of Corne* he quotes the first book of the *Georgics* where Virgil suggests steeping (or at least mixing at the time of sowing) seed with the lees of black oil and nitre. Plat comments that these ingredients are not cheaply available in England but the passage prompts him to consider native alternatives. This time, six years after the publication of his previous remarks, he is more favourably disposed towards seed steeps.[25]

Before presenting his own ideas on steeps, Plat quotes at length from G.B. della Porta's *Magiae Naturalis*. Porta describes how to increase the yield of corn in allegorical language, summarized by Plat as 'the

---

24 Plat, *Diverse new sorts of Soyle*, pp. 35–38.
25 Plat, *The new and admirable Arte of setting of Corne*, 1600.

mysticall marriage betweene God *Bacchus* and the Goddesse *Ceres* (at the solemnization whereof onely *Vulcan* and *Luna* were present, as though neither *Saturne, Iupiter, Mars, Sol, Venus*, nor *Mercury*, nor the rest of that celestiall crewe were neither worthie to dine nor daunce at the Wedding)'. As far as Plat is concerned, Porta is simply advocating a seed steep of dung and wine. Plat thinks new ale a better steep for corn. He notes that Barnaby Googe, in his translation of Heresbach, advocates a water and saltpetre steep for beans, modifying Virgil to local circumstances. 'But,' continues Plat, 'why should we spend these costly liquors that are fitter for Tavernes & Alehouses, then for rusticall imbibitions? when as with common water and the dung of cattell, especially of Oxen, Kine, and Sheepe, or Pigeons dung (wherof more quantity, with a great deale lesse charge (being not much inferior in effects) may so easily be had & obtained)'.[26]

Acknowledging that earlier authors have advocated steeps, he points out that they have been vague about 'the strength of the liquors, the time of imbibation,[and] the proportion between the liquor & graine'. (Understandably, since most earlier authors simply quoted the vague remarks of Virgil.) Plat provides detailed advice on the strength of steeps (roughly two parts water to one of dung) and full directions as to how the steep should be prepared. 'And as for the time of imbibition, it is a rule in naturall Philosophie, that every thing hath his stomacke' so seed corn will absorb all it can bear without breaking but it is only by trial and error that the correct steeping time will be discovered. As to the ratio of steep to corn, he sensibly suggests just covering corn with the steep and topping up if need be as the liquor is absorbed. Concluding this 'true use of all imbibitions,' he again notes that steeps have 'hitherto been confirmed by some, and condemned by others, each severall man reporting according to his owne experience'. That steeps were a subject of much experimentation is confirmed in Plat's manuscripts where he frequently notes possible liquors: corn in ale, beer or wine, water and chalk, water and lime, water and sandiver (a saline liquid by-product of glassmaking), a lee of soap-ashes, soot, dung and urine. His proposals for steeps, often in combination with other ideas for fertilizing ground, are sophisticated:

26 Barnaby Googe, *Foure Bookes of Husbandry*, 1577; G.B. della Porta, *Magiae Naturalis*, Naples, 1558; Plat, *The new and admirable Arte of setting of Corne*, 1600.

How to have a rich cropp of Corn uppon a barren soile.

C infuse rayne water or other water uppon some fallow grounde that have not ben donged, let this liquor run through 2 or 3 severall earths, imbibe corn therin some 24, some 48 & som 72 howres, then whilst it is moist rowle the same in the fine powder of sleakt lime & it will gather a white coate or crust uppon it. this hath ben proved good in somme grounde but the liquor was only such as distilled from a muck heap, I take the same to be best in cold & moist grounds but prove others. qre of coating your corn wth pigeons or sheepes donge, first dried qre of searced ashes. & if it will not stick to the corn moisten the corn after it is steped and dried a litle, in somme strong gome water. prove this in beanes & pease

and:

To have a rich Cropp of Corne uppon arable groundes by infusing the greine in stale and Horsedonge, Cowdunge, Sheepes dung, pigeons dunge, &c or in rayne water that hath run through 2 or 3 severall Earthes that are good mold & have not ben lately donged. or in theinfusion of liquor uppon soote, or in a brine made wth the salt that is gotten in peterhowses wch I take to be the salt of nature or in a lee made stronge wth fearne ashes. & such like.[27]

His Oxfordshire acquaintance Bostock experimented with steeping:

Stepe barlie, wheate, or pease in the deepe collor red water that descendeth from a mucke heape or rather from an ox stall for 2 dayes & 3 nights in a tubb, drive away the water, and in one day after sow your greine in ground apte in nature for the samme, & by reason of the swelling of your corne you shall sow it much thinner, becawse half a handfull maketh almost a handfull, & you shall reape almost a doble cropp in respect of other men. and every corne will yeald a greate many eares vz.20.15.25. &30 as in settinge. probat per Mr Bostock oxfordshire, to his exceeding greate profit. beanes and rie, and otes may bee used in this manner.

27   BL, SL 2245, ff. 82v–83, 104.

In a rare glimpse of his own farming, Plat notes,

> 15 of november at night, I sowed wheat at the north end of a peece of ground that had ben stiped in a liquor wch had run through 3 earths before (wch earth had lien fallow & had not ben dunged in many yeeres before) for 48 howres, and after lay 12 howres to drie. The 16 of november in the morning I sowed wheat that had ben steped 3 dayes & three nights in an infusion uppon soote: this I sowed nigh to the former wheat and next to this I sowed more wheat prepared as in the first manner, save only the wheat lay.3. dayes & 3 nights in infusion./ as this was turned under furrow after it was sowed in Taylors ground the ground wherin I sowed was utterly spent and barren./ but the rest of the field was dunged/[28]

One gains the impression that many gentlemen tried out a variety of seed steeps in Plat's time and there was much sensible discussion of the benefits. (Early in the reign of James I, and during Plat's lifetime, an attempt was made to gain a patent on one such steep.) Maybe this enthusiasm was helped by the relative cheapness of such experiments – one bucket or barrel of steep and half a bushel of corn was enough for a trial sowing and if barren land was used, so much the better. The vague passage from Virgil, giving classical credence to the idea but imprecise instructions, may also have been a spur to literate gentlemen.[29]

On the subject of ordinary dung Plat had little to say, explaining, 'I will passe over the triviall uses of Cow-dung, Horse dung, folding of sheepe, Hogs dung, Pigeons dung, and such like, for that they are aready knowne, and common in this land with every Country *Coridon*' (and, he could have added, they had been covered by the classical writers on husbandry). He did, however, strongly recommend covering dung heaps to prevent rain leaching out the goodness. He advocated a cheap thatched or boarded roof sliding up or down on corner posts, 'as they use in the low Countries to make their barnes, (a patterne whereof standeth to be seene nere unto St Albones not far from Parkmill; in the backside of one of my tenaunts howses there)'. Is it possible this structure was erected as

28   BL, SL 2189, f. 11; SL 2210, f. 89; SL 2245, f. 104.
29   *Agrarian History of England and Wales*, vol. iv, Cambridge, 1967, p. 167.

an experiment by Plat? The heaps should be carefully constructed, with alternating layers of dung and soil to improve putrefaction.[30]

On one type of fertilizer Plat was deliberately evasive. This was his own secret compost, 'A new and extraordinary meanes for the inriching of arable grounds'. All the information he was willing to publish was set out as an advertisement at the end of his pamphlet on famine of 1596:

> And because the multiplying of corn is not greatly abhorring from our purpose, and seeing the greatest part of Dearth, must of necessitie begin from scarcitie of graine, I will here (without praying in aide of M. AIAX,[31] or of his stale marginal notes, whose reformation hath already more offended the eares of Honorable persons, then his first salts could over offende their noses) make a publike offer to all those Gentlemen and Farmers of England, whoe dwell in such partes of this Realme, as doe neither yield any store of Marle, or other common and ordinarie dung or soile, how they shall bee sufficientlie furnished, with a newe and plentifull *Compost*, and whereof there have beene already sundrie and rich trials made, whose quantitie shall not exceede eight bushels, whose yearlie charge shal not amount to xviii. Pence the acre *communibus Annis*, one year with another, and whose nature is so transmuted and disguised, as that one neighbour, yea M. *Aiax* himselfe, though he were present at the disposing or scattering thereof, shall not be able to discerne what his next neighbour hath done to his ground. In which secret, all those whome the author shal finde willing and worthie of the same, may upon reasonable composition, become owners of the skill, aswell for their owne as for the good of their countrey. Neither doe I know any iust obiection, why the same should not inrich aswell pasture ground as arable.

This invention may have been based on nightsoil (material dug from privies), as Harington alleged, but Plat implied it was not. Selling the

30   Plat, *Diverse new sorts of Soyle*, pp. 33–5.
31   'M. AIAX' and his 'stale marginal notes' are references to Sir John Harington and his jibes at Plat in *The Metamorphosis of Ajax* published also in 1596. Harington attacked Plat's attempts to gain a patent of monopoly on the manufacture of 'coleballs' and his consequent secrecy over the process, speculating that Plat mixed coal dust and dung to produce the balls.

compost nationally was not practical and instead he hoped to make a profit thus:

> C iff there were an acte of parliament made that whosoever shold disclose any secrete for thenriching off ground shold have 2d uppon every acre throughowt England of all such as shold putt the same in practize for a certain tyme, & the 10 th part of the proffitt reserved to the crowne, then this secrete were worth the publishing & like to prove very gaynfull.
>
> C But now I see not how the Author shold reape any proffitt answerable to his discovery, unlesse somme greate person or somme whole shire wold compound for the same.[32]

Although apparently unsuccessful in making any money from his ideas on fertilizers, Plat did have the satisfaction of knowing that his ideas were taken seriously by gentry readers, for he writes, 'A Western Gentleman by direction of my Book of Husbandry [i.e. the 1594 treatise on soils], steeped two years together Barley for twelve hours in the Sea-water, and then sowed the same, Anno 1595. and 1596. and had a very plentiful crop.'[33]

In *The new and admirable Arte of setting of Corne* published in 1600, Plat set out other ideas to improve cereal yields.[34] This work, of eight short chapters, advocated digging corn-lands rather than ploughing, setting corn by hand in prepared holes in neat rows, carefully choosing seed grain, and steeping and manuring prior to sowing. I have found very few manuscript notes on the two most important parts of the work, digging and setting, but it is obvious from the text that some thought, and consultation, had gone into this pamphlet. Plat refers to his own experiments and to many gentlemen who had told him of their successful trials setting corn and digging land, 'if I should report all the several trials that have beene made by several persons, aswel of the Nobility as others, within these two yeares, I should both weary myselfe with recording, and you with reading such infinit numbers of practises as I could produce'. The immediate cause of its publication may have been the good harvest in 1599 which drove wheat

32 Donald Woodward, 'Recycling in Pre-Industrial England', *Economic History Review*, 2nd Ser. XXXVIII, 1985, pp. 188–9; BL, SL 2197, f. 19v.
33 Plat, *The second part of the garden of Eden*, para. 118.
34 Plat, *The new and admirable Arte of setting of Corne*, 1600.

prices down to a six-year low and led MPs to question the wisdom of the laws to maintain tillage passed a few years earlier when grain prices were very high. Lower grain prices, and low wages, lead Plat to the conclusion that by using the spade and setting each seed by hand (i.e. employing less seed and more labour) higher yields would overcome any increased costs and be of overall benefit to the husbandman.[35]

Digging was an operation too well known for Plat to describe in detail 'not a fit subject for a Scolers penne' but he did advise against digging too deep: eight to ten inches was usually sufficient. He referred to John Taverner's *Certaine Experiments Concerning Fish and Fruit* of 1600 which identified topsoil as that part of the earth where fertility resided, cautioning against penetrating the barren subsoil below. Initial ploughing, followed by digging, was a way to avoid excessive labour charges.

Setting seeds individually, a more novel art, *was* discussed in depth. Plat runs through various setting instruments: simply a hole made with a finger; a small board with pins and a handle like a rake to thrust into the ground to make several holes at once; a board three feet long with holes at appropriate distances that a man could thrust a dibber through; a similar board with fixed pins which, with the aid of a gardener's line, might allow corn to be set in rows. The latter device, although not perfect, Plat considered the best. He provided detailed dimensions, recommending that two men, one to make holes and the other to set the corn, worked most effectively with it. He had been informed that setting wheat three inches deep at three inches apart had yielded 30 quarters an acre but only experiment would find the optimum depth and space in any piece of ground. Plat thinks it best not to put dung in each hole with the seed but, as we have discussed, he debates at length the virtues of seed steeps prior to sowing.[36]

The benefit of all this careful work in digging and setting was greatly increased yields. Plat reports that normal wheat seed to yield ratios were between six and eight to one, with ten to one considered exceptional. With spade and setting board, he has heard of a yield of up to 1000 to one achieved by an Essex JP (who was inspired by a conversation with a

35   BL, SL 2189, f. 174.
36   Plat, *The new and admirable Arte of setting of Corne*, ff. A3–A4, Ch. 4, 5.

lawyer at the Court of Requests to carry out the experiment). A Surrey gentleman 'reaped 16 bushels out of one pint of wheat which he set'.

> I have heard also by sundry reports of 20.30.&32. quarters of wheat upon an acre, & of 15. quarters of barly upon an acre; yea there have been some which have reported, that they have had 15.quarters of wheat upon one acre by this manner of setting, the ground being spent and out of heart by often plowing before.

Such dramatic yields were possible because, he claimed, set corn produced many ears on one stalk. In contrast, broadcast winter corn was attacked by harsh winter weather and, even when it took root it,

> hath taken so slender holde in the grounde, eyther the inward Balsamum is washed awaie with moysture, or nipped with extreme colde, that it cannot possiblie send forth so manie spiring stalkes and eares as naturallie it would: besides, the earth being full of clods, and not sufficientlie broken into fyne moulde with the Plough, the Corne cannot so easilie and plentifully by this attractive nature drawe for his owne nourishment such store of that vegetative salt from the earth as it desireth.[37]

Plat was the first of many English agricultural writers to discuss the benefits to field-cultivation of digging and setting in rows, essentially gardening techniques. The debate continued for two centuries and it is significant that in 1799 William Marshall wrote of experiments he had carried out: 'It would not, perhaps, in the present state of things, be eligible to reduce each field to a garden; but I would wish to see an arable field and a kitchen garden bear some resemblance'. I have argued in detail elsewhere that innovative farmers for many years sought the goal of making ploughing as efficient as digging and mechanized seed-setting a reality, taking their cue from commercial kitchen gardeners.[38]

As we have seen, Plat made much of the use of dung and seed steeps to improve corn yields. Following one manuscript note on these topics, he suggested an alternative way to 'doble the profitte yeerely wch our english corne grownd doth nowe yeald'. Botanically unsound and hopelessly

37 Plat, *The new and admirable Arte of setting of Corne*, f. A3.
38 Thick, *The Neat House Gardens*, Totnes, 1998, pp. 68–76.

optimistic, the idea certainly displays his ingenuity and, one might argue from the preamble, some arrogance. Readers were requested to 'meditate uppon this Invention wch iff yt fall owte trew, then let Englande confesse & testifie, that by this one Invention, the Author of this booke hath well deserved to bee registered amongst the best Benefactors that ever brittaine bredd'.

Plat thought grain plants could be induced to produce a succession of crops by repeated mowing throughout the year, the harvested crops being used for fodder or liquor production:

> Sow one or 2 acres off grownde parte wth wheate, part wth rie, part wth barly & part wth otes, cutt of these severall croppes, when they are comme to somme reasonable groath, and see yff they will not eare againe, for soo, thow haste the firste cropp for nothinge, wherof thow maiste drie somme & other somme to make triall off beinge greene by infusinge the same in water, till by arte thow haste procured a heady liquor, either to serve for a common drinke or els for viniger or Aqua vite. those that can not prove this woorkmanshipp for want of skill in natures woork, let them content themselves wth the first cropp to feede their cattell. iff thow wilte draw this conclusion further, then mow or Cutt downe as often as thow shalt see any resonable cropp to comme, and use it as before, & perhapps yt shall not bee amisse, to leave yt in the ende (when thow canst have no more Croppes) to grow to eare & so to bringe foorth Corne, wch iff yt prove well, then shalt thow have for the same coste and labor wch ordinary fermors bestow uppon their grownd yeerly, diverse cropps wherof I thinck everie one employed uppon aqua vite, beere or viniger, will bee as rich as the best cropp wee were wonte to have in corne.[39]

Apart from this fanciful scheme Plat had little to say on conventional arable crops and his notes on the husbandry of the most common farm animals, sheep, pigs and cattle, are not extensive. He had some ideas on curing two of the most common animal diseases, 'rot' in sheep (seemingly a general term, in Plat's day, for any disease of the internal organs) and measles in pigs. For measles he recommends mixing human urine with pig-

---

39   BL, SL 2210, ff. 41–42.

feed, a mixture of wine and a few grains of antimony, or juniper berries and rye. Various ingredients for cures were put forward for rot: a treacle, woad, briers, broom, or dry food for some days. Alternatively, 'some commend the barck which Mr Winter brought owte of the Streights and therby is called cortex winterianus'. One Doctor Jordan brought a recipe from Italy for curing rot which consisted of bay salt, wild thyme and Arsmart (water pepper), although he 'did of late confesse unto mee, yt all other additions added unto the salt…were but only to obscure ye secrete'.[40]

To improve milk yields in cows Plat also recommends doses of salt. Alternatively, sow trefoil in pasture once every five to seven years 'as they do in Italy,' or feed them boiled hay in winter (an idea from the Low Countries). To put flesh on cattle he recommended a handful of brimstone and salt mixed with food two or three times a week. Fattening hogs could be achieved with tangle (sea kelp), woad roots, rotten apples, peascods, or garden stuff: 'C plante infinite stoare of pompions & feede them thwerwth, or els sow carretts, parsnippes or turnippes & when they are ripe turne in your hoggs, & let them feede theyre bellies full. qre iff you may not feede them wth Carrett Toppes only, & make other use of your rootes. Note that your pompions muste bee sodden to papp either in fayre water or powder beoff broath.' He added that 'kyne feeding uppon Carret Toppes give excellent milcke.' Sheep might be pastured without doing harm in orchards or gardens as long as they were restricted to grassy areas. Plat recounts seeing sheep on short tethers attached by a horn ring to long lines, allowing them to go up and down 'the gardein at the Rolles in aprill 98'.[41]

As a Londoner, Plat had little to do with farm animals: at Bethnal Green he 'never kept but twoe kine, in any one summer'. This London perspective conditioned the livestock (and crops) which did attract his attention. We have seen that he was interested in the market gardening around London and this interest spread to other forms of intensive, suburban, food production.[42] Commercial rabbit production, in rural

40   BL, SL 2210, ff. 158v, 166; SL 2244, ff. 31v, 67v, 76v; SL 2189, ff. 159–160; Gerard, *Herball*, 1633, pp. 445–6. 'The Streights' are in south-east Asia.
41   BL, SL 2244, f. 32, 57, 67v; SL 2210, f. 143; SL 2189, f. 148. The garden of the Master of the Rolls was in Chancery Lane, on the site of the old PRO.
42   Plat, *Jewell House of Art and Nature*, pt. 3, pp. 10–12.

warrens, was well established by Plat's time; warrens flourished in various parts of the country in the thirteenth century and the number of place names in England incorporating 'warren' or 'conney' is an indication of once widespread rabbit exploitation. Joan Thirsk has identified rabbit production as one element of 'alternative agriculture' – marginal rural activity in times of buoyant prices for mainstream agricultural produce but of increasing importance in times of agricultural depression.[43]

Plat describes not a rural warren, with high banks, low scrub and grass, and only the most rudimentary burrows and internal control of the rabbits, but he talks of an intensive operation in the suburbs, run by an old friend of his, 'Mr. Parsons my scholefellow being an Apothecarie'. With much detail, he outlines the structure and operation of the warren. To keep 10 or 20 breeders with 2 bucks, a grass plot of at least one rood (1210 square yards) was enclosed by a brick wall. In part of this enclosure a pit five feet deep and 15 or 20 yards square was sunk, brick-lined but with some bricks missing at the bottom to afford the rabbits room to make their earths. Gravel round the walls will prevent the rabbits tunnelling too vigorously. A long wooden chest was placed in this vault, with a hinged door-flap at one end and a hinged roof-flap. The rabbits were fed by a large rack with little partitions so that each animal could feed without disturbance from its neighbour. Plat recommends the rack be made of wire to prevent the rabbits nibbling it. A stepped slope out of the vault allowed the rabbits access to the grass enclosure. A board hung on each wall could be let down suddenly to stop the animals entering their earths. 'This vault or pitt must be covered with a fraime wherin you may have either a banquettinge house or somme other roome of necessitie or pleasure' and looking through windows into the vault, 'you may with no small pleasure behold them feeding at theyre racke and skipping upp and downe'. An alternative which some breeders preferred to the vault was a wooden shed divided into many small hutches.[44]

A little bundle of hay hung on a string in the vault gave some food to the rabbits: in the rack they were provided with bran and parched oats, 'I

43 John Sheail, *Rabbits and their History*, 1972, pp. 35–8; Thirsk, *Alternative Agriculture*, pp. 17, 38–9.
44 BL, SL 2216, ff. 50v–52.

have heard parched otes greatly commended to make the doe to take Buck the sooner'. The rabbits could chew on the grass in the enclosure, 'and it will not bee amisse to throw them now and then somme greene meate or Carrett toppes mallowes, dandelion etc.'

Every morning the rabbits were to be let out of the vault, when the floor must be swept clean. To count or sort the rabbits, the side boards were let down to stop them entering their earths, the door flap of the chest or bin was opened, and by clapping hands the rabbits were scared into it. The door was closed and the roof flap opened, 'and you shall see them all like a flock of sheepe together. Then take or lett goe as you please.' Surplus bucks (above the ratio of 1 buck to 10 does) were castrated when three or four months old, 'this buck will grow in a short tyme to be exceeding fatt, very white of flesh, and of a good leane.'

As to the gains to be made from a warren, Plat advised, 'Now because this enclosure will be costly, I hold yt best to keepe only such Bucks and Does as bee of silver coullor. By this manner of keeping Mr Parsons affirmed unto me that of one Buck and 2 does bee hadd in one yeere 33 Rabbets whose skinns be sold for 2s and 3s a peece.' Silver-grey rabbits, thought by some to be a distinct species, were much prized for their fur, even though they were considered weaker than common rabbits, needing better food and breeding less prolifically. Half a century later, Adolphus Speed was uninhibited in his enthusiasm for rabbit production near London, estimating that 500 acres of land near London could be taken at a rent of 8 shillings an acre, producing rabbits yielding a gross return of £5 6s. 8d. per acre. He claimed the acquaintance of two London gentlemen who had made £10 an acre from five-acre suburban warrens. After a plethora of further statistics, Speed recommended rabbits as 'the most profitable creatures of England, rightly used'.[45] In the 1720s, Richard Bradley described another suburban London warren, possibly in Hammersmith. Four rods in area, with 10 couple of does and four bucks, this warren shared many of the characteristics of Plat's – the brick sides, feeding racks, hutches for does with inspection flaps, all echo Plat's advice.

45 Sheail, *Rabbits and their History*, pp. 25–6; Adolphus Speed, *Adam out of Eden*, 1659, pp. 1–13.

And their feed of oats, bran, hay and garden waste is the same as in Plat's notes.[46]

Poultry-keeping was another suburban occupation noticed by Plat (and others – Leonard Mascall published the first English book on the subject in 1581 and a second, translated from the French, appeared in 1584). Plat favoured intensive poultry-keeping, battery-hens in fact: 'Let everie hen have speedely streight roome onlie to turne in, & he will fatten quicklie'. Various feeds were recommended to him by friends – barley half-sodden, boiled carrots, turnips, parsnips, apples, or pumpkins. More imaginatively, he thought a noxious by-product of London's butchers, blood, might be recycled as feed, either by mixing blood well with grain and bran or 'make square litle shallow pitts of brick putt bloodd therin cover them wth plates of yron full of litle holes, that the flies may gett in to blow and engender maggotts & feed your fowle wth them'. Cracked acorns were commended to Plat by his uncle to feed turkeys.[47]

Plat made notes on efficient dove-keeping and he also recorded a commercial venture to store salt-water fish on the Isle of Sheppey in north Kent which might be duplicated close to the capital:

> Sr Edward Hobby hath stoared certein dikes in the yle of sheppey wth sea fish of sundry sortes, into wch dike by sluces hee doth let in from tyme to tyme chaunge of Sea water to norish them. qre iff this may bee don in ponds of fresh water by putting one 1/18 or 1/20th part of bay salt in them to make them of an equall saltnes wth the Sea water that is brackysh.[48]

Plat made notes on liquorice, a plant whose roots were important to apothecaries and comfit-makers for medicines and sweetmeats. It was imported from Spain but also grown in England where deep and preferably sandy soils were available. Near London, liquorice-growing thrived at Croydon by the mid-eighteenth century and some acres were grown at

46 Richard Bradley, *General treatise of husbandry and gardening*, 1723, vol. II, pp. 354–358.
47 BL, SL 2210, ff. 137v, 138, 152v, 160; G.E. Fussell, *Old English Farming Books*, I, pp. 9–10, Collieston, 1978.
48 BL, SL 2210, f. 184. Sir Edward Hoby was a courtier and MP who Plat may well have met at Court.

the Neat House gardens in Westminster in the 1660s. The impetus to try this difficult and time-consuming crop around London may have come from immigrant gardeners from the Low Countries in the second half of the sixteenth century and it is significant that Plat, concluding instructions on growing the crop, mentions that his friend Thomas Gascoigne 'was credebly enformed that a stranger made an infinite profitte therby in a few yeeres.'[49]

Another agricultural product important to Londoners and of vital importance to the prosperity of Plat himself was the hop. Hops were introduced from the Low Countries at the beginning of the sixteenth century and steadily became the main flavouring and preservative in English beer. Plat provides detailed notes on the planting, tending and drying of hops which were not culled from the book on the subject by Reginald Scot which was published in 1574 but came from talks with Jefferies, Royal Gardener at Sheen and Sir Henry Constable, a north-country gentleman and MP. Aside from these workmanlike notes, there is an interesting aside that 'In essex they have vessells as big as hoggshedds owte of wch they squirte water uppon the woormes wch otherwise wold destroy theyre hoppes.' This engine seems to have been akin to the 'great Squirt' made of tin illustrated in Thomas Hill's *Gardener's Labyrinth* of 1577 and used as a watering device. As usual, Plat is keen to improve on existing techniques. Rather than straight, upright poles for the hop bines, he suggests 'short poles of a yarde longe, & then laying other poles alonge, & not uprighte, as wee doo use in england to make our frames for vines' or even 'strong packthreede waxed or tanned…insteede of poles for the hoppes to run uppon', presaging the wires used for today's hops.[50] Plat believed that green, undried hops had more strength than the dried flowers (a fact denied by Scot) and he speculated that costly drying and storing of hops might be avoided by making 'a strong decoction' of the green hops 'even to a sirup'. The syrup would be much easier to use in brewing.[51]

49   BL, SL 2210, f. 37v; Thick, *The Neat House Gardens*, pp. 48, 108.
50   BL, SL 2210, ff. 30–31v; SL 2189, ff. 85a–86a. Reginald Scot, *A Perfite platforme of a hoppe garden*, 1574; Thomas Hill, *The gardener's labyrinth* [1577], ed. Richard Mabey, 1987, pp. 80, 83.
51   BL, SL 2189, ff. 85v–86v.

AGRICULTURE

Drawing together his thoughts on new crops, efficient and profitable husbandry, and the provision of good and diverse foods, Plat sets out his plan for a small estate which we may envisage as a retirement home for a London gentleman. This is an ideal farm, combining, in a phrase much used in seventeenth-century horticultural works, profit and pleasure. To his original idea Plat added many brief notes as more ideas occurred to him. Here it is in full:

C How a Country Gentleman may wth a small cost in a few yeares procure a proffitable & sufficient revenew him selfe uppon 40 or 50 acres off good grownde to mayne teyne himselfe & his family.

C Havinge a convenient howse wth 40 or 50 acres on the backside off good meadow or pasture, or rather some of either, I cold wish these points of Husbandry ensuing to be put in practize. You may plant one acre or two wth pompions, plowinge up the earth in borders for them, & makinge evrie border 10 or 12 yardes or more distant the one from the other, wth these pompions you may both feed yr familie, & if yr store bee over greate, you may feed sheepe & hogges thereon or slicing 2 or 3 of them evrie day you may seeth them in your powder beoff broth to the thickness of a pappe and thereof feed ducks geese hens & such other fowle as you shall finde to like them well. Quaere if you may not keepe them longer by drieyinge them in a stove or oven that is but gentlie heated and quarae iff the same beinge sliced and well dried in the same iff after they will not serve to make breade off./Next you may plante either a wood of willowes (if yr grounde be watery) or doble ranke of them to make convenient walk (as in ye backside of St Johns Colledge in cambridge) & to loppe specially for fire./ Also plante 10 or 12 acres wth apple trees or peare trees each distance from other at the least 12 or 16 yeards so shall you save the grass till yr trees beare fruite, & after your fruit will bee very rich, let yr trees bee very large when you plante them you may fatten hoggs wherewth soundely./ and iff you have greate store of fruite you may practise the profitable conclusion of making of cider & perry ...... also plant 8 or 10 acres wth hoppes according to the order & excellent manner sett downe in this book./ also iff the grownde serve make a Connygrth, sowe the same wth silver heard connies onlie killing all that come of any other coller make a dich within a yarde of your enclosure, of 2 fote broade and a foot & ½ deepe, evrie morninge looke about the Connaygrth, and iff you percieve any to draw or undermine, wherby they shellbe likelie to get owte of the Conneygar, then stoppe the hole wth a litel cowdunge & they will goe no further, here you must have care specially for the killing of vermin./ quere iff in yr orchard or wodd of willows you may not putt sheepe, to eate upp the grass, for so there will bee no losse. & to bee ye better assured herof, you must learne whether sheepe will barke yr trees or no./ also for the makinge off your orchard speedily, see all the secretes thereof in this booke./ iff you graft in a feild full of ashes, you shall have store of good walnut trees speedily ./ also you may keepe all sorts of fowle & poultrie ware you cheape sicut hic no.368./ also you may buy all kindes of fowle when they are good cheape & so keepe them either in their feathers or wthout theire feathers. sicut hic no.344./ a dovehouse used sicut hic no.287. will be also profitable./ put also younge frie of fish into ponds to store your fishponds wch in 6 or 7 yere will bee very profitable some commend sowing of onions seede for a very proffitable matter./ also practise the making of salt or swete water if the place doo serve./ pursue all wayes for thenriching of your corne grounde./ iff pigeons will feede either of the pompions ante, or of the blood and gredynes, or gentills, in no.368; then may you keepe tame pigeons that are good breeders good cheape/. iff you have great store of fruites, you may drie them in embers ut alibi paret. quare how profitable yt will bee to prevent the first bearing in all fruites and seedes for perhapps one cherrie tree in september will be worth 20 in June/ some hold opinion that an acre of onions seede is worth at least 100L./ quer of sowing Aaron, & the seedes of Roses in an acre or 2 of grounde/ parre slicing of yr pompions in 4 or 8 peece and then either sonning of them or stoving of them./ graft store of Apricock Trees for theyre fruite is very profitable/ Sow Carretts or parsnipps, & when they are ripe turne in yr hoggs to fatten prove the same in Turnipps./ or lett your kine feed uppon the Topps of the Carretts as they grow, & they will give greate store of excellent milke, this practize in Colchester is saide to make an acre of grownd woorth 10L yerely by a successive kinde of plantinge. plant Saffron./ Sow linseede plant liquoris if the ground serve./

oyle of radysh seede is profitable [*crossed through* – to make oyle]:/iff you plant yr trees of willows or fruite at somme distance, you may also in the samme borders plant either roses, pompions or muske mellons, & yet not loose the grasse of your feilde./ Sowinge of linseede, rice, and planting of saffron are most profitable/ make woad of beane leaffes, marigold do. & of amber & indico./ make Kuchrenells / artificial wynes, butter of a new taste/ sow artichoke seedes, reasons of the son and Curreins./ sow Annis seedes & Canarie seedes/ Coliander. 6. acres of corne./ Sow thyme, sweet margorum, ysopp, winter savorie &c. and sell the plants in the springe tyme in the markett. / qre if one may plant Lavender of seedes./ sow alpistoe seedes/ sow the seedes of coloquintida./[52]

On this farm Plat envisages growing a wide variety of garden and field produce to supply a gentleman's family with a surplus for the market or for animal fodder. He suggests pumpkins, onions, carrots, parsnips and turnips for the pot, as well as the more delicate fare of artichokes and melons. He includes ingredients for flavouring savoury dishes or sweetmeats such as saffron, liquorice, aniseed, coriander, marigolds and roses, as well as herbs like sweet marjoram, hyssop, winter savory and lavender. Somewhat unrealistically, Plat wanted to try exotic plants such as rice, currants and raisins, or the bitter gourd coloquintida, used as a purgative which, according to Gerard 'commeth to perfection in hot regions'. He wondered about sowing 'Aron' (cuckoo-pint) for starch. Meat and eggs are provided by a variety of fowls, pigeons, sheep and pigs while rabbits yield meat and fur. Fresh and salt-water fish add variety to the diet. Orchards provide fruit for desserts, pies, cider and perry. Willows provide pleasant walks, fuel and timber. The 'corne ground' contributes bread grain and barley for malt which, together with hops, provide the household with beer. Running through this piece are Plat's novel ideas on food and gardening, many just cross-references to earlier notes: flavoured butter, artificial wine, dried fruit, dried pumpkins and bread-flour made thereof, new ways to preserve meat, and holding back fruit trees to have fruit out of season.[53]

52 BL, SL 2210, ff. 29v–33.
53 Gerard, *Herball*, pp. 914–5.

SIR HUGH PLAT

Plat wrote his plans for an ideal farm in the 1580s or 1590s, an early attempt to sketch out a retreat where a gentleman might recreate an earthly paradise. In 1659, Adolphus Speed published a similarly sketchy and hurried summary in chapter 16 of *Adam out of Eden*, a book of brief notes on agricultural improvement. We can ignore the fantastic size of Speed's scheme (500 acres) and the inflated gross returns and regard Speed's notes as another ideal list of novel and potentially profitable produce which might be tried on farms near London.

A Discovery of the benefit might be made with part of 500 acres formerly limited; and so accordingly by other parcels.

Impr. Rape-seed 20 acres.	0400*l*.
Mustard-seed 20 acres.	0400
Licquorish 20 acres.	1600
Hops and pumpions 20 acres.	1000
Saffron 20 acres.	0400
Flax 20 acres.	0400
Clover grass 20 acres	1000
Roman Beans 20 acres	1000
Tobacco 20 acres	0600
Improvement of Ground for Orchards, woade 20 acres.	
Madder 20 acres	0400
French Beans 20 acres	0400
Pumpions and Cabbages 40 acres.	0400
Tassels 2 acres	0050
Osiers 2 acres	0125
Clove July Flowers, and red Roses 5 acres	0100
Milch kine	0100
Feeding Cattel	0100
Rabbits in Hutches	0500
Malt	
Poultry and Fowl	0020
Pidgeons	0020
Swine	0020
Bees	0100

AGRICULTURE

Silk worms   0500
A Decoy   0040

Oats hath been made 300*l*. per acre, the times being still seasonable therefore, Coriander seed need no Dung, and will grow very well here.
Anniseeds will grow very well in England, and so Canary seed.
Aspragus once sown will last twelve years.
These before specified, one with another, will afford twenty pound an acre yearly.⁵⁴

In 1670 another and much more detailed ideal gentry farm was described by John Smith as 'Directions set down how to plant 200 acres of Land as well for Pleasure as Profit wherein shall be pleasant Walks with Timber-trees and groves of under-woods, and several Orchards and Gardens, with Fruit, Flowers and Herbs both for Food and Physick, variety of Fowl, Bees, Silk-worms, Bucks, Does, Hares, and other Creatures of several kinds: And a short account of the Charges and Profit of keeping a thousand Doe-Conies in Hutches, the profit amounting to 450 pounds per annum; Also Fish-ponds and Streams of water stored with many kinds of Fish, and stocked with Decoy-Ducks; And the Use and vertues of all the Plants in the garden of Pleasure.' This lengthy description only covers some of the produce of this earthly paradise, which also contained 40 acres of pasture for milking cows, 20 acres of corn and peas in rotation, a vineyard, a dove house, space for liquorice, asparagus, and tobacco, and a full kitchen garden. Smith added an overtly spiritual dimension to such ideal farms, this was a 'Theatre of Nature' for 'Speculation and Contemplation'. In the summer venison, fowl, fish and small game were available to eat, 'your Orchards and Gardens will afford you several sorts of Fruit and pleasant Flowers both for colours and scent; and in the Groves and Woods most stately Trees and pleasant Walks: What shal I say, a thousand pleasant Delights are attendant in this Pleasant Land'.⁵⁵

54   Speed, *Adam out of Eden*, 1659, pp. 142–4.
55   John Smith, *England's Improvement Reviv'd*, 1673, pp. 160, 176–77, 191–93. I have taken the quotations from an article which deals with Smith's ideas in some detail: Peter Goodchild, 'John Smith's Paradise and Theatre of Nature', *Garden History*, vol. 24, 1, 1996.

We can trace political and social differences in the way Plat, Speed and Smith describe their ideal farms: profit and the maintenance of family in pleasant surroundings with good food motivate Plat. Speed is writing in the age of Hartlib and the Commonwealth and economic advancement is his preoccupation, whereas Smith is a post-Restoration gentleman, acquainted with Evelyn and other members of the Royal Society, able to combine profits with aesthetics in his ideal farm. But many of the crops and animals recommended are the same and the fundamental aims of self-sufficiency and variety of food run through all three. We should perhaps be not surprised that in times when a monotonous diet was common and dearth an ever-present danger, even well-off gentlemen had recurring dreams of Cockaigne.

Plat's worth as a writer on agricultural matters, especially his thoughts on soil fertility and manures, was recognized by many later commentators. Gabriel Plattes concluded in 1639, 'I would wish every well-willer to the publicke weale, to be diligent in the furtherance of it; and to reade Master Markham, Master Googe, Master Tusser, Sir Hugh Platt, and others, who have manifested their good will, by publishing their knowledge in this behalfe', linking Plat with three other important agricultural authors of the time.[56]

Plat came to the notice of Samuel Hartlib and his circle, who were diligent in searching earlier writers for useful material. Hartlib published a long letter from Robert Child in his *Legacie* in 1655. Written in 1651, the letter includes Plat amongst the 'learned Philosophers' who have studied soil fertility. Another associate of Hartlib, Adolphus Speed, made several references to Plat in 1659 and borrowed without acknowledgement from his publications, considering him 'a man of the greatest judgement and experience in Husbandry'.[57]

After the Civil War Plat was well thought of by members of the Royal Society; I have quoted Evelyn's approval of Plat on soil science. In the course of a review of Blagrave's *Epitome of the Art of Husbandry* in the *Transactions of the Royal Society* of May 1675 the anonymous writer heaps praise on Plat's publications:

56   Gabriel Plattes, *A discovery of infinite treasure*, 1639, p. 63.
57   *Hartlib's Legacie*, p. 38; Speed, *Adam out of Eden*, 1659, p. 119.

AGRICULTURE

It is but half an age or in fresh memory since improvements in husbandry began to have any name or to bear any credit in England. Sir Hugh Platt had a long and tedious task and spent many years in pleas, defence, apologies, solicitations, printing and reprinting many tracts before the husbandry would stir. But by importunities and perseverance at least he prevailed so far that in most countries they were convinced and began to see, and taste and enjoy the public benefit.

In the same year John Worlidge, another Royal Society member, quoted Plat on soils in his book on husbandry and yet another member, John Mortimer, in 1716 referred with approval to Plat's recommendation of soap-ashes. John Laurence also mentions Plat in connection with soap-ashes in 1727 and commends his advice on ploughing and digging. That Plat was regarded as relevant after almost one and a half centuries must be some testimony to his ability as a writer on agricultural matters.[58]

---

58 *Philosophical Transactions*, vol. x, 1675, no. 114; John Worlidge, *Systema Agriculturae*, 1765, p. 63; John Mortimer, *Whole Art of Husbandry*, 1716, vol. I, pp. 98, 121; John Laurence, *A new system of agriculture and gardening*, Dublin, 1727, p. 55.

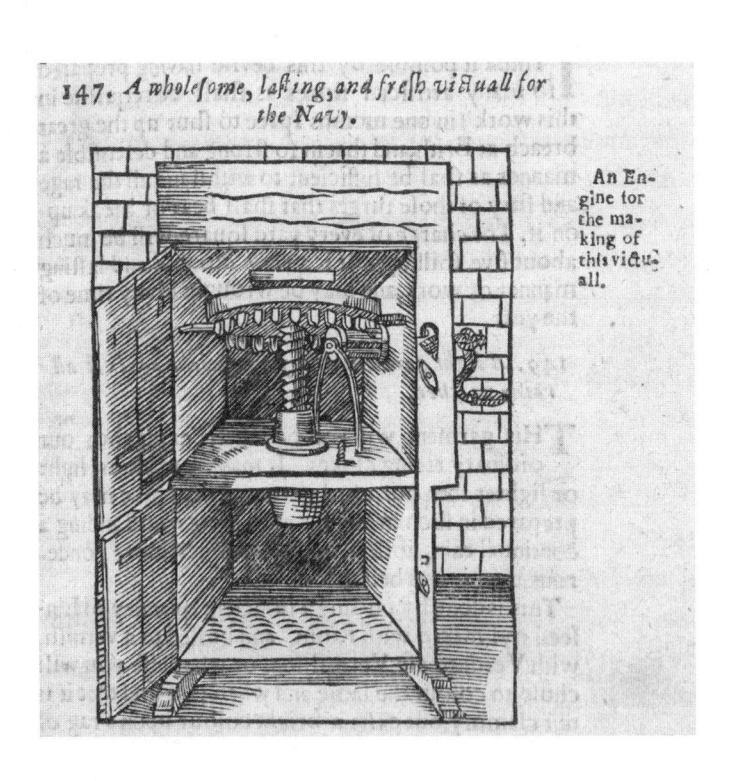

Figure 4. Plat's pasta machine, from The Jewell House of Art and Nature, 1594.

CHAPTER FOUR

Military Food and Medicine

Sir Hugh Plat was never, as far as we know, a sailor or a soldier. He did, however, talk at Court with the military leaders of his day, especially Drake and Raleigh, and a relative was a captain under Drake. He also encountered at Court men involved in the campaigns against Irish rebels and he was aware of the war in the Low Countries against Spain. Like many gentlemen in London or at Court, he was excited by the growing confidence of England as a naval power and proud of the exploits of the seamen who were Court heroes. He knew John Dee, who may have told him of his ideas of imperial expansion and he would have known merchants in London heavily involved in commercial ventures overseas.

Some of his military contacts discussed the problems of feeding ever larger armies and navies abroad and at sea and Plat considered the matter. He was in any case interested in food and drink preservation: many recipes in his cookery book *Delightes for Ladies* are for 'conserving', as are numerous ideas and recipes in his manuscripts. His alchemical experiments set him thinking about chemicals which could be added to food and drink as preservatives and his skill as a physician reminded him of the importance of diet to the health of soldiers and sailors. If these men fell ill, Plat had a range of medicines to cure them.

Throughout his publishing career Plat put forward ideas to improve military food and drink. *The Jewell House of Art and Nature* of 1594 contains detailed suggestions on keeping meat in brine and how to preserve water fresh at sea, as well as introducing his 'New Invention' of pasta as a victual for the navy. A brief prospectus of new inventions published in the following year announced a way of preserving meat for the sea without salt but gave no hint of the secret. In *Sundrie new and Artificiall remedies against Famine*, a pamphlet rushed out in 1596 to suggest ways to cope with major food shortages, Plat re-introduced pasta (macaroni) as a sea-victual, listing its virtues and offering to supply it in bulk. *Delightes for Ladies* of 1600 once more dwelt at some length on keeping meat in brine at sea and put forward a recipe for keeping orange & lemon juice for a year. Finally, in about 1607 (there is no publication date), a single broadside sheet, *Certaine philosophical preparations of foode and beverage for sea-men*, brought together macaroni, a secret preservative to keep water and other liquids fresh a long time, and a number of medicines which Plat said he could supply for use at sea.[1]

Plat's suggestions for improved naval and army victualling were occasioned by the needs of the time. Food for soldiers and sailors was becoming more expensive because long-term English food prices were rising throughout the second half of the sixteenth century, mostly because population rose faster than food production. In some years bad harvests caused sharp increases in prices – the 1580s and 1590s experienced particularly bad years. In the spring and summer of 1586 the Crown struggled to provision the army in Ireland, forces sent to the Netherlands and the garrison in the Channel Islands in the face of diminishing food stocks and popular resentment of state procurement.[2]

A more immediate problem occurred in 1585 when the government was shocked by notice from Edward Baeshe, the long-standing General Surveyor of the Queen's Victuals for Sea Causes (in effect head of naval procurement), that he intended to end his contract in six months because the fixed rates per capita he was paid for feeding the navy were so much lower than the prices of food he procured that he was losing over £500 a

[1] Plat, *Jewell House*, 1594; Plat, *A Discoverie of Certaine English Wants*, 1595; Plat, *Sundrie new and Artificiall remedies against Famine*, 1596; Plat, *Delightes for Ladies*, 1602; Plat, *Certaine philosophical preparations of foode and beverage for sea-men*, c. 1607. See below for manuscript references.

[2] Brian Pearce, 'Elizabethan Food Policy and the Armed Forces', *EcHR*, vol. 12, 1942, pp. 39–40; *Agrarian History of England and Wales*, vol. iv, ed. Joan Thirsk, 1967, pp. 848–9.

quarter. For well over 30 years Baeshe had provisioned the navy by using his bargaining skills to buy food cheaply in the market; his resignation in these difficult times would mean a return to widespread purveyance (compulsory acquisition of food at a fixed price) with potential civil unrest in the market places of England. The government's immediate response was temporarily to increase rates paid to Baeshe, persuading him to stay on for another two years. More generally, the government sought to increase the regulation of food markets, intervening directly on occasion to obtain victuals for the army, trying to balance the need for provisions with the threat of unrest from people who found their markets suddenly denuded of basic foodstuffs by purveyors. The policy was not very successful. Baeshe finally went, exhausted by his office, in 1587 and thereafter complaints about the quality of naval victuals became more frequent.[3]

The resignation of Baeshe was symptom of the general problem of provisioning military and commercial expeditions overseas. Small expeditions to distant lands could get by carrying food for only the first part of the voyage on the expectation of revictualling en route: fishing at anchor, hunting on shore, buying or stealing food from indigenous peoples or existing European colonies. Larger and longer expeditions to unknown parts needed more careful provisioning to avoid the not-infrequent cases of crews exterminated by starvation or disease brought about by malnutrition. Voyages to tropical regions posed particular problems – the climate rotted and wasted the beer, biscuit, bread, salt meat and salt fish which comprised the normal fare of English seamen. Moreover, longer voyages meant more valuable space devoted to food, 'to victual a crew of 200 for three months required 150 tons of stowage', more than half the capacity of a 300-ton ship.[4]

English naval expeditions expanded in size and range towards the end of the sixteenth century. In addition to the sailors, the Cadiz raid of

[3] Pearce, 'Elizabethan Food Policy and the Armed Forces', pp. 40–43; David Loades, *The Tudor Navy*, 1992, p. 206; M. Oppenheim, *A History of the Administration of the Royal Navy*, 1896, pp. 140–2.

[4] Richard Hakluyt, *Voyages*, 1962, vol. 8, pp. 107–32, This description of Edward Fenton's voyage in 1582 to Africa and South America typifies the hand-to-mouth existence of smaller expeditions; Loades, *The Tudor Navy*, p. 207.

1587 included 6000 troops; 12000 troops were carried on the Portugal expedition of 1589. The large fleet assembled to oppose the Armada in 1588 was provisioned for six weeks at a time in the hottest part of the year. Towards the end of each period the food deteriorated. These decaying victuals may have caused widespread, and often fatal, food poisoning. The fleet of Drake's last voyage to the West Indies in 1595–6 embarked with 2500 men. Inadequate victualling soon became a problem. Such a force on such a long voyage, sailing against an enemy forewarned of its approach, could no longer victual for the first part of the expedition only, thereafter relying on food scavenged en route. Scammell summed up the problem, 'From the late 1500s ships were expected, as a matter of course, to make longer voyages and to implement strategies demanding months rather than weeks in commission. But the successful execution of such schemes was heavily dependent on the endurance – not to say survival – of crews, and this in turn was to a considerable degree governed by adequate victualling.... as a rule the length of a cruise was decided by how long provisions would hold out.' More cynically, James Williamson pointed out that until the 1580s the English navy operated only in European waters, a few days' sail from a friendly port: 'No doubt ... swarms of men consumed victuals at inordinate speed and themselves died off rapidly of burning fevers and other penalties upon dirt and overcrowding. It did not matter; they were always close to England, whence beef, beer, and sailors could be had.'[5]

At the same time as the navy faced novel victualling problems, the food requirements of English armies overseas were a cause of concern. After some years without any significant military activity, the Netherlands expedition in the 1580s drew food from a wide area of England. Over a still longer period, there was need to supply an army of fluctuating size in Ireland. Although Ireland was a country with a rich and diverse agriculture, English armies found it difficult to live off the land. An occupying garrison of 2000, the norm when times were quiet, was a feasible

5   K.R. Andrews, *Drake's Voyages*, 1967, pp. 164–6; John J. Keevil, *Medicine and the Navy*, 1957, vol. 1, pp. 66–76; Cyril Falls, *Elizabeth's Irish Wars*, 1996, pp. 49, 196; G. Scammell, 'The sinews of war: manning and provisioning English fighting ships c. 1550–1650', *Mariners' Mirror*, vol. 73, no. 4, 1987, p. 365; James A. Williamson, *Hawkins of Plymouth*, 1969, p. 243.

proposition; but, the historian Cyril Falls observed, 'at the height of Tyrone's rebellion, with 15,000 troops or more in the country, with that country largely devastated, it was another matter.' He continued, 'It may appear strange that in a country stocked with enormous herds of cattle, sheep, and pigs, and growing a considerable amount of corn, food should have had to be imported for the troops, but so it was all through. In 1568, when the force in Ireland numbered only about 1700 men, the Crown entered into a contract with a purveyor named Thomas Mighte to victual the army.... Mighte was given permission to import 25000 lb butter, and 50000 lb of cheese annually, and licence to bring in malt and wheat in time of scarcity.'[6]

The ability of the Irish to move themselves and their cattle into the hills deprived the English troops of meat. Irish arable farming was crude, and not up to supplying a substantial army, especially as the disruption and devastation of a long campaign curtailed production. The terrain made transport by wheeled vehicle difficult, so the army was often supplied by sea, entailing its own problems of prearranged coastal rendezvous.[7]

Plat put forward his ideas against this background of a growing need for substantial quantities of long-lasting food for military expeditions on sea and land. Broadly they were for either new types of food or improved methods of preserving food and drink. Plat's most original new food for sea and land forces, which he advocated persistently and with some modest success, was pasta (which he usually called macaroni). In 1596 he set out the case for this victual as follows:

> 1. C First, it is durable, for I have kept the same both sweet and sound, by the space of 3 yeares, and it agreeth best with heat, which is the principal destroyer of Sea victuall.
>
> 2. It is exceeding light: for which qualitie Sir Frances Drake did highly esteeme therof, one man may carie upon any occasion of land service, so much thereof, as will be sufficient to relieve two hundred men a day.
>
> 3. It is speedily dressed, for in one halfe houre, it is sufficientlie

[6] Pearce, 'Elizabethan Food Policy and the Armed Forces', pp. 39–40; Falls, *Elizabeth's Irish Wars*, pp. 33, 63.

[7] Falls, *Elizabeth's Irish Wars*, pp. 32–3, 46, 64.

sodden, by which property it may also save much fewell and fiering, which occupieth no small roome in a ship.

4. It is fresh, and therby very pleasing unto the Mariner in the midst of his salt meats.

5. It is cheape, for in this dearth of corne, I dare undertake to feed one man sufficientlie, for 2 pence a meale.

6. It serveth both in steede of bread and meate, wherby it performeth a double service.

7. Not being spent it may be laide up in store for a second voyage.

8. It may be made as delicate as you please, by the addition of oyle, butter, suger, and such like.

9. There is sufficient matter to bee hadde al the yeare long, for the composition thereof.[8]

This is a powerful list of qualities, addressing many of the needs of the time: durability and resistance to hot weather (a growing problem on long voyages to the tropics); absence of salt (the ubiquity of which in much sea-victual was a source of concern); ease of cooking in the cramped conditions of a galley; the possibility of making the macaroni 'delicate' for officers to eat (although too much pasta might be monotonous, nevertheless 'It may be used now & then for change of diet'); use as a food for seamen to replace both meat and bread or biscuit; ease of carriage when raiding ashore or for armies stationed abroad. Macaroni could save valuable space at sea. If, as Plat proposed, pasta replaced some of the sailors' allowance of beef, butter, cheese, and salt fish, the navy would need less than the usual allocation of one-sixteenth of a ton of stowage space on board ship for a month's solid food per mariner. As pasta used less fuel in cooking, further space was saved.[9] Cost was an obvious consideration, both to merchants outfitting private voyages and to naval victuallers. In 1594 Plat estimated that when corn was 20 shillings a quarter he could supply pasta at 8 ounces a penny, enough for a reasonable meal. By 1596 – a famine year when available statistics indicate wheat prices some 87 per cent higher than two

[8] Plat, *Sundrie new and Artificiall remedies against Famine*.
[9] Oppenheim, *A History of the Administration of the Royal Navy*, pp. 140, 144; Plat, *Sundrie new and Artificiall remedies against Famine*.

years earlier – Plat's costs rise proportionately: 'in this dearth of corne, I dare undertake to feed one man sufficiently for 2. pence a meale'. In a manuscript price-list of about the same time he thought of charging 6d. a pound, while in 1589 he quoted 1d. for 3 ounces. By substituting pasta for some of the more expensive elements of the navy's usual diet, Plat could save the government money.[10]

Plat's pasta was 'called by the name of Macarone amongst the Italians'. Plat never published the raw ingredients of this food although he assured his readers that, 'There is sufficient matter to bee hadde al the yeare long, for the composition thereof'. Most contemporary readers must have guessed from the description of the finished product and the recommended mode of cooking and serving that it was made from a flour-based paste. Some Londoners might have seen the food prepared by members the Italian community and a few might have read of pasta in the English translation of an Italian cookery book, *Epulario or the Italian Banquet*, published in 1598 which included a clear and simple recipe for macaroni.[11]

In his unpublished notes Plat is explicit about the composition of the macaroni of the same wheat flour as for white bread, with or without the addition of beaten egg. This was just the basic recipe. Plat, as soon as he was introduced to pasta and recognized its potential as a military victual, thought of many variations on the theme of macaroni to add colour, taste and extra goodness, making it more palatable and nutritious. To the basic paste he considered adding: saffron; cream or new milk; sugar; sack, muscatel, spirits of wine or aqua composita; dried and powdered cucumber, carrot, parsnip or beef; essence of cinnamon, cloves, ginger, aniseed or pepper. These thoughts, crammed into the margins of his notebook, are typical of the way he returned to an idea again and again, linking it with his other interests.[12]

10   Plat, *Jewell House*, pt. 3 p. 74; Plat, *Sundrie new and Artificiall remedies against Famine*; BL, SL 2172 f. 5; Oppenheim, *A History of the Administration of the Royal Navy*, p. 140.
11   Plat, *Jewell House*, pt. 3, p. 74; Plat, *A Discoverie of Certaine English Wants*, 1595; Plat, *Sundrie new and Artificiall remedies against Famine*; *Epulario, or the Italian Banquet*, 1598.
12   BL, SL 2244, f. 29v; SL 2189, f. 118v; SL 2210, f. 134–134v.

Two basic shapes of pasta were put forward by Plat. Either flat: 'in shape like wafers ... cut ... owte either in losenges, or longe narow peeces', which would pack together close for storage, or macaroni-shaped, 'in the forme of hollow pipes'. Initially, he seemed more inclined to favour flat sheets, which were prepared simply with a rolling-pin. Once he resolved to furnish the navy with this victual himself, he advocated the 'hollow-pipes' because, sometime before 1594, Plat acquired a macaroni press, illustrated in *The Jewell House* as 'An Engine for the making of this victual'. The crude woodcut bears strong similarities to early Italian macaroni presses which extruded the wheat-flour paste through a die filled with round holes in which wires were centred to produce the hollow macaroni, which was then cut off in lengths and dried. Plat describes the process in 1590 probably before he acquired a machine, 'put the same paste into an apte presse that is full of holes in the bottomme, and as the same spirteth owte of the holes, cut every peece of it a reasonable length/ these wil be like strawes, & you may make them hollow also by putting in of wires at the holes.' A later note confirms that to make macaroni he would 'drive it into hollow pipes in my new engin, drie them well in the wind uppon canvas frames'. In 1594 Plat hoped to interest an entrepreneur in the machine – either to buy outright or to acquire the secret of how it worked. From 1596 onwards he offered to produce macaroni himself, 'Al those which are willing to victual their ships therewith, if they repaire unto me, I wil upon reasonable warning, furnish them therewith to their good contentment.' A similar offer for land forces was made in 1598, and in 1607 he claimed macaroni 'may be upon reasonable warning provided in any quantity'. His maximum production, with the aid of the machine, was one hundredweight in a day. As dried pasta, this is about 450 four-ounce servings.[13]

Plat's detailed knowledge of macaroni suggests he had strong Italian connections. He refers in his works to some Italian authors and quotes recipes from 'Maister Bartholomeus Scappius, the maister cooke of Pope Pius Quintus, his privie kitchen'. Scappi was only available in Italian, so Plat knew the language. Plat may have read the pasta recipes in *Epulario*

13 Plat, *Jewell House*, pt. 3, pp. 74–5; Plat, *Sundrie new and Artificiall remedies against Famine*, 1596; Giuseppe Prezzolini, *Spaghetti Dinner*, New York, NY, 1955, p. 51; BL, SL 2210, ff. 134–134v; SL 2244, f. 29v; SL 2172, ff. 5, 21.

or the Italian banquet in one of the Italian editions which preceded the English translation of 1598. He knew something of the Italian community in London: in 1594 he mentioned their method of drying walnuts.[14] Plat's main informant on macaroni was a Spanish gentleman, identified only as 'Sr Romero' who provided him with secrets in military matters, food and cookery, medicine and general knowledge such as the principles of the diving bell and how to engrave in glass. Who he was – diplomat, merchant, or religious exile – I have not established, but he knew a great deal about Naples, then under Spanish rule, and probably lived there prior to his arrival in England. On Christmas Eve 1590 he met Plat and told him in detail about macaroni in Naples: how it was eaten, how it was made in mechanical presses, then 'honge upp, uppon foddes to drie presently after they are made like candells' (a method familiar to nineteenth-century travellers in the region, who saw spaghetti hanging to dry from racks), and that it was sold 'by the l. in Naples, where (as Sr Romero assureth mee uppon his creditt) that more then 100 persons doo live very richly by making of them.' Naples was the area of Italy most associated with macaroni (by Plat's time a generic name for pasta), where one was most likely to encounter pasta dried and sold in shops rather than freshly produced at home. Sicily and southern Italy had been areas of dried pasta production since the twelfth century, makers in Naples were selling it by 1295, and it was also made in Morocco at that time. Moreover, some Mediterranean ships were provisioned with dried pasta instead of ships' biscuit by the fourteenth century.[15]

Macaroni was originally made by hand, wrapping dough round a wire which was then extracted. Then a syringe tipped with a plate with several holes was invented to extrude small quantities of the pasta for domestic consumption. By the end of the sixteenth century, this idea had

---

14   Plat, *Delightes for Ladies*, pp. 26, 32; Plat, *The Jewell House*, pt. 1, p. 38; Bartolomeo Scappi, *Opera Di M. Bartolomeo Scappi*, Venice, 1570; *Epulario, or the Italian Banquet*, 1598; Denis E. Rhodes, 'The Italian Banquet, 1598 and its origins,' *Italian Studies*, vol. 27, 1972, pp. 60–3.

15   Silvano Serventi and Françoise Sabban, *Pasta: the story of a universal food*, New York, NY, 2002, pp. 42, 52–3; BL, SL 2210, ff. 87, 131v, 134–134v; Prezzolini, *Spaghetti Dinner*, pp. 16–29, 89.

been adapted for commercial use: machines based on the wine press were developed to extrude macaroni through holes in a die. They were used by macaroni manufacturers in Naples. In 1579, a charter of the vermicelli-makers' guild of Naples laid down that each shop should have an extrusion press. The likelihood is that Plat's 'Engine' for making macaroni was either constructed under Sr Romero's supervision, or imported from Naples. The illustration of this machine in the *Jewell House* in 1594 maybe the earliest depiction anywhere of an Italian extrusion machine (see figure 000, above). Plat knew that macaroni was eaten in Italy 'by the best and most honorable gentlemen of all the Countrie' but Romero also told him of the mass production in Naples and he may also have mentioned its use as victuals on ships trading in the Mediterranean.[16]

Another food used as a sea-victual in that region was couscous. Plat mentioned it only once in his published works – in an aside in 1607 when he described macaroni as 'not unlike (save onely in forme) to the Cus-cus in Barbary'. Several notes he wrote about this food in his manuscripts show that he was attracted to it as a military victual because it had the same advantages as pasta of portability, ease of storage and preparation, and palatability. He may have learned of is use at sea from English sailors who traded in the Mediterranean. He did not, however, have an informant to tell him how it was prepared. Couscous was a food of North Africa, part of the world not much traded with or visited by Englishmen. Some was produced in Sicily but this was not, apparently, known to Señor Romero. Nevertheless, Plat made a good guess as to how it was made:

> C Qre how they make Cuscusow in Barbary. per adventure after the manner as they corne gonpowder first making a paste of an apt temper. per adventure by strawinge of flower upon a flatt streyner of tyn plates and sprinkling water wth a fine brush uppon it, and after searcing the flower upp and downe, and then reiterate your worke till you have spent all your flower.

A modern description of couscous preparation is:

16   Serventi and Sabban, *Pasta: the story of a universal food*, pp. 60, 62–3, 82–87.

> A bowl of [wheat] flour is sprinkled intermittently with salted water as the fingers of the right hand rake through it in sweeping, circular movements, causing balls of dough to coagulate. The granules are also rubbed between the palms or against the side of the bowl to shape them, and when complete they are dried. The contents of the bowl are sieved several times to obtain granules of uniform size.

The process covers the perishable bran and germ with starch, especially when couscous is made from freshly ground whole grains, thereby preserving the food for months or even years.[17]

The resemblance of couscous to 'corned' gunpowder (an improvement on finely powdered gunpowder much used at the time) led Plat to speculate whether it could be disguised as white gunpowder, although quite why is unclear. Surely, upon capture, the victors would be as likely to confiscate gunpowder as food? Maybe Plat considered it beneficial for officers to have a secret store of food unknown to the men to prevent pilfering. An additional curiosity is a small entry in the 1596 accounts for Drake and Hawkins' final voyage which reads 'Cusq'us viii *l.* the Tonne v Tonne xl *l*' (Couscous £8 the ton, 5 ton £40). Plat, as outlined below, contributed victuals to this voyage but, as he did not know how to make couscous, he was unlikely to have been the supplier. Maybe Drake or Hawkins introduced it to *him*.[18]

Plat also worked on a victual based on oatmeal for army use. One variant was to 'infuse otemeale in water till the water have gotten owte all the white substaunce & harte therof, divide yt, make starch therof if yt bee white enough, or els somme foode for soldiers if yt will serve'. He finally settled on the following:

> Take half a pinte of hole cleaned otemeale weying about 9oz ½, steipe the same 10 or 12 houres, then take 4 oz of butter, clarifie the same & boyle your otemeale therewith (having first drayned away all the water) over

17 BL, SL 2189, ff. 118a, 127a; SL 2210, f. 134a; SL 2244, f. 29a; Barbara Santich, *The Original Mediterranean Cuisine*, 1995, p. 14; Charles Perry, 'Couscous and its Cousins', Staple Foods, Oxford Symposium on Food and Cookery 1989, Totnes, 1990, pp. 176–7.
18 Perry, ibid.; BL, SL 2189, f. 127v; TNA, E351/2233.

a gentle fier, stirring it continually till it grow dry then mix half an oz of powdered suger therwith & it will tast very pleasingly./ you shall have 18 oz in all, wherof 4 oz are water as I ghesse. prove if you can evaporate the same wholly away, that the victuall may last you longer, remember to add a litle salte.

He added, 'this I devised for the Irish service. January 94'. This invention was probably the 'Victual for Warr' described by him elsewhere as 'A portable, delicate, light & cheape victuall & lasting longe. Each soldier may carrie a moneths provision wth him, it is most fitt for the yrish service & it exceedeth all ordinary foode in nourishment.'[19] Proposed for the army, this oatmeal mixture was similar to the oatcakes widely eaten in northern England and was also a dried and portable version of a recipe popular with seamen. In 1615 Gervase Markham praised the naval version fulsomely,

nay, if a man be at sea in any long travel, he cannot eat a more wholesome and pleasant meat than these whole grits boiled in water till they burst, and then mixed with butter, and so eaten with spoons: which although seamen call simply by the name of loblolly, yet there is not any meat how significant soever the name be, that is more toothsome or wholesome.[20]

'Gelly', meat broth boiled down to a thick and dry jelly, in later years called 'portable soup', was a military food which appeared several times in Plat's manuscripts, the idea being refined over time in Plat's usual way but never aired in his published works. The idea was good, taken up successfully after Plat's death and he was probably reluctant to allude to it in print for fear of revealing a valuable secret. Plat's basic recipe was to boil the feet or legs of beef cattle for a long time to make 'a good broath' which was then strained and boiled down 'to a strong & stiff gelly'. This in turn was dried on clean cloths in a windy place out of the sun, cut with wire into pieces, powdered with flour to stop the pieces sticking (much like Turkish delight is powdered with sugar today), and stored in wooden

19   BL, SL 2189, f. 112v; SL 2244, f. 31; SL 2247, f. 2.
20   Gervase Markham, *The English Housewife* [1615], ed. M.R. Best, Montreal, 1986, p. 203.

boxes. Made in March, it would 'keepe all the yeere'. Alternatively, the dry jelly could be 'stamped' into shape with a wooden die, like the 'Genoa Paste' of quinces familiar to Plat and other cooks of the time. He instructed that no sugar or salt should be added because their taste would be concentrated by the boiling process, although he speculated that saffron might add colour and rosewater could also be added at this stage. He wondered whether baked flour or grated bread could be incorporated to make the jelly 'serve as bread and meate the better' and whether the addition of isinglass would make it stiffer.[21]

The jelly was at different times described as a 'Victual for Warr', 'dry gelly carried to the sea', and a food for soldiers on the march. Plat envisaged using this jelly either as the base for soup or 'neat' as a concentrated food. Reconstitution as soup simply involved dissolving a piece of the jelly in hot water to make ' good broth' and, because jelly and water alone would be rather bland, adding such flavourings as were available or to taste – sugar, salt, liquorice, aniseed, or other 'convenient spice'. Plat emphasized the utility of the jelly as field rations, for 'a soldier may satisfie his hunger herwith, whilst hee is in his march'. He recommended that a leg of beef or veal be boiled with every 6 or 8 neats feet to produce a jelly which would more easily dissolve in the mouth. As field rations, the jelly would have been of interest to those involved in victualling troops in Ireland.[22]

Meat and fish jellies were not uncommon, both in medieval kitchens and in cookery books of Plat's time, but they were for immediate consumption, not dried out for keeping. John Evelyn, writing in the middle of the seventeenth century, gives a recipe using calves feet for a 'Strong Gelly', meaning one which was fairly stiff.

In 1736 Richard Bradley produced a recipe for what he described as 'Veal-Glue, or Cake-Soup, to be carried in the Pocket' which was, in effect, Plat's jelly. The flesh of a leg of veal was boiled until the liquor was 'very strong,' then:

> strain the Liquor thro' a Sieve, and let it settle; then provide a large Stew-pan with Water, and some China-Cups, or glazed Earthen-Ware; fill these

21 BL, SL 2244, f. 29v; SL 2189, f. 119–119v.
22 BL, SL 2244, f. 29v; SL 2189, f. 119–119v. 'A Neat' denotes any bovine animal.

Cups with the Jelly taken clear from the Settling, and set them in the Stew-pan of Water, and let the Water boil gently till the Jelly becomes thick as Glue: after which, let them stand to cool, and then turn out the Glue upon a piece of a new Flannel, which will draw out the Moisture; turn them six or eight Hours, and put them upon a fresh Flannel, and so continue to do till they are quite dry, and keep it in a dry warm Place: this will harden so much, that it will be stiff and hard as Glue in a little time, and may be carry'd in the Pocket without Inconvenience. We are to use this by boiling about a Pint of Water, and pouring it upon a piece of Glue or Cake, of the bigness of a small Walnut, and stirring it with a Spoon till the Cake dissolves, which will make very strong good Broth.[23]

By Bradley's time this jelly had become one of the earliest commercially produced processed foods. 'Some of this sort of Cake-Gravey has lately been sold, as I am inform'd, at some of the Taverns near Temple-Bar, where, I suppose, it may now be had'. A century later the manufacture of such jelly caused the chemist Liebig concern. He doubted the nutritional value of commercially produced cakes of gelatine. 'By degrees people took to the gelatinising matter for the true soup ... [and] manufacturers found that the best meat did not yield the finest jelly tablets, but that tendons, feet, cartilage, bones, ivory and hartshorn yielded the most beautiful and transparent jelly tablets.[24]

Plat wondered about the nutritive value of his jelly. He gathered tales of men who could not eat solids:

Mr Pill of Exeter sometimes a draper in bristow but now dwelling in exeter, hee hath kept shopp in both places. being a very old man at the least of 80 yeeres of age, and hath for the space of these 10 yeeres at the least lived only of the iuyce of his meate and by chewing of his bread

---

23 Constance B. Hieatt, *An Ordinance of Pottage*, 1988, pp. 184–5; Markham, *The English Housewife*, ed. M.R. Best, p. 111; Hilary Spurling, *Elinor Fettiplace's Receipt Book*, 1986, p. 206; *John Evelyn Cook*, ed. Christopher Driver, Totnes, 1997, p. 65; Richard Bradley, *The Country Housewife and Lady's Director*, 1736, ed. Caroline Davidson, 1980, pp. 58–60.
24 Bradley, *Country Housewife*, 1736, pp. 58–60; E.S. Dallas, *Kettner's Book of the Table*, 1877, p. 216.

> wthout suffring any of the gros parts to goe downe his boddie and that in perfect health, hee is of a faire complexion & a leane man, hee hath his dogg to eate that which hee hath chewed and casteth from him, hee doth forbeare to swallow any thing becawse hee hath no passage for it, hee was alive wthin these 4 dayes, and liveth at this present of his owne revenew. you may heere of him by somme of his neighbors at the star or 3 Cupps in breadstreate. this Note I had of Spenser the Travayler and Coster the 28 of May.97. I have set downe this example to satisffie men that they may live sufficiently of my gelly … which is nothing else but the Iuyce of meate and therby neede not to feare the emptines of their guttes or the clinging of them together for want of repletion.

Plat later heard from a friend of a Warwickshire gentleman, Mr Woodward, who had lived for 10 or 12 years on meat juices alone, as had a man in Enfield, 'all which examples comend my victuall & prove that there doth not neede any greate repletion of the gutts'.

As with pasta, Plat thought his jelly would be cheap to produce and therefore save the Crown money. He modified domestic recipes for jelly, replacing sugar with liquorice as a sweetener, and substituting a proportion of beef suet and legs of beef for the more expensive neats feet. The beef was, in any case 'cheap enough to the Queene whoe giveth away the leggs in fees' (possibly a reference to a patent granted for 'Ox Shin-bones' between 1597 and 1601).[25]

The proposals so far outlined by no means exhaust Plat's ideas on new foods for military use. Some were foreign, such as a suggestion from his Spanish friend Romero 'used by the moores and in Turckie', of 'how to keepe butter very sweete for the Navy, a longe tyme'. The majority however, were new recipes, although some, like the following, he must have discarded as impractical (or unpalatable): 'Qre of incorporating flower crumbs of bread, isinglass &c, wth stockfish, ling saltfish, codd &c. for a new victuall.' Drawing on existing techniques of preservation, he suggested slicing and drying cheese, then powdering it; or drying and powdering parsnips and carrots, or making a 'marmalade' of them with

---

25   BL, SL 2189, ff. 119–119v, 126v; William Hyde Price, *The English Patents of Monopoly*, 1906, Appendix E.

sugar and baked flour. Eggs could be hard-boiled and barrelled in salt or butter, 'they are cheaper in Skoteland' he added helpfully. Many of his ideas, like the oatmeal recipe, involved adding 'goodness' to flours, pastes, or bland vegetables: chestnuts or carrots boiled in sugar and wine; pastes mixed with 'restorative broths' and baked; parched peas or beans steeped in sugared water, whiskey or other spirit and dried again.[26]

Soldiers and sailors expected to be provided with meat and potable drink and Plat put much thought into keeping these fresh and palatable. To judge by his many notes on the matter, he gave much thought to preserving meat, only publishing his best ideas. He tried to find ways to minimize salt, moving away from the usual techniques of packing meat with a good deal of salt in barrels then preparing it for eating by soaking and boiling. After a time such meat went bad and stank. At best it was hard, flavourless and salty, difficult to digest, monotonous to eat, and by many considered dangerous to health in hot climates. As one answer to the problem Plat proposed using brine rather than salt crystals as a preservative. The idea came originally from a seaman: 'I heard a Maister of a Shipp affirme confidently that hys beoff doth eate much better being kept in bryne then in drie salt, only here must a care bee hadd that the vessells bee kept full. And herby I gather that the Sea water wold bee an excellent secrete for this purpose, either to be used at the first, or at the Least after the Beoff hath lien 5 or 6 dayes in theyre ordinary brine.' Plat thought about this, considering that the first brine should be discarded after 24 hours to remove blood, and a fresh brine used. He wondered what strength the brine should be, how large the pieces of meat could be and how long the beef would remain soft in brine, eventually concluding that seawater alone could preserve meat if flowed constantly over the meat. Excusing the intrusion of an unladylike passage into his 'Delightes for Ladies' in 1602 he set out his idea in full:

> A conceipt of the Authors, how Beefe may be carried at the Sea, without that strong & violent impression of salt, which is usually purchased by long and extreame powdering. Heare, with the good leave and favour of those courteous Gentle-women, for whom I did principally, if not only,

26   BL, SL 2244, ff. 29v, 30v, 51v; SL 2189, ff. 119, 127v.

intend this little Treatise; I will make bold to lanch a little from the shoare, and try what may bee done in the vast and wide Ocean, and in long and dangerous voyages, for the better preservation of such usuall victualls, as for want of this skill doe oftentimes meerly perish, or else, by the extreme piercing of the sale, do lose even their nutritive strength and vertue: and if any future experience doe happen to controul my present conceipt, let this excuse a scholar, quod in magnis est voluisse satis. But now to our purpose: let all the bloud be first well gotten out of the Beefe, by leaving the same some nine or ten dayes in our usuall brine: then barrell up all the peeces in vessels full of holes, fastening them with ropes at the sterne of the ship, and so dragging them thorow the salt sea water (which, by his infinite change and succession of water, will suffer no putrefaction, as I suppose): you may happely finde your Beef both sweet & savory enough, when you come to spend the same.[27]

Salting briefly and then drying in air or smoke was an alternative proposal, of use at home and on tropical voyages.

> Make a strong Brine, so as the Water bee overglutted with salt, and being scalding hote, perboyle therein the fowle or flesh which you would preserve some reasonable time, that is to say, according to the greatnes and grossnesse therof, then hang it up in a convenient place, and it will last a sufficient time without any bad or oversaltish taste, as I can testifie of mine own experience. This I thought good to publish both for the better preservation of mutton, Veale, and Venison, whereof a great deale in this lande is yearely lost, in hote and unseasonable Sommers, as also for the benefite of our English Mariners, which are forced sometimes to vittaile themselves in such intemperate Clymates, where no flesh will last sweet foure and twentie houres together, by reason that they have no meanes to make the same to take salt, which without all question will enter this way and make penetration verie speedily by reason of the hote and fierie spirite of salt thus prepared.[28]

27 Plat, *Delightes for Ladies*, c. 20; BL, SL 2189, ff. 116, 118v.
28 Plat, *The Jewell House*, pt. 1, p. 8.

He also suggested,

> mince or shredd into small pieces, any manner of flesh that is leane, and then let it drie thoroughly and barrell it upp in drie casks. this will soone bee sodden, and therfore if they shall bee forced at the Sea to seeth it in Salt water for want of fresh, I thinke it will not eate very salt, becawse it is so soone dressed, it may bee served in with oyle or butter and somme curreins, or without curreins after it is sodde.

He went on to consider how it should best be dried, and whether the dried mincemeat might be mixed with flour and made into a meat macaroni.[29]

Plat experimented with vinegar as a preservative for meat, fetching out the blood with brine as before, then sousing in vinegar and sugar and packing into barrels. But he was most pleased with his recipe for 'potting' roasted meat with butter. He alluded to this recipe in print in 1595 and 1600, and in a petition to the government in 1598, but such was Plat's confidence that this secret was valuable and might be saleable that he never published it in detail:

> But of all other wayes I hold this the finest sweetest and cheapest vz. first clarify your butter in balneo casting away the foote therof, then bake byrds, fowle, or other ioynts of meate (heere I think a little parboyling wold doo well to take away the blood in some water & salt) in pewter coffins and mixing good stoare of butter, pepper, salt & cloves in the baking, then put your meate into new stoane potts well leaded, but first cover the bottom of your potts half an inch with clarified butter wherin there is both salt and pepper & cloves, let yt coole and when yr baked meates bee almost cold, then pile them in your potts or wodden vessells one by one as clos as you can packe them, then cover them an ynch thicke wth butter prepared as before powring the butter in at a temperate heate. this way I have knowne baked meates kepte 3 monthes very sweete. qre how long they will last. I thinke that roaste and sodd meats being layd upp in the same manner

---

29   BL, SL 2216, f. 52v.

whilst it is warme and covered wth butter clarified & dissolved in balneo will kepe likewise. & so all Seamen may have roaste, baked and sodd meates wth them to the Sea ready dressed. this secret saveth ye carriage of much wood and coale. and the butter will serve either a second tyme for ye same purpose or else yt may bee converted to diverse other good uses.[30]

The provision of fresh water at sea was of extreme importance and Plat must have heard from his seafaring friends tales of thirst and death when water ran out or became undrinkable. The most pressing problem was to keep water, shipped in barrels, from putrefying. Unless the barrels were very sound and the water and barrels particularly clean, putrefaction because of organic impurities was likely sooner or later. Heat accelerated the problem. Plat put forward thoughtful solutions, wondering if boiled water, distilled water, or snow water (melted snow) should be tried. Rainwater was known to be 'pure' and Plat had heard that rainwater which fell in August would never putrefy. He wrote in 1594 that water might be preserved ' by the addition of some small proportion of the oyle of Sulphur with it, incorporating them both together, whereof I have long since made a sufficient triall. Some commende the oile of Vitriol to the same end'. He made a note of his 'sufficient triall': 'I didd put to 5 or 6 gall of thames water, i oz of oyle of sulphur, & yt was very sweete & cler 2 monethes after, & tasted like rennish wyne qre how longe more yt laiste.'[31]

Another solution he considered was to let water putrefy, once or several times, until it would degenerate no more, and then barrel it up. In Plat's words :

> Let the owner, Marchant, or mariner, having sufficient leisure to make his provision of fresh water, before hee beginne his voiage, prepare his water in this manner. First let him fill eyther some Renish Wine fattes, sack buts, or White wine pipes, such as have tubs having tapholes within three inches of the bottome, at the which after the water hath passed his first putrifaction, and is become sweet againe, he may then drawe it from his

30 BL, SL 2210, pp. 43v–44.; Plat, *A Discoverie of Certaine English Wants*, 1595; Plat, *Delightes for Ladies*, 1602; BL, SL 2172, f. 21; SL 2189, f. 127.
31 BL, SL 2210, f. 42v–43; Plat, *Jewell House*, 1594, pt. 1, p. 9.

residence into a cleane caske, and by this meanes it will last much longer at the Sea than other wise, and yet if there were but two or three handfulls of salt dissolved in a pipe of the same water (which would not much offend either the tast or stomach) it would preserve it much more then the bare preparation of the water will doe in the aforesaid manner.[32]

Once the almost inevitable putrefaction occurred, Plat had various suggestions for rendering the stinking water palatable. He had conversations with his naval relative, Anthony Plat and *his* admiral Sir Francis Drake who both gave him the benefit of their experience:

Sir fraunces Drake, ... who hath sought for all the helpes which he might, either in his water or his victual, for the better comfort and reliefe of his Mariners, in one of the last conferences which I had with him, did assure me that the most putrified and offensive water that could happen at sea, would by 24 howers agitation rolling up and downe, become sweet and good beverage. And Captaine Plat in whome sir Frances Drake for his good partes did alwaie repose great trust and confidence, did usuallie carrie certein long and thicke peeces of sheet lead with him, which he would cause to bee hanged by lines at the bung-hole even to the verie center of the vessel, wherby he did attract much of the fecicall part of the water, and the Leades would become very slimie therwith. This he did with often change and iteration, alwaies clensing the leades as they grew filthie, and so with much adoo, he found the water a great deale more pleasing then before.[33]

To make water more palatable at sea, Plat advised that if a captain of a ship carried '4 or 5 Tonn of water ... and a little vessell of aqua vite, hee may have sufficient good drincke for himself and his ffreindes for a long tyme' by mixing the water with some of the alcohol. Alternatively, aqua composita or Irish whiskey could be used as an additive. These mixtures were for officers. For the men a recipe borrowed from the Spaniards was recommended '⅛ part of malassoes maketh good beverage wth 7 parte

32 Plat, *Jewell House*, 1594, pt. 1, p. 9.
33 Plat, *Jewell House*, 1594, pt. 1, pp. 9–10.

water in whott places.' Drake gave Plat another healthy mixture for seamen on long hot voyages:

> Sir Francis Drake hath found by good experience that one vessell of viniger will make greate store of good beverage, and as I remember he spake of one parte of viniger to 20 partes of water which hee comended greatly for his cooling and refreshing Nature, when his Mariners wth much eatinge of salt beoff and other salt victualls (which is usuall at the sea) were inwardly enflamed, and then iff they shold drinke either wine or any other stronge drinck it wold drie and burne them upp.[34]

An expensive but innovative way for sailors to make fresh water at sea was 'by the helpe of some distillatorie vessell (wherin as also in divers other of the same kind and qualitie. I have found maister Sergeant Gowthrowse, the moste exquisite and painfull pracizer and performer of our times) they can make separation of the freshe part thereof on Ship-boord.'[35] Ideas on preserving water at sea came from a variety of sources but Plat was particularly beholden to his friends and acquaintances for suggestions. As well as help from Sir Francis Drake and his kinsman Anthony Plat, Thomas Lodge, playwright, sometime seaman and medical practitioner, told him of the molasses and water drink, the judge Sir Julius Caesar suggested the addition of aqua composita, and his Spanish friend Señor Romero advocated mixing oil of vitriol with water as a preservative. One has a picture of Plat identifying a problem, consulting his library and his notebooks, then drawing on advice from a wide circle of friends and acquaintances.[36]

In his c. 1607 printed broadside advertising food, beverages and medicines for the seafarer Plat boldly put forward a secret ingredient, a 'Fire of Nature' or 'Philosophical Fire' with the potential to solve the whole problem of preserving water, and a good deal else:

---

34   BL, SL 2210, f. 43; SL 2244, f. 29v; SL 2216, f. 50v.
35   Plat, *Jewell House*, 1594, pt. 1, pp. 9–10. 'Sergeant Gowthrowse' was William Goodrus, the Queen's surgeon.
36   BL, SL 2210, f. 43; SL 2189, f. 135v.

> All the water, which to that purpose shall bee thought needefull to be carried to sea, will bee warranted to last sweete, good and without any intention to putrefaction, for 2, 3, or 4 yeeres together. This is performed by a Philosophical fire, being of a sympatheticall nature with all plants and Animals. In the space of one moneth, the Author wil prepare so many Tunnes therof, as shall be reasonably required at his hands. By this meanes also both Wine, Perrie, Sider, Beere, Ale, and Vineger, may be safely kept at Sea, for any long voyage, without feare of growing dead, sowre or mustie.

The 'fire' would also preserve liquid medicines and distilled waters.[37]

The nature of this miraculous substance was never revealed in print by Plat but some historians thought cooking was involved, possibly simply a bottling and boiling process, producing a sterile liquid. I doubt these explanations. Plat's Fire of Nature, or Inward Fire of Nature, was a product of his interest in alchemy. The philosophy behind the 'fire': an essence within plants which could be isolated and used as a universal preservative, sits well with the alchemical thinking of the time. But Plat was a practical, as well as theoretical, alchemist and he produced this 'fire' in his laboratory. The 'fire' was of two types, an essence within all liquids which could be induced to work as a preservative on the particular liquid involved, and a universal 'fire' which could be added to whatever liquid one wished to preserve. It was this second 'fire' which Plat claimed he could manufacture and supply.[38]

As to the first 'fire', this, Plat wrote, could 'preserve any liquid, as water the iuyce of any hearbe or flower &c from putrifyinge' and was 'donn per rotationem philosophicam v[iz] by stirrynge upp the fire of Nature wthout fire'. Going into more detail, we are told the method for preserving vegetable juices:

> Expresse the uiyce of the herbe, put the same in a round glas that hath a neck of some length, leave a third part vacuum, stopp the glas wth a corke shake the same violently upp & downe in your hande for the span of one

37 Plat, *Certaine philosophical preparations of foode and beverage for sea-men*, c. 1607.
38 John J. Keevil, *Medicine and the Navy*, 1957, vol. 1, p. 108; Margaret Pelling, *The Common Lot*, 1998, p. 53.

howre, opening the corke now & then to let owt the wold & venemous spirits. stopp your glas, & once evry 3 days make agitation only for ½ howre. (qre if once a weeke will not serve) & so you shall keepe the same wthout putrefaction, as long as you please. I have provd this in the uiyce of Cowslipps, wch is a hard uiyce to preserve any longe time. also this woorke serveth to exalt & spiritualize the uiyce so as after .6. months preserving, I thinke by a gentle Balneo it will yeald a Spirit.

Plat thought this first 'fire' was released simply by vigorously shaking liquid in a closed vessel but it was of little commercial use, needing small and expensive glass vessels to enclose each liquid to be effective.[39]

The second fire is more difficult to identify, although Plat tantalizingly provides many clues. A manuscript chapter headed 'Severall uses of the inward fier of Nature partly approved & partly imagined' lists 54 uses of the 'fire'. First its general properties are listed: 'it is of the Nature of simple fier not combustible … it augmenteth the fier of nature in evry animal and vegeteble wth wch it is mixed … it digesteth all his Components … it helpeth the weak stomach of ye ostridge upon a surfet … it turneth the inside of nature owtward'. These enigmatic phrases are alchemical, meaning something to the initiate but tell us nothing of its real nature. Many uses as a preservative are listed: 'this fier preserveth water from putrefaction … it kepeth any white wine or renish from decay or descoloring … keepe white & yolck of egg beaten together wth this fier for long voyages … qre if brothes, collesses, &c may be kept this way for long voiages'. It would also preserve ink, milk, citrus peel, cider and perry, and all distilled waters, as well as retaining the 'naturall smell & tast of violet, cowslip, primrose' etc., if they were soaked in it for a time. Then 'if you wold preserve any thing wth the fier of nature, fist fill your glas or vessell therwth and then put in your water, wine vineger &c. do this once a month & you may keepe it as long as you please. qre if 3 or 4 iterations will not serve for a long time'. I presume Plat is here suggesting swilling the vessel out with the 'fire' before pouring in your liquid, and repeating the dose regularly. To keep water from putrefaction he speculated that '24 oz of fier to 8 gall of water' might suffice but wondered 'whether often iteratid be not requisite'.

39 BL, SL 2210, f. 58; SL 2245 ff. 34–34v.

A medley of other uses were put forward for the 'fire', some to 'strengthen' other liquids such as brine, oil of sulphur or 'dissolved metals'. Many of his recipes are associated with strengthening or improving alcoholic drink: preserving white wine from decay or discolouration; strengthening 'any decoction of dull drinke'; improving cider when mixed with its must; mixing with water used to make beer, making it 'a mightie drinke' and swilling round beer casks before they are filled. Going further, he speculates that water in which the 'fire' has been mixed might, 'serve for beere & ale. or at the least beinge strengthened wth a little Corne, or fruit, liquorice, annis seedes &c.' or might after 3 or 4 months yield (presumably by distillation) 'great store of spirit'. Alternatively, mixed with water, sugar & honey, it might make wine in a couple of months. Hay soaked in water with the 'fire' will yield a spirit.[40]

A liquid which preserves the taste and colour of juices and strong waters, strengthens and prolongs the life of beer and wine, and turns water into wine must be strongly alcoholic and I conclude that Plat's mysterious 'fire of Nature' was a strong, clear, tasteless spirit. As proof, I put forward the description of the effect the 'fire' has in preserving lemon juice according to the *c.* 1607 broadside. Plat is familiar with the 'normal' way of keeping the juice – letting it ferment naturally, with consequent loss of colour and change of taste (which he suspected might also destroy its effectiveness against scurvy). The addition of the 'fire' would avoid this – fresh lemon juice would be simply preserved in alcohol, much the same way as fruit is still preserved in brandy or kirsch for use at Christmas.

> And though their iuice will, by natural working and fermenting, in the end so spiritualize itselfe, as that it will keepe and last either simply of itself, or by the help of a sweete olive oyle supernatent yet this Author is not ignorant, that it hath lost much of his first manifest nature, which it had whilest it was conteined within his owne pulp and fruit (as is evident in the like example of wine, after it hath wrought long, which differeth exceedingly both in taste and nature from the grape out of which it was expressed) whereas being strengthened with this philosophicall fire, It

40 BL, SL 2245 ff. 95v–101.

retaineth still both the naturall taste, race, and verdure, that it had in the first expression: and so likewise of the Orange.

I doubt whether the doses of the 'fire' Plat suggested for water were strong enough to preserve water at sea. If it was 80 per cent alcohol, 24 ounces to 8 gallons would produce a liquor of 1.5 per cent alcoholic strength. If the dose was repeated frequently, strength would rise significantly: about 10 per cent by volume would be required to protect against putrefaction. Some of the other published ideas of Plat on water preservation remind one of the secret 'fires' – Drake's suggestion of rolling putrid barrels has an echo of stirring up the first type of 'fire' and the addition of aqua vitae or Irish whiskey is a crude version of the second.[41]

Plat envisaged preserving medicines (many of which he could supply) at sea with his 'fire of Nature'. As the chapter on his medical activities will reveal, Plat practised as a physician and possessed the recipes for many medicines. To separate food and medicine in the sixteenth and seventeenth centuries is difficult; diet was seen then by both doctors and patients as a prerequisite to good health. Nevertheless, Plat makes the distinction in his 1607 broadside, with lemon juice falling into the medicines, whereas in earlier publications he discussed its culinary use. He viewed the juice as a medicine because it 'hath of late been found by that worthy Knight Sir James Lancaster to be an assured remedy in the scurvy'. In 1601 Lancaster 'brought to Sea with him certaine Bottles of the Juice of Limons, which hee gave to each one, as long as it would last, three spoonesfull every morning fasting'. In contrast to the fate of sailors on other ships in the expedition who succumbed, Lancaster's crew did not get scurvy. There is no evidence that Plat knew of the use of lemon juice against scurvy prior to Lancaster's success, or that he supplied the lemon juice to him, but the prospect of seamen needing a supply of well-preserved juice with a good taste and colour (i.e. 'fresh') prompted Plat to add lemon juice to the medicines preserved by his 'fire of Nature'.[42]

41   Plat, *Certaine philosophical preparations of foode and beverage for sea-men*, c. 1607.
42   Pelling, *The Common Lot*, 1998, pp. 1–16; Keevil, *Medicine and the Navy*, 1957, vol. 1, p. 112; Plat, *Certaine philosophical preparations of foode and beverage for sea-men*, c. 1607. See BL, SL 2209 for surviving notes of Plat's medical practice.

Significantly, the broadside advertised 'Philosophical Preparations' of food and drink and 'Hermeticall medecines and Antidotes', proclaiming the author's alchemical approach to medicine. More specifically, his medical thought was influenced by the theories of Paracelsus and his followers, placing an emphasis on distilled essences, metals and other inorganic substances, and simple, strong medicines which would cure a number of diseases such as the 'antidote powder' which, if taken as soon as the disease appeared, would cure soldiers or mariners of the plague, burning fever, smallpox, measles, any kind of poison, swooning, or 'sodaine passion of the heart'.[43]

Like many others interested in medicine at the time, Plat culled his cures from a wide range of friends and acquaintances, as well as from his own laboratory and practice. From a great store of medical recipes, he produced what he considered was most needed by those going on long voyages. Agues, fevers, headache, scurvy, rheumatic diseases, fluxes, and the pox were the main ailments mentioned in the broadside and in a fuller manuscript list of 'Waters, oyles, Electuaries & other compositions necessary for the Sea'. Plat also undertook to supply (in the manuscript list) plaisters and ointments for wounds and burns; a wide range of distilled waters, including Dr Stevens Water – a sharp-tasting water sometimes recommended for scurvy – an ointment, syrup, and lotions for scurvy; several conserves and syrups – roses, barberries, ginger, violets, quinces, cherries – pleasant vehicles for vile-tasting medicines and a comfort for the sick; as well as medicines for a variety of other diseases. The list also includes oil of vitriol and Irish whiskey for mixing with water for a cooling drink (see above), as well as macaroni, ink powder, ship-piercing musket balls, and an anti-rust treatment for armour. By the standards of the day, Plat's medicines would have been regarded as a good foundation for a sea-surgeon's chest.[44]

Plat spent much time, effort and money on assembling ideas for better food, drink, and medicines for the army and navy, drawing from

43 BL, SL 2242; SL 2209; Plat, *Certaine philosophical preparations of foode and beverage for sea-men*, c. 1607.
44 Plat, *Certaine philosophical preparations of foode and beverage for sea-men*, 1607; BL, SL 2171, f. 30.

books in his possession, his own experiments, and a number of friends and acquaintans. Some of the new products he eventually offered to manufacture in commercial quantities, lobbying and advertising to drum up custom. His links with prominent seamen such as Drake and Raleigh have been mentioned. He was also acquainted with Sir John Hawkins, Treasurer of the Navy between 1577 and 1595, probably also his son Richard, and other merchants and seamen involved in long-distance expeditions overseas. Plat's contacts gave him a chance to promote his new foods and medicines, as well as 'picking their brains'. He planned to use one 'Bradbury, to harken after all the Captains and officers of long voiages. for the streights merchants etc.' to promote his 'macaroni for the sea'. The successful placement of new food or drink on the vessel of a prominent seaman was vigorously exploited as 'product endorsement' in subsequent publications. Drake's shipping of his pasta was repeatedly mentioned in Plat's books and pamphlets as he sought to associate himself with a national hero, 'that Spanish scourge & Magna spes altera Troie'.[45]

Plat's publications abound with more general advertisements for new types of victuals. In *Sundrie new and Artificiall remedies against Famine*, the hastily prepared pamphlet with suggestions to relieve the dearth of corn in 1596, he inserted a detailed advertisement for pasta as a sea-victual, listing all its virtues, mentioning the link with Drake, and successful placement on *The Bear*, one of a small fleet sent on a trading expedition to China in that year. Despite the claim that ' if I might once finde any good incouragement therein, I would not doubt but to deliver the same prepared in such sort, as that without anye farther dressing thereof, it should bee both pleasing, and of good nourishment unto a hungrey stomach', pasta, made from wheat, was no use in alleviating a hunger caused by the failure of the grain harvest in that year. The real reason for the advertisement was to encourage 'Al those which are willing to victual their ships therewith, if they repair unto me, I wil upon reasonable warning, furnish them therewith to their good contentment'. In similar vein, Plat advertised his new method of keeping meat without salt for long voyages in a *Discoverie*

45 BL, SL 2172, f. 14; Plat, *Jewell House*, pt. 1, pp. 8–9; Plat, *Sundrie new and Artificiall remedies against Famine*, 1596; Plat, *Certaine philosophical preparations of foode and beverage for sea-men*, c. 1607.

of certaine English wants, a pamphlet of 1595, offering to disclose details of the secret only for reasonable reward.

The culmination of the published advertisements for sea-victuals came with the single broadside sheet of about 1607, *Certaine philosophical preparations of foode and beverage for sea-men, in their long voyages: with some necessary, approved, and Hermeticall medicines and Antidotes, fit to be had in readinesse at sea, for prevention or cure of divers diseases*. This is a shop-card, a 'flyer', printed to be distributed to merchants, captains, and officials with an interest in provisioning private and naval vessels (and probably posted on quays and docks), in which Plat promised 'if it shall receive entertainemnt according to the worth therof and my iust expectation, I may happily be encouraged to prie a little further into Natures Cabinet, and to dispense some of her most secret Iewels, which she hath long time so carefully kept, onely for the use of her dearest children'.

Just how well prepared Plat was to supply the goods advertised is revealed by a single manuscript sheet, probably either a draft for another printed flyer or a note circulated to potential customers. Headed: 'Waters, Oyles Electuaries & other compositions necessary for the Sea', this paper, neatly written in Plat's best hand and logically organized, set out some 40 medicines for use at sea with prices per ounce, pint, pound etc., followed by a price list for pasta, bullets, anti-rust ointment for armour, and an ink powder, all inventions of Plat himself.[46]

Direct lobbying was, he recognized, essential when trying to place victuals and medicines with the army or navy, or to obtain reward from the state either directly by payment or in the form of pensions, privileges or honours. Approaches to the right people were of some concern; his manuscripts contain several lists of inventions with thoughts on how to bring them to the attention of powerful men. For instance, three abbreviated lists jotted on the fly-leaf of a notebook all contain, amongst other things, ideas on military victualling. That headed 'L. Tho Arundell', mentions couscous, flesh preserved without salt and beverage for the sea; under the heading 'my L. Threserer', he mentions some of the foregoing plus macaroni; and under 'Sr Wat. Raylegh', we find additionally flesh

---

[46] BL, SL 2172, ff. 13, 14, 30. See Appendix IV.

kept dry, flesh kept moist, bottled drink, and a variety of strong waters. Thomas Arundell, first Baron Arundell of Wardour was a noted soldier who fought the Turks in Hungary, a favourite of Queen Elizabeth and also favoured by James I who made him baron in 1605. The Lord Treasurer was almost certainly Thomas Sackville, Baron Buckhurst, a powerful and well-respected Privy Councillor who held high office under both Elizabeth and James I. Raleigh, sometime favourite of Elizabeth and a maritime adventurer, needs little biographical background. All were prominent courtiers, potentially useful to an ambitious suitor for royal favour. Another list, prosaically headed 'gayninge of money', includes a plan to lobby 'Sr John Fortescew to get the victuallyng of the Queenes ships' and then, 'seeke a pension for victuall for the navy'. Fortescue, Chancellor of the Exchequer from 1589, was, like Buckhurst, an important and well-respected government minister in charge of a department which handled royal finances including payments for the provision of the navy. Plat was, therefore, trying to interest the right people.[47]

Plat made a direct approach to Lord Buckhurst by a letter delivered in the Michaelmas law term of 1598. Headed, 'An offer of certen light, fresh, lastinge & portable victualls for the service of Ireland wch the author will, undertake eyther to make for the said service or uppon reasonable Composition to discover unto suche as shalbe thought meet for the same', it set out the virtues of macaroni, cooked meat which would keep fresh without salt for 3 months, and a third victual which was coyly described as a 'kinde of victaull the author doth preferr in extencion of norishment before anye other victuall whatsoever thoughe taken in a duble proporcion to the same. It serveth in stead both of bread & flesh & performeth theffect of both in the bodie'. (Either this was the 'portable soup' or the cooked oatmeal which Plat said he perfected for Irish service in January 1594.) He made the usual reference to Drake's use of macaroni, quoted prices for the pasta and the unnamed victual per ounce, and undertook to supply the foods as required. 'And because the author will not be thought to seek his owne good wth the hazard of the life or health of anye of the meanest of her ma.ties subiects by theis newe kinde of victaulls he will become bound

[47] BL, SL 2189, IV; SL 2172, ff. 13–14a; TNA, E. 351/2233. See also entries in the *ODNB* for these men.

himselfe to live one whole moneth or Sixe weeks together upon eyther of thaforesaid foods & wth the proportions aforesaid onelie havinge beere allowed him att his meales in stead of water.'[48]

Were these efforts successful? In the most obvious sense, that neither the English navy nor army immediately adopted pasta or any of the other new foods Plat proposed, we may conclude he made little impact on victualling. Even if he had convinced those in charge he would have come up against the conservatism of the ordinary sailors. Towards the end of the seventeenth century, they remained 'so besotted in their beef and pork that they had rather adventure on all the calentures and scurvies in the world than to be weaned from their customary diet, or lose the least bit of it'. They would not live like Spaniards and Italians, who existed 'upon rice, oatmeal, biscuits, figs, olives, oil and the like', or the French and Dutch who balanced meat consumption with dairy-produce, grains and pulses.[49]

Plat deserves recognition for the sensible way in which he considered the problems of victualling and suggested possible solutions. He understood that food preservation over ever longer periods and the protection of food from decay in hot climates were the most important concerns of the time. The use of 'fire of nature', new ways to preserve meat, concentrated meat jelly, and pasta were all attempts to address these matters. Pasta, 'gelly', and the Irish oatmeal recipe were all concentrated or dried foods, space-savers on long voyages and lighter for land-carriage. His advocacy of a still to produce fresh water at sea was potentially the most complete solution to both preservation and storage of fresh water. The medicines he offered were aimed at the major diseases of seamen and if we smile at the 'antidote powder' which, he claimed, was effective against the pox, plague, burning fevers, measles, and any kind of poison, such a powerful remedy against known diseases and unknown poisons and infections was required for voyages into strange lands.

Moreover, there is evidence that Plat's ideas were considered and his foods used. He claimed that his method of preserving roast beef in vinegar 'was fully proved in that honourable voyage unto Cales' (the Cadiz raid of 1587). Macaroni was used at sea on two expeditions in 1596 : 'With this

48 BL, SL 2172, f. 21.
49 'Boteler's Dialogues', ed. W.G. Perrin, *Navy Records Society*, 1929, p. 65.

MILITARY FOOD AND MEDICINE

food I furnished sir Francis Drake in his laste voyage which hath beene well approved and commended by sundry of his followers upon their return for England, whereby I was the more encouraged to make a second triall thereof in the Beare which went latelie for China'. Plat's claims are substantiated by an entry in the accounts for Drake and Hawkins' final voyage for a 'Lastinge victuelle … A kinde of victuells for the Sea service devised by Mr Hugh Platte'. Four barrels were bought at a cost of £4 10s. each, giving Plat the not inconsiderable sum of £18. Did he sell more pasta to victuallers for which no records have survived?[50]

Plat may well have provided pasta for an earlier expedition. We have encountered Captain Anthony Plot or Plat, almost certainly a relative of Sir Hugh. He was commander of a company of soldiers on Drake's voyage to the Indies in 1585 and was in overall charge of land forces on the Cadiz raid of 1587. In 1588 he married the sister of another of Drake's captains and leased a house owned by Drake in Plymouth. He was given the task of commandeering ships and loading them with victuals to provision the fleet led by Sir John Norris and Drake which raided Spain and Portugal in 1589. Anthony Plat worked slowly and Captain Crosse was dispatched to Plymouth by Lord Burghley in May 1589 to hasten matters. He 'inquired how it came to pass that these ships were so far behindhand…. Answer was made me that Captain Platt had the looking unto it and he had a preparation of a quantity of victual to be carried to the fleet, which he thinks to make gain of. And because he took it up among his friends and acquaintances in the country, it was so much the longer ere it could be ready and therefore he made no haste to make ready the shipping.' It is likely that one of Anthony's circle referred to was Sir Hugh, desperately manufacturing pasta in London and sending it to Plymouth. [51]

Mention has been made of Plat's recommendation of a still to provide fresh water at sea, an idea repeated with approval by the captain and admini-

50 Plat, *Sundrie new and Artificiall remedies against Famine*, 1596; Plat, *Delightes for Ladies*, 1602; TNA, E351/2233.
51 Plat, *The Jewell House*, pt. 1, pp. 10–11; G.M. Thomson, *Sir Francis Drake*, 1988, p. 199; J.S. Corbett, *Drake and the Tudor Navy*, 1898, vol. II, p. 14n; *The expedition of Sir John Norris and Sir Francis Drake to Spain and Portugal, 1589*, ed. R.B. Wernham, 1988, p. 163; John Sugden, *Sir Francis Drake*, 1990, p. 82.

strator Nathaniel Boteler some years later. While the expense of fuel and small capacity of such equipment might seem to make this an outlandish suggestion, circumstantial evidence suggests that Richard Hawkins, son of Sir John, may well have heeded Plat or even been supplied with a still by him. In 1593, on a voyage to Brazil, 'our fresh water had failed us many dayes (before we saw the shore) by reason of our long Navigation, without touching any land ... yet with an invention I had in my Ship, I easily drew out of the water of the Sea sufficient quantitie of fresh water to sustaine my people, with little expence of fewell, for with foure billets I stilled a hogshead of water, and therewith dressed the meate for the sicke and whole. The water so distilled we found to be wholesome and nourishing.' On the same voyage, for the first time Richard Hawkins took 'Dr Stevens' water (one of the medicines Plat advertised) with him against scurvy.[52]

The letter to Lord Buckhurst about provisioning the army in Ireland made some impression. Buckhurst wrote a paper in January 1599/1600 headed 'Considerations touching Ireland causes', about the forthcoming military campaign. He noted 'That Mr plott's device for victual of continuance may be considered by my Lord Mountjoy and by some of the Captains, especially for the keeping of forts and sudden services.'[53]

Despite the receptivity to Plat's ideas of private seafarers and those with influence over the navy and the army in Ireland, the income he sought from these and other inventions did not match his expectations. The final paragraph of the 1607 broadside threatened that if no interest was shown in buying his wares, 'I doe here free and enlarge my selfe from mine owne fetters: purposing to content my spirits, with my place and dignitie, and in likelyhood proove also more profitable in the ende, then if I had thankelesly devoted my selfe to Bonum Publicum. In which course, happy men are sometimes rewarded with good words: but few or none, in these dayes, with any reall recompense.' In other words, failing more tangible reward, he would console himself with the knighthood recently bestowed upon him.[54]

52 Keevil, *Medicine and the Navy*, 1957, vol. I, pp. 101, 103; 'Boteler's Dialogues', ed. W.G. Perrin, *Navy Records Society*, 1929, p. 65.
53 *Calendar of State Papers, Ireland*, vol. CCVII, 1599–1600, p. 377.
54 Plat, *Certaine philosophical preparations of foode and beverage for sea-men*, c. 1607.

CHAPTER FIVE

The Writing of Delightes for Ladies *and* Sundrie new and Artificiall remedies against Famine

This chapter looks closely at how Plat composed his two works largely devoted to food and drink, *Sundrie new and Artificiall remedies against Famine*, (1596) and *Delightes for Ladies* (1600 or 1602).[1] Though he devoted much of his intellectual energy to examining these matters, to judge by his published output and surviving notes, Plat was no epicure, albeit he could appreciate a well-made cheese or delicately flavoured butter and enjoy the taste and colour of a good wine. His cookery knowledge (as I argue below) was not great. His primary interest in all of the natural science he investigated was technological improvement – new and better ways of doing things, be they industrial processes such as dyeing, moulding or saltpetre production, or the machines he proudly exhibited at his London house. And this love of technology is reflected in his writings on food and drink: improvements in brewing and making and keeping wines; a new machine to sift flour; new types of food for use

1 Part of this chapter is drawn from Malcolm Thick, 'A Close Look at the Composition of Sir Hugh Plat's *Delightes for Ladies*', Chapter 3 of *The English Cookery Book, Historical Essays*, ed. Eileen White, Totnes, 2004.

at sea; and many more. One has the feeling, for instance, that in his long and detailed description of making 'sugar-plate, and casting thereof in moulds', his interest has been captured by the difficulty of making a perfect paste, casting it cleanly and gilding the cast with gold-leaf. The fact that the cast has anything to do with dining is incidental. Above all, Plat was interested in the technology of food preservation. 'The Arte of Preserving, Conserving, Candying &c.' forms the first and longest section of *Delightes for Ladies* and preserving fruits and flowers also begins *The Jewell House*. Plat was not the only writer to take an interest in preserving at this time. Joan Thirsk has drawn attention to a general upsurge of writing on fruit and flower preservation in the late sixteenth century, due partly to fashion, and partly to a desire to enjoy the widening range of fruit then grown.[2]

Delightes for Ladies was, to judge from its publishing history, a successful book. Entered into the Stationers' Company records in October 1599, it may have first appeared in 1600. There certainly was an edition in 1602, four more by 1610, and at least 16 further editions by 1656.[3] It is undoubtedly a pleasing book to handle. The small pages of early editions are each surrounded by an ornate border. An introductory poem by Plat, much better than most doggerel found at the beginning of popular books of the time, summarizes the contents and also Plat's publications to date. Indeed, the poem serves as an excellent and elegant abstract of the work. (Maybe learned journals today should require abstracts in verse.) A decent index leads one into the text, arranged by short numbered chapters in four sections: 'The Art of Preserving, Conserving, Candying &c.'; 'Secrets in Distillation'; 'Cookery and Huswifery'; and ' Sweet Powders, oyntments, beauties, &c.' While not the first book published in England to tackle these subjects, it is better arranged than most of its predecessors, with less evidence of recipes culled uncritically from the 'books of secrets' then popular. Pleasing to the eye and easy to use, the book's success can be readily understood.

2   BL, SL 2189 f. 65–65v; Joan Thirsk, *Food in Early Modern England*, 2007, pp. 50–55.
3   Bent Juel-Jensen, 'Some Uncollected Authors XIX, Sir Hugh Plat ?1552–?1611', *Book Collector*, 1959, pp. 64–66; Plat, *Delightes for Ladies*, Intr. G.E. & K.R. Fussell, 1948, p. vi.

As I confine myself in this chapter to Plat's food writing, I will say little of his last section on cosmetics, save to record that there are 37 recipes, some for perfumed soap and washing water, and many for improving the appearance by means such as colouring hair, removing facial blemishes or cleaning teeth: a fair representation of the preoccupations of the time, especially in the competitive world of the Court which Plat knew well. Some gentlewomen (and men) were tempted to go to extreme lengths to improve their looks. Plat warns gentlewomen to be careful when barbers clean their teeth with aqua fortis (nitric acid), 'for unlesse the same be well allayed, and carefully applied, shee may bee forced to borrow a ranke of teeth to eat her dinner, unlesse her gummes do help her the better.'[4]

On reading the text, it is clear that 'Cookery and Huswifery' is the weakest section. Only 35 pages long, it consists of several chapters [5] on preserving, some interesting and detailed passages on flavoured butters, a few miscellaneous chapters on such topics as candle-making, table-salt, flavoured vinegars, bottling beer and oddities like how to keep flies from oil paintings. The recipes do not provide even a basic guide to the main cooking techniques – only sousing and boiling are included. Many of the recipes themselves are poor, missing out vital stages of the cooking process. Take this:

To boil Sparrows or Larks
Take two ladles full of Mutton broth, a little whole mace: put into it a peece of sweet butter, a handfull of Parsly being picked: season it with sugar, veriuice, and a little pepper.

---

[4] Plat, *Delightes for Ladies*, 1628, d. 26. Note: unless otherwise specified, I refer throughout to the 1628 edition. There is no pagination. I have followed the index to this edition, which allocates a letter to the four sections of the book: a. *The Art of Preserving, Conserving, Candying, &c.*; b. *Secrets in Distillation*; c. *Cookery and Huswifery*; d. *Sweet Powders, oyntments, beauties, &c*. Chapters are numbered within each section.

[5] I refer to the numbered recipes as 'chapters' and the parts of the book dealing with preserving, cookery etc., as the 'sections'.

This one-sentence recipe does not specify a cooking vessel, gives no instructions for cooking or serving the birds – in fact, no birds are mentioned at all![6]

Why, when most of his recipes and instructions are detailed and comprehensive, are the cookery ones relatively poor? Part of the answer is that Plat did not have an interest in cookery as such – there are no purely cookery recipes in his own hand in the pages of manuscripts on food which have survived. Moreover, the cookery recipes in *Delightes for Ladies* were not, in fact, written by Plat. In a notebook in the British Library, now Sloane MS 2189, we find the origin of the bulk of them. These manuscript notes, headed 'Divers receipts of cookery', are not in Plat's hand or usual spelling: they are not his work. He simply placed them in the cookery section, with all faults, to form the backbone of this part of the book.

Plat's borrowing from this source did not end with cookery. Twenty-seven out of the seventy-three chapters in the preserving section (the largest section of the book) are from this source. And these are not minor recipes used to fill out the text: one is an eleven-page treatise, 'The art of comfet-making, teaching how to cover all kinds of seeds, fruits or spices with suger.'[7] This detailed and practical chapter takes one through the whole process of manufacturing comfits, beginning with a list of the utensils required, then moving from basic comfit-making to variations of flavour, shape and colour. The other 'borrowed' preserving recipes cover sugar plate, 'jumbolls', 'marchpane', gingerbreads, 'bisket-breads', puff pastry, jellies, quince pastes, marmalades, preserving citrus fruits, and candying with hard-rock candy. The recipe 'To make sucket of Lettuce stalks', which has been commented upon as one of Plat's more unusual ideas, comes from the borrowed source, as does that for gingerbread 'used at the Court, and in all Gentlemens houses at festivall times' and that for 'Quidini of

6   Contrast this recipe with that 'To Stue Sparrowes or larkes' in A.W., *A Booke of Cookrye*, 1591, p. 11: 'Take the best of Mutton broth, and put it in a Pipkin, and put to it a little whole Mace, whole Pepper, Claret wine, Marigolde leaves, Barberies, Rosewater Vergious, Suger, and Marrowe, or els sweet Butter. Perboile the Larkes before and then boyle them in the same broth and lay them uppon Sops.'
7   Plat, *Delightes for Ladies*, a. 54 (which in some editions is misprinted as a. 24).

Quinces'. In short, many recipes from *Delightes for Ladies* which have been quoted by food historians over the years as the work of Plat are by someone else.[8]

Who was the author to whom Plat was indebted? Manuscript BL, SL 2189 is composed of two notebooks bound together. The original first leaf of the longer one was written by our unknown author: instructions to make putty.[9] The putty recipe concludes, 'The first that ever I made proved T.T.' Numerous other recipes 'proved' by 'T.T.' in the manuscript provide us with the initials of the author.

A gap of a page after the putty recipes is followed by two and a half pages of bread-making advice in the unknown hand, followed by recipes written by Plat. This pattern of recipes by T.T. with additions by Plat is repeated throughout the notebook, with sections on preserving and cookery, bread, beer, gardening, candle-making, dyeing, leather, ink, paints, and organ pipes. T.T. thus wrote his notes before Plat and Sir Hugh used them as a template around which he entered his own observations, a conjecture confirmed by the older spelling used by T.T. and a couple of dates in his manuscript: a candle wax recipe dated 1559 and a small comfit recipe of 1561. Two recipes out of the *Widowes Treasure* of 1585 were, I think, added by another hand, so Plat probably acquired the papers after that date but before 1588, the earliest dated recipe in Plat's hand in the manuscript.

In *Floraes Paradise*, his book on gardening, Plat reproduced many of the notes T.T. wrote on gardening, attributing one to 'T.T. a Parson'.[10] T.T.'s observations on the repair of organ pipes and his detailed notes on candle-making support this profession, in particular, a recipe for a Candlemas candle 'for a trendell on Candlemas day'. A *trental* was a set of 30 successive masses for the dead, a Catholic rite. I believe that T.T. may have been a priest in Mary's reign and much of his manuscript may date from as early as the 1550s. He made frequent mention of the urban occupations of apothecaries and brewers, making it probable that he

8   Plat, *Delightes for Ladies*, a. 22, a. 28, a. 32; Laura Mason, *Sugar-Plums and Sherbet*, Totnes, 1998, p. 229.
9   BL, SL 2189, f. 55.
10 Plat, *Floraes Paradise*, 1608, p. 17.

Figure 5 (above). The title-page of the 1628 edition of Delights for Ladies.

Figure 6 (below). Pages from Delights for Ladies *with decorated border and recipes attributed to 'T.T.'*

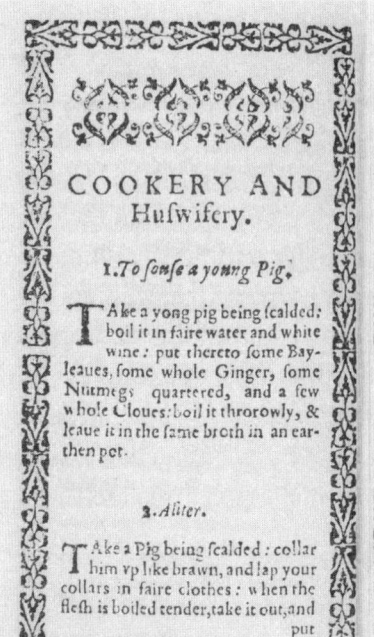

was a Londoner. His cookery recipes are not good but he had a close acquaintance with comfit-making and banqueting stuff, as evidenced by the detail of some of the recipes Plat published.[11]

Only two men can I find who might fit the few facts known of T.T. Thomas Thurland, appointed master of the Savoy Hospital around 1557, who died in 1574, was a priest but also a mining entrepreneur, foreign traveller and an embezzler, described as 'a papist of scandalous life'. A letter written from a prison cell when he was arrested for debt, however, is not in the same handwriting as that of T.T. in Plat's manuscript. The other possibility, Thomas Tymme, Rector of St Antholin in Budge Row, London, was also an unusual clergyman, who turned from writing devotional works to books on alchemy, a subject of great interest to Plat. He, however, was probably too young to have written Plat's cookery notes: he was not ordained until 1566 and he lived until at least 1620. His publications show no interest in cookery. While it is unlikely that a debtor in gaol would employ an amanuensis to write his letters, Thurland is the likeliest author of the notes by T.T.[12]

I believe that the manuscript by T.T. was the starting point of *Delightes for Ladies*. As noted above, there is a dated recipe by Plat from 1588, the next is in 1595, and a scattering of dates occur between then and 1599, the year the book was entered in the Stationers' Company records. The position in the manuscript of these later dated recipes indicates that Plat added many recipes in the few years before *Delightes for Ladies* was published, building around T.T.'s notes a corpus of ideas from which he could write his new book. He also drew some recipes from another notebook written up in the 1590s[13] and took material from his earlier book of 1594, *The Jewell House of Art and Nature*. These sources account for most of the recipes in the food sections of *Delightes for Ladies*.

11 Plat, *Delightes for Ladies*, a. 54.
12 Robert Somerville, *The Savoy*, 1960, p. 42; M.B. Donald, *Elizabethan Copper*, 1955, pp. 15–35; G.L. Hennessy, *Novum repertorium ecclesiasticum parochiale Londinense*, 1898, pp. 295, 302; TNA, State Papers, 12/36; Bruce Janacek, 'Thomas Tymme and Natural Philosophy', *Sixteenth Century Journal*, vol. XXX, no. 4, 1999, pp. 987–1077.
13 BL, SL 2216.

About a dozen recipes are copied exactly or paraphrased from *The Jewell House of Art and Nature* but he was not merely recycling them unmediated.[14] Some have been copied without passages later considered irrelevant – he removed a long criticism of vintners for adulteration from the wormwood wine chapter, and recipes for drying rose leaves and extracting spirit of roses are cut down.[15] The chapter 'To keep the iuyce of Orenges and Lemmons all the yeare for sauce...' is a good example of a rambling entry in *The Jewell House* which Plat edited and revised, adding for his London lady-readers a useful money-saving tip:

> And because that profit and skill united do grace each other, if (curious Ladies) you will lend eares, and follow my direction, I will heer furnish a great number of you (I would I could furnish you all) with the iuyce of the best Civill Orenges at an easy price. About Allhollantide, or soone after, you may buy the inward palp of Civill Orenges, wherein the iuyce resteth, of the comfit-makers for a small matter, who doe onely or principally respect their rindes; to preserve and make Orengadoes withall, this iuyce you may prepare and reserve as before.[16]

There is a chapter on purifying salad oil in both books, but they are entirely different, and although both *The Jewell House* and *Delightes for Ladies* have instructions for keeping oysters, he rethinks his earlier reliance on sea water or brine as a preservative, adding vinegar to produce a more reliable liquor. In the case of flavoured butter, Plat provides a cross-reference to the earlier book, as he does in the detailed chapter on moulding and casting animals in sugar-paste, where he speculates that the wax moulds described in *The Jewell House* might be used for these elaborate banqueting conceits.[17]

14 The Fussells exaggerated when they wrote 'Many .... recipes were later included in Delightes for Ladies', G.E. & K.R. Fussell, *Delightes for Ladies*, 1948, Introduction, p. xvi.
15 Plat, *The Jewell House*, pt. 1, p. 15; pt. 3, pp. 34, 42; Plat, *Delightes for Ladies*, a. 36, b. 17, c. 33.
16 Plat, *Jewell House*, pt. 3, pp. 36; Plat, *Delightes for Ladies*, c. 35.
17 Plat, *Delightes for Ladies*, a. 10, c. 21.

THE WRITING OF *DELIGHTES FOR LADIES* &c.

The only other publication acknowledged by Plat as source for material in *Delightes* is a treatise on cookery by Bartolomeo Scappi published in Venice in 1570. Plat mentions this book by 'the maister cooke of Pope Pius Quintus, his privie kitchen' both in *The Jewell House* and in *Delightes for Ladies* and he was clearly impressed by it, as indeed he was by many aspects of Italian cuisine.[18]

Although few names of informants are printed in *Delightes for Ladies*, we can augment our knowledge of Plat's sources by going back to his manuscripts and by looking for allusions in the printed text. In particular, he includes many references to London tradesmen involved in supplying food, drink and medicine. Thus he suggests using lead-lined earthenware pots 'such as the Gold finers call their hookers, and serve to receive their Aqua fortis' for storage of rosewater and dried rose leaves. Other London traders supplied him with materials such as the orange pulp from comfit-makers we have already mentioned and the wine lees used to preserve quinces which he obtained from taverners or wine merchants.[19]

Sometimes he claims that he has improved on the traders' own practices: so his wormwood wine was 'a more neat and wholesome wine for your body, than that which is sold at the Stillyard' (presumably imported from Germany); by following the recipe for syrup of violets 'you may gain one quarter of syrup, more than diverse Apothecaries doe'; he promised bottled ale 'that should farre exceede all the Ale that mother Bunch made in her lifetime'.[20] He was afraid that his method of preserving lobsters and other crustaceans would hurt fishmongers' businesses 'who, onely in respect of their speedy decay, doe now and then afford a penny worth in them'. But, if they took note of his way of preserving salmon, 'Vintners and Cookes may make profit thereof when it is scarce in the markets and salmon, thus prepared, may be profitably brought out of Ireland, and sold in London or elswhere'. He was scathing about ready-made mustard sold in London: 'I thought it very necessary to publish this manner of making your sawce, because our mustard which we buy from the chandlers at this day, is many

18 *Opera Di M. Bartolomeo Scappi*, Venice, 1570.
19 Plat, *Delightes for Ladies*, a. 3, a. 67, c. 35.
20 Plat, *Delightes for Ladies*, a. 4, c. 30, c. 33; 'Mother Bunch' was a generic name for London ale-wives.

times made up with vile and filthy vinegar, such as our stomacks would abhorre, if we should see it before the mixing therof with the seeds.'[21]

From apothecaries he got both recipes and ideas: he was on good terms with the regular suppliers of materials and medicines he used in his medical practice.[22] One Crosley provided a number of recipes, mostly medicinal, including one interesting preparation of fine powdered steel for the bloody flux, a powerful chemical and metallic medicine like those advocated by Paracelsus, the Swiss-German doctor and alchemist. Plat patronized apothecaries who supplied Paracelsian medicines, such as 'Maister Kemmish' a well-known apothecary who provided Plat with his recipe for Doctor Stevens Aqua Composita and to whom Plat recommended readers wanting oils distilled from herbs. They might 'repair to Maister Kemish, that aunciant and expert Chimist dwelling neere the glashouse, at whose hands they may buy any of the aforsaid oiles in a most reasonable manner'.[23] A third apothecary, Mr Parsons (a schoolfriend), was commended as an 'honest and painefull Practicer in his profession'. He provided Plat with two of the preserving recipes in *Delightes for Ladies*.[24]

Another acquaintance whose profession was germane to the contents of Plat's book was a Mr Webber who told Plat how to make 'A most delicate and stiff suger past, wherof to cast Rabbets, pigeons, or any other little byrd or beast, either from the lief, or carved in moldes'. With this paste one could mould fowls and other animals, cover them 'with crums of bread, cinnamon and suger boiled together: and so they will seem as

---

21   Plat, *Delightes for Ladies*, c. 16, c. 25, c. 29.
22   Notes of his medical practice are to be found in BL, SL 2209. See also Chapter 7.
23   Plat, *Delightes for Ladies*, a. 53; BL, SL 2189, ff. 155–156v; SL 2203, f. 213v; SL 2189, f. 156v. George Baker, who edited a translation of a work by Conrad Gesner concerned with the new chemical medicines, published as *The Newe Jewell of Health* in 1576, in the preface listed those who sold such preparations in London, including 'one mayster Kemech an Englishe man dwelling in Lothburie'. Quoted in Paul H. Kocher, 'Paracelsan medicine in England', *Journal of History of Medicine*, Autumn 1947, p. 460. Plat was also a friend of Master Jacob, who ran the glasshouse, and they talked of ways to use waste heat from glassmaking to keep plants warm indoors. Plat, *Floraes Paradise*, 1608, pp. 38–9.
24   Plat, *Delightes for Ladies*, a. 69, a. 70; BL, SL 2209 f. 2; SL 2216 ff. 50v–52.

if they were rosted and breaded ... By this meanes, a banquet may bee presented in the forme of a supper, being a very rare and strange device.'[25] He was probably a leading expert on banqueting conceits being, when Plat met him in December 1595, 'one of her Ma[jes]ties Privie Kichin'.[26]

Outside the food, drink and medical suppliers, Plat had a wide circle who provided him with recipes and ideas – Master Rich of Lee (probably in Essex), 'Nepper the Skott' (possibly John Napier, the inventor of the logarithm), Thomas Gascoigne who had an old parchment book from which Plat took recipes, and Mr Hill, an Exchequer Auditor who provided recipes for Aqua Rubea and drying rose leaves.[27] Some of Plat's informants were foreigners who brought with them experience of strange lands. One was the Spaniard from Naples Sr Romero, an intriguing figure who was the source of just one recipe in *Delightes for Ladies* ('Chestnuts kept all the yeere') but, from the evidence of Plat's manuscripts, provided him with many ideas on food, wine, medicine, military technology.[28] Some of the ideas for *Delightes for Ladies* came not from abroad but from Plat's own household, more specifically, his wife. A recipe not included in the book but found in manuscript for 'sallets of flowers all ye yeere', involving reconstituted dried flowers, was 'my wiffes invention'. In the book she is associated with dairy produce, traditionally a housewife's domain.[29]

Plat was very taken with recipes to flavour and colour butter and cheese. Many are to be found in his manuscripts and they appear first in *The Jewell House*, to be developed further in *Delightes for Ladies*. In the earlier publication, after explaining his method of extracting oils from herbs and spices, he describes 'How to make sundry sorts of dainty butter

25 Plat, *Delightes for Ladies*, a. 10. A banquet in Plat's time was the dessert course of a formal meal, often served in a separate room at a table decorated with elaborate sugar sculptures.
26 BL, SL 2189, f. 71v. Webber was skilled at moulding for he also told Plat how to make papier mâché into 'frontages, antiques, or other beastes to abide the weather' which 'the Italians garnishe the owtsides sides of theyre howses wth'.
27 BL, SL 2189, f. 155v; CSPD, 1595–7, pp. 68, 157, 167. These men occur in *Delightes for Ladies* as follows: Rich: a. 2, a. 3, BL, SL 2189, f. 71; Nepper: b. 21, BL, SL 2189, f. 151a; Gascoigne: a. 45, a. 46, BL, SL 2216, ff. 230–231; Hill: b. 7.
28 BL, SL 2210, ff. 87, 131v, 134–134v.
29 BL, SL 2189, f. 38v.

with the saide oils' providing us with a glimpse of his family's breakfast table at Bishop's Hall, Bethnal Green: 'In the moneth of may it is usuall with us to eat some of the smallest, and youngest, sage leaves with butter in a morning'. He suggests using oils with butter instead for 'a far more lively and penetrative taste then can be presently had out of the greene herbe'. He continues, 'This last sommer I did entertaine divers of my friends with this kinde of butter amongst other country dishes, as also with cinnamon, mace, and clove butter … and I knew not whether I did please them with this new found dish, or offend them by denying the secret unto them'. Despite his expertise with butter, he confesses himself no countryman, for he 'never kept but twoe kine, in any one summer'.[30]

In *The Jewell House* Plat suggests a flavoured cheese made by mixing vegetable and spice oils with the curds. He continues by describing 'a tricke in the making of a cheese' by subjecting it to a gentle pressing only, thereby expelling only a thin whey 'and so your cheese wil bee much bigger, and better than otherwise it would be.' Not going into details, he explains, 'I would be loath to offend a Gentlewoman that presumeth of a great secret herein' … But *I think I have given light sufficient to a good dairie Woman to find out al the circumstances therof in time*.'[31]

In *Delightes for Ladies* Plat is much more expansive about this new type of cheese, giving precise quantities, the design of a press, and the exact handling required. The object, it appears, is to produce a close-textured but full-cream cheese, 'your ordinarie Cheeses are more spongious and full of eyes by reason of the violent pressing of them; whereas these cheeses setling gently and by degrees, doe cut as close and as firme as Marmalade.'[32] The gentle pressing yields, he estimates, about a quarter more cheese than by ordinary pressing. Displaying a knowledge of Continental cheeses, Plat speculates, 'I suppose that Angelores in France may bee made in this manner in small baskets, and so likewise of the Parmeesan.'[33]

30   Plat, *Jewell House*, pt. 3, pp. 10–12.
31   Plat, *Jewell House*, pt. 3, p. 13.
32   Plat, *Delightes for Ladies*, c. 22.
33   According to an application to list Livarot cheese as an EC regional speciality: 'Livarot is one of Normandy's oldest cheeses; like Pont-l'Evêque, it claims to be the descendant of the Angelot cheese referred to in the *Roman de la Rose*, written

The anonymous gentlewoman of 1594 whose secret he guarded is now revealed as his wife. 'I have robbed my wives Dairy of this secret, who hath hitherto refused all recompences that have beene offered her by gentlewomen for the same: and had I loved a Cheese myself so well as I like the receipt, I thinke I should not so easily have imparted the same at this time.' To impress his readers with the sacrifice he has made (albeit of his wife's secret) he adds: 'And yet I must needs confesse, that for the better gracing of the Title where-with I have fronted this pamphlet, I have been willing to publish this with some other secrets of worth, for the which I have many times refused good store both of crowns and angels. And therefore let no Gentlewoman think this Booke too deare, at what price soever it shall be valued upon the sale therof: neither can I esteem the work to be of lesse than twenty yeeres gathering.'[34]

From what I have written so far of *Delightes for Ladies*, the reader may be wondering if Plat had many ideas of his own. He did include recipes which were clearly derived from his own experience. From these, we can learn something of his main interests in food and medicine. Plat often tells us that a recipe is his own, or at least that he has used it regularly, with phrases such as, 'This way I have often proved excellent'; 'accept [this] as a new conclusion'; 'I know by mine own experience'; 'This an approoved Secret, easie and cheap'. Talking of preserving artichokes, he digresses into another glimpse of his household:

---

by Guillaume de Lorris in 1260. Thomas Corneille, in his *Dictionnaire Universel Géographique et Historique* of 1708, recites its praises. In the 19th century it was consumed in greater quantities than any other cheese by the people of Normandy. In April 1970, the producers of Livarot formed an association and applied for the Appellation d'Origine, which they obtained in December 1975 (Regulation of 17 December). Method of production: The curds, after being cut into large cubes, are left to stand then cut again, pressed, allowed to settle and put into moulds. After being turned several times and drained in a drying area, the cheese is salted with cooking salt or in brine. During its time in the cellar, it is turned over and washed at least three times with water, to which annatto is added for the final wash. It can be eaten after a ripening period of at least three weeks.' *Council Regulation (EEC) No. 2081/92 Application for registration.*

34 Plat, *Delightes for Ladies*, c. 22.

In a milde and warm winter, about a moneth or three weeks before Christmas, I caused great store of Artichokes to bee gathered with their stalkes in their full length as they grew: and, making first a good thicke Lay of Artichoke-leaves in the bottome of a great and large vessell, I placed my Artichokes, one upon another, as close as I could couch them, covering them over, of a pretty thicknesse with Artichoke-leaves: these Artichokes were served-in at my Table all the Lent after, the apples being red and sound, onely the tops of the leaves a little vaded, which I did cut away.[35]

Whether or not recipes were his originally, he put considerable effort into perfecting them. In the preserving section is a very short chapter, one sentence long, giving directions for sweet cakes made from pounded dried parsnips and flour. Plat's original manuscript note of this recipe is beset with inserts and marginal notes: 'qre of stammpinge these rootes, iff so they will bee made into a paste to make cakes of'; 'qre iff they will not grinde'; 'qre of dried pompions'; 'qre of dried chestnutts./ skirret rootes, & the cakes of almonds after the oyle is expressed will make excellent bread'; 'qre what the seedes of pompions wolde doo in any victuall'. This simple recipe has generated several ideas and variations in his mind. He concludes by confirming that 'this I did both Invente and prove in Cakes'. The cakes first appeared in his pamphlet on famine relief published in 1596 and, despite their origin as a desperate measure to eke out flour in hard times, Plat took a liking to them, eating them 'diverse times in my own house'.[36]

A project Plat worked on for many years was the improvement of victuals for the navy and the army abroad (described in the preceding chapter).[37] In *Delightes for Ladies* are hints of this project such as: 'Trosses for the Sea' (cough sweets); 'How to keepe rosted Beefe a long time sweet and wholesome' (the recipe involving wine vinegar which was tried on Drake's Cadiz expedition); and the detailed 'Conceipt of the Authors, how Beefe may be carried at Sea, without any strong & violent impression of salt, which is usually purchased by long and extreame powdering.' This last,

35 Plat, *Delightes for Ladies*, a. 69, b. 11, b. 23, c. 24, c. 29.
36 Plat, *Sundrie new and Artificiall remedies against Famine*, 1596; BL, SL 2210, f. 31.
37 See Chapter 5.

by which beef was to be preserved in a box full of holes, dragged behind a ship using the flow of sea water as preservative, is uneasily inserted into the text with an apology to his 'courteous Gentlewomen' readers for his sudden launching 'a little from the shoare'. Plat clearly wants to take the opportunity of a new publication to publicize further his ideas for naval victuals, promising to follow up with a new way of storing food at sea and another way of keeping roast meat of all kinds at sea for up to six months: 'And this I hold to be a most singular and necessary Secret for all our English Navie; which at all times, upon reasonable terms, I will bee ready to disclose for the good of my countrey.'[38]

The substantial manuscript of T.T. which forms a backbone to *Delightes for Ladies*, plus the many recipes and ideas from friends, relatives, and acquaintances acknowledged by Plat, may leave us disappointed that so much of what the book has to say on food and drink did not originate with him. Cookery, however, is rarely truly original; recipes and methods are variations on themes suggested by others; indeed subtle changes may transform a recipe into a far superior dish. Cooks throughout the ages have kept notes and some have turned these into books with or without revealing the sources of recipes copied from others. Close examination of the material in Gervase Markham's *The English Housewife*, published in 1615 only a few years after *Delightes for Ladies* first appeared, reveals parallels between the two works (and indeed between the writers). Markham, like Plat, spent time on the fringes of the Court and, after early flirtations with poetry, applied himself to writing on many topics but mainly agriculture, horticulture, and household management. Like Plat, he was a gentleman who was keen to find ways of making money. And *The English Housewife* was based on the work of others: the printer of the first edition announced 'this is no collection of his whose name is prefixed to this work, but an approved manuscript which he hath happily light on, belonging sometime to an honourable personage of this kingdom'. A recent editor has established that Markham based the work on several manuscripts. Markham also took recipes from printed books, including one from *Delightes for Ladies*. Sir Hugh Plat, therefore, was not exceptional

---

38   Plat, *Delightes for Ladies*, a. 39, c. 18, c. 20.

in his reliance on others, only in his leaving manuscripts from which we may discern how he arrived at his published text.[39]

We should not belittle his achievement. Apart from the insertion of T.T.'s manuscript, the work has freshness and originality – it is not merely a rehash of recipes culled from books of secrets. Plat's ideas came, in the main, from contemporary experience, his own, his friends' and those of tradespeople whose living depended on the success of the recipes they divulged to him. The idea for a 'history of trades' which runs through the seventeenth century from Bacon, though Samuel Hartlib to Evelyn and the Royal Society, can be seen at work in Plat's questioning of apothecaries, comfit-makers, and professional cooks. This book is the most frequently reprinted of all his works, republished at least 23 times up to 1656.[40]

******

Plat's other publication devoted to food, *Sundrie new and Artificiall remedies against Famine. Written by H.P. Esq. upon thoccasion of this present Dearth*, of 1596 is also partly composed of material taken from a manuscript not written by Plat (as well as much that he has put together from other published sources). The reason for such a jumble was haste. This is a pamphlet (of 33 pages) full of advice on how to cope with a major food shortage. It was published in the middle of what one historian has described as 'The Great Famine [which] extended over nearly all Europe, lasting for some three years.' 1596 was England's worst year. Despite a 'temperate winter' which 'made mens hearts to leape for joy, and the Barnes, as it were, to enlarge themselves for the receipt of this promised plentie', unseasonable weather, especially at harvest time, with 'never ceasing raine' and 'tempestuous winds which choake out the corne when it would have been shorne', destroyed early hopes and left the standing corn 'utterlie rotted and corrupted'. The shock of the dearth was intensified by contrast

39 Markham, *The English Housewife*, ed. Michael R. Best, pp. xii–xiv, xvi–xxii, lvi–lviii.
40 Michael Hunter, 'John Evelyn in the 1650s: A Virtuoso in quest of a Role', pp. 86–91, in *John Evelyn's "Elysium Britannicum" and European Gardening*, Washington, USA, 1998; Bent Juel-Jensen, 'Some uncollected authors XIX', *The Book Collector*, 1959, pp. 60–68.

with earlier abundance: harvests of 1591, 1592 and 1593 were all good. The bad weather affected all major bread grains, while alternatives for the poor such as pulses and dairy produce also rose significantly in price. Cattle and sheep prices rose less steeply, but such meats were usually beyond the pockets of the poor. Mortality in London, as measured by burials, rose in 1597. Unless there was an outbreak of some unrelated illness, some inhabitants must have died from malnutrition or hunger.[41]

Government response to the crisis came from the Privy Council which issued 'The Book of Orders' (a comprehensive set of instructions on action to be taken in times of dearth), proclamations, and exhortations to local authorities. An attempt was made to even out shortages by persuading areas with more grain to send some to places suffering a lack. The Privy Council condemned speculators, whom it blamed for keeping prices high. It commented in 1597 that, although the harvest was better, 'yet there are seene and fownde a number of wicked people in condicions more lyke to wolves or cormorants then to naturall men, that doe moste covetously seeke to holde up the late great prices of corne and all other victuells by ingrossing the same into their private hands'. The Council banned exports of grain, encouraged imports, sanctioned hijacking of foreign grain ships in the Channel, and regulated malting, brewing and starch-making to maximize bread grains. Lastly, moral suasion was tried: archbishops were ordered to tell clergy to preach against gluttony, encourage charity, and urge the poor to accept their lot; the Lord Mayor of London was ordered to prevent excessive feasting in the city.[42]

With some 200,000 inhabitants, London posed an especial problem. The poor, and many of the rich, had no gardens and depended entirely

41 W.G. Hoskins, 'Harvest fluctuations and English economic history, 1480–1619', *Ag. Hist. Rev.*, xvi, I, 1968, pp. 37–38; Gustaf Utterstrom, 'Climatic Fluctuations and Population Problems in Early Modern History', *The Scandanavian Economic History Review*, 3, 1955, pp. 27–28; W. Barlow, trans., *Three Christian sermons made by Ludovike Lavatore, Minister of Zuricke in Helvetia, of Famine and Dearth of Victualls*, 1596; Henry Arthington, *Provision for the poore, now in penurie*, 1597; A.B. Appleby, 'Nutrition and Disease: the case of London, 1550–1750', *Journal of Interdisciplinary History*, VI, I, 1975, p. 5; A.B. Appleby, *Famine in Tudor and Stuart England*, 1978, pp. 138–9.

42 Joan Thirsk, ed., *The Agrarian History of England and Wales*, vol. iv, 1967, pp. 581–2: Appleby, *Famine in Tudor and Stuart England*, pp. 140–5.

on the markets. The Privy Council had frequently to order provincial authorities to export food to London although members understood the problems of both sides: 'wee have founde exceeding great difficulty to reconcile the wantes of the citty and countrie, the one requiring great supply, the other not so able in these as in other tymes to affoarde such stoare'. One side-effect of the dearth was to persuade many poor Londoners to substitute roots and other garden vegetables for grain in their diets. These 'new' foods were supplied by gardeners in the London suburbs and immigrant gardeners settled in East Anglia and Kent.[43]

This great national crisis engendered several pamphlets offering advice and comfort. Many were religious and moralistic in tone, blaming dearth on sinfulness. A Wakefield gentleman, Henry Arthington, thought 'the hand of God is heavy upon us, in most places in this Realme of England', but as well as a religious solution his pamphlet put forward practical measures such as abolition of customs duties on grain exports, official encouragement of imports, and action against engrossers of grain. Two pamphlets stand out as urgent attempts to provide practical answers, Richard Gardiner's *Profitable instructions for the manuring, sowing and planting of kitchen gardens. Very profitable for the common wealth and greatly for the helpe and comfort of poore people* of 1599, and Plat's *Sundrie new and Artificiall remedies against Famine*. The first concentrated on gardening as a remedy for hunger, giving very detailed instructions for growing vegetables, based on the author's experience of market-gardening in Shrewsbury. Like Plat's, the book shows signs of being produced in haste. The first 20 or so pages read like the start of a comprehensive manual on raising vegetables from seed: the remainder are full of urgent advice on avoiding hunger.[44]

---

43 *APC*, 1596, p. 269; 1597–98, p. 291–2; Malcolm Thick, 'Roots and other garden vegetables in the diet of Londoners, *c.* 1550–1650, and some responses to harvest failures in the 1590s', *Staple Foods, Oxford Symposium on Food and Cookery 1989*, 1990, pp. 228–235.

44 Arthington, *Provision for the poore, now in penurie*, 1597; Richard Gardiner, *Profitable instructions for the manuring, sowing and planting of kitchen gardens. Very profitable for the common wealth and greatly for the helpe and comfort of poore people*, 1599; 'Richard Gardiner's "Profitable Instructions", 1603', ed. Dr. Calvert, *Shropshire Arch. Nat. Hist. Soc.*, series II, vol. 4, 1892, pp. 241–2.

THE WRITING OF *DELIGHTES FOR LADIES* &c.

Plat's pamphlet was the earlier of the two, coming out at the height of the dearth. Plat recognized that his was a preliminary response to the crisis and he encouraged others to put their minds to the problem:

> yet I will bid the base to those choice, and delicate wits of England, who if they would either associate themselves unto me at the first, or second me, when I have begun this proud attempt, I would not doubt, but that by these our ioynt labours we should frustrate the greatest parte of these covetous complots, and by new, and artificial discoveries of strange bread, drinke, and food, in matter and preparation to full of variety, to worke some alteration and change in this great and dangerous dearth. Nevertheles (though I do only break the yce, for those that shal follow me in this kind) yet according to that poore talent of mine, I will trie mine owne strength and confer as well my conceipt, as knowledge herein: which though it bee neither such as I could wish, nor as these urgent times require, yet I wil be bold (in the fulnes of mine affection) to prefer and present the same to the view of the well disposed Reader, whose courteous acceptation hereof, may one daie peradventure wring from me some matter of higher reach, and farther service than as yet I see either iust cause to promise or reason to speake of.[45]

There is much internal evidence that this pamphlet was a quick response to the food crisis. As we will examine shortly, the disparate nature of the various pieces which make up the pamphlet indicate haste and Plat confirms this directly in the quotation above. Later, discussing possible sources of starch, he comments, 'It may bee that at my better leisure I may handle this subiect more at large, but now the present times enforce me to deliver that knowledge which I have'.

He advocates untested ideas: a general section of proposals taken from various sources is prefaced with the warning, 'These as I dare not warrant, so yet because I have received them either from good Authors, or from the credible report of men of worth, I will deliver them as faithfully as I have received them'. Describing making bread flour from the starch

[45] Plat, *Sundrie new and Artificiall remedies against Famine*, A3–4. 'Bid the base' means to challenge to a game of some kind.

of cuckoopint roots he comments, 'This carieth some good sence and likelihood of truth with it, for we finde by dailie experience yt it maketh as faire, if not a fairer starch, then our wheat.' He *has* extracted oil from beechmast but not solved the problem of removing the leathery hull of the nuts before boiling them for flour-substitute.[46]

Friends and acquaintances provide suggestions. Parson Batemen (Bateman) of Newington tells him of a tale told by Dr Grindal, when Archbishop of Canterbury, about English prisoners in Turkey who survived a sentence of starvation to death because their gaoler secretly gave them each a small piece of alum to suck which sustained them for 30 days. The Spanish ambassador 'Mendozza himselfe' told Plat he had relieved dearth by mixing ground wheat straw with wheat meal to make bread go further. After the conversation Plat was 'farther informed, that the same practise hath beene usuall in harde yeares in some partes of England'. He experimented, 'and for mine owne better satisfaction, I caused some of the flower to bee kneaded into bread, but it was verie browne in colour; and verie grettie in the mouth, and therefore it shoulde seeme that our stones be not apt for the grinding of it, & I have heard some affirme, that the same cannot wel be ground but in a steele mill, or hand mil.' He was assured by 'divers Gentlemen of good credit' that a drink could be made for travellers from 'aqua composite' and spices mixed with water, and of a drink for the poore made from ling flowers and water he comments, 'And this liquor is commended unto mee, by one of the most sufficient professours of Physicke of our times, and that upon his owne and often experience'. These ideas however, are not found in his notebooks, they were probably gathered from a quick visit to Court or to his legal haunts and put straight into this treatise. In fact there is no thematic section in his notebooks on famine relief, a further indication that he had not planned to write on it.[47]

Sundrie new and Artificiall remedies against Famine is a rushed pamphlet with a hotchpotch of material. Plat draws on major works of his own time: putting a turf on one's stomach to relieve hunger pangs comes from Paracelsus' *Archidoxis*; and Plat also draws inspiration from della

46   Plat, *Sundrie new and Artificiall remedies against Famine*, B1, B3–4.
47   Plat, *Sundrie new and Artificiall remedies against Famine*, B2, C2–3.

Porta's *Natural Magic*. Without acknowledgement Plat paraphrases the English translation of Nicholas Monardes' book on plants and medicines from America: 'And the east Indians, as I have read, do use to make little balles of the iuice of Tabaco, and the ashes of cockle shells wrought up together, and dryed in the shadowe, and in their travaile they place one of these balles betweene their neather lip, and their teeth, sucking the same continually, and letting downe moisture, and it kepeth them both from hunger and thirst for the spece of three or foure daies.'[48]

Plat also raids his own publications. Better-fertilized land will bring forth more food crops, and 'Here I thinke it not impertinent to the purpose, which I have in hand to wish a better survey to be made of my booke of Husbandry, being a parcel of the Iewel house of Art and Nature printed an.1594. Wherin sundry new sorts of Marle are familiarlie set down, and published for the good of our English farmers'. He helpfully adds, 'The book is to be had at the Greyhound in Paules churchyard.' (This is an advertisement! And not the only one in the pamphlet – at the end are advertisements for coalballs and a new compost.) In similar vein, Plat takes up two pages to extol the virtues of macaroni as a victual for seamen, listing ten points in its favour (with the excuse that dried pasta is an appropriate food to mention here), concluding 'Al those which are willing to victual their ships therewith, if they repair unto me, I wil upon reasonable warning, furnish them therewith to their good contentment.'[49]

Towards the end of the pamphlet, after a printer's decoration, comes a new section of six pages headed, 'An abstract of certaine frugall notes, or observations in a time of Dearth or famine, concerning bread, drink, and meate, with some other circumstances belonging to the same, taken out of a Latin writer, intituling his booke, Anchora Famis & sitis.' Forty short paragraphs of advice follow, mostly covering emergency measures to combat hunger and thirst: man can live on milk alone; aniseed steeped in water makes a wholesome drink; toasted bread soaked in wine is very

48 John Frampton, trans., *Joyfull newes out of the Newe Founde Worlde*, Nicholas Monardes [1577], intr. Stephen Gaselee, 1925, Book II, pp. 90–91. Plat quotes this book (on the uses of snow) in BL, SL 2210, so he may have owned a copy.
49 Plat, *Sundrie new and Artificiall remedies against Famine*, C1–2.

filling. Many ideas are for bread-flour substitutes: rice, Indian millet, Turkish wheat, lentils, wheat starch, bean flour and the like. This treatise named *Anchora famis, sitis & valetudinis* (in Plat's notes, 'The anchor of life, good health and hunger') was written by a German physician, Joachim Strupp, and published in 1573. Strupp worked in various German courts, eventually becoming personal physician to the Landgrave of Hesse. Plat had links with this court, which may explain how he acquired the book. More extensive transcriptions of this treatise by Plat are in BL, SL 2210 ff. 19–21. The original Plat worked from was at least 120 folios long and he only extracted items relating to food and drink.[50]

Despite all the borrowed material, there are plenty of original ideas from Plat himself. He has experimented with ways to 'to take awaie a great part of that ranke and unsavoury tast of Beanes, Pease, Beechmast, Chestnuttes, Acornes, Veches, and such like'. He devotes some pages to the use of wormwood instead of hops to flavour beer, and the use of a mechanical bolter in cornmills, returning bolted bran immediately to the mill to make bread go further. And he gives a succinct recipe for parsnip cakes, assuring his readers, 'I have eaten of these cakes divers times in mine owne house Quaere, what may be done in carots, turneps, and such like rootes after this maner.' Plat recognized that, even if the pamphlet was a hasty compilation, it must focus on the immediate problem of dearth. In a paragraph meant probably to precede *Anchora famis* but inserted several pages early by the printer (another sign of haste) he writes, 'And because in the treatise following my Author hath raunged over all manner of trees, plants, roots, greene pulse and herbs, out of which hee might by any p[r]obabilitie draw out any kind of sustenance for the reliefe of man, I will onely content myselfe with the handling or preparation of some of these particulars which are most plentifull in their quantitie, least offensive in their nature and most familiar with our soile and bodies, so as thier offensive taste being first removed by arte, they may serve us in a far better manner and to our greater liking then nowe they do, either for bread, drinke or food.'[51]

50 BL, SL 2210 ff. 19–21; *Wikipedia* sv Strupp, Joachim (accessed 17. 4. 2010).
51 Plat, *Sundrie new and Artificiall remedies against Famine*, A4, C4.

CHAPTER SIX

Alchemy

Alchemy is a big subject, made larger by the mystery with which alchemists surrounded it and its intrusion into literature, philosophy, and the popular imagination. We still use the word figuratively to suggest the indefinable – the alchemy between two disparate individuals whose combined efforts outshine all others, the incomprehensible luck of some politicians, etc. We also speak of successful cooks as 'alchemists' emphasizing the practical aspects of the subject and the act of creation involved in an alchemical process. In late sixteenth-century England it would be fair to say that almost any man with some education and leisure would have had an opinion about alchemy and most would have some interest in it. Plat was one of a minority of gentlemen of his time with a deep knowledge of alchemy, both as a reader and collector of alchemical texts and as a practical alchemist who, over many years, toiled in his own laboratory.

General and particular interest in alchemy towards the end of the sixteenth century was stimulated by the publication of several books on the subject, reprints of ancient Latin texts and new translations into English of hitherto obscure works. The influence in some medical circles of the Swiss-German doctor Paracelsus, whose theories were much associated with alchemy, may have brought some English readers to the subject. More generally, alchemy was but one intellectual pursuit which filled the time of gentlemen increasingly unburdened by the obligations of military service, central government administration and, in the case of Plat and other 'London' gentry, free from the administration of large country estates. For some, alchemy held out the hope of religious and political progress, sidestepping the great rift of the Reformation, bringing the prospect of a return to a natural, unified interpretation of life. The world was explained as a vast series of chemical reactions initiated by God, discovered with the help of ancient texts which predated Christianity. Bruce Janacek, reviewing the works of Thomas Tymme, an Anglican minister who became deeply interested in alchemy, summed up Tymme's reasons thus:

> Tymme turned to alchemy sometime in the last years of the sixteenth century, when alchemy was increasingly associated with the restitution of nature. Alchemy was thought to be a redemptive process for the fallen natural world; it was therefore not merely a search for individual wealth, but a pious pursuit of the "philosophers' stone", the stone that could prolong life, salve injured bodies, purify individual's souls, and heal political and religious divisions. Acting as priests of natural philosophy, alchemists believed they were redeeming corrupted matter and therefore possibly transforming and redeeming the entire natural world, returning it to its pristine perfection.[1]

The serious studies of the likes of Plat and Tymme shaded into a popular recognition of the subject amongst literate men and women, reflected (and reinforced) by the literature of the time. Ben Jonson's satirical play *The Alchemist* (1610) springs most readily to mind. Not so much a satire of alchemy as one which uses alchemy and its rich language to ridicule human greed and credulity, the play reveals that Jonson had studied many alchemical books, including works by Arnald of Villanova, Nicholas Flamel, George Ripley, Sendivogius, the compendium *Theatrum Chemicum* of 1602 and translations and interpretations of Paracelsus. John Donne, William Shakespeare, and many lesser-known writers also brought alchemy into literature.[2]

The goal of the serious alchemist was transmutation – changing one substance into another by chemical operations. Transmutation was

[1] Bruce Janacek, 'Thomas Tymme and Natural Philosophy', *Sixteenth Century Journal*, XXX, 4, 1999, p. 990.
[2] Charles Nicholl, *The Chemical Theatre*, 1980, pp. 98–102; Ben Jonson, *The Alchemist*, ed. F.H. Mares, 1997, p. xxxi.

possible according to Aristotelian theory: all substances were composed of four elements – earth, fire, water and air in various proportions – changing these would change the substance. The four elements were in turn each composed of two primary qualities – earth, dry and cold, air wet and hot, etc. Changing these components changed the element and thus the substance of which the element was a part. Aristotle hinted that a combination of earth and water composed metals and that, imprisoned in the ground, varying compositions of these elements produced the metals buried there. Arab and later Western European alchemists refined these theories. The 'earthy smoke and watery vapour' which produced metals became philosophical mercury and sulphur, pure versions of the ordinary elements. Purity, and precise combinations, determined the metals produced: perfect purity and correct combination produced gold.

Alchemists sought the means to speed up and harness natural transmutation, to produce pure metals or other substances at will. To do this they sought the philosophers' stone or the elixir, a catalyst which would bring about transmutation at will. Because of its power to purify, the stone also cured disease and prolonged life. An awesome substance, producing it was an extremely hazardous operation, demanding dedication and purity on the part of the alchemist. Searching for the stone was akin to a religious experience – some texts emphasize the divine aid necessary for success and that ultimately the achievement of the stone is only possible by divine revelation.[3]

Charles Nicholl defines the modern image of alchemy as tending in two directions: alchemy as early chemistry, a scientific study, or alchemy as magic, an occult art. The scientific explanation emphasizes that alchemists stumbled on basic chemical truths and techniques whilst pursuing an illusory quest. This illusory quest, the search for the stone, links the scientific with the occult. Modern uncertainty over what alchemy was is mirrored in the historiography of the subject, which is almost as complicated as the alchemical texts on which it relies. Chemists in the eighteenth century sought to divorce their 'new' chemistry from alchemy and painted it as an obscure and often fraudulent practice. The mystery

---

3   Nicholas H. Clulee, *John Dee's Natural Philosophy*, 1988, pp. 96–101.

of alchemy was emphasized by some secret societies founded in that century which adopted alchemical symbolism and terminology. Much was written in the nineteenth century on the spiritual and mystical side of alchemy: practical experimentation was interpreted as a series of spiritual exercises and some writers denied that any chemical operations had taken place. These writings informed Jung's research into alchemy. He concluded that alchemists 'projected' their inner thoughts on the experiments they performed, allowing them to 'see' the bizarre images they reported. Consequently, the experiments themselves were meaningless in practical terms. The nineteenth-century spiritual interpretation, together with that of Jung, has influenced more recent writers to downplay the practical side, but Principe and Newman have recently produced cogent arguments that alchemy was essentially a practical activity, despite the flowery language of the texts. They point out that many alchemical operations have been decoded and repeated in modern laboratories, 'we have shown clearly how extravagant alchemical imagery was consciously contracted to hide actual laboratory operations and how the very same alchemists who penned bizarre allegorical descriptions in print were able to express their knowledge in clear, unambiguous "chemical" terms in private communications.'[4]

Alchemists also had to deal with public scepticism in their own time. Catch-penny pamphlets told lurid tales of false alchemists and more detailed criticism came from authors such as Reginald Scott, who devoted a chapter to exposing false alchemists in his *Discovery of Witchcraft* of 1584. Criticism is also found in Samuel Harsnett's *Declaration of Egregious Popishe Impostures* of 1603. Interestingly, Harsnett was the censor at the Stationers' Company who passed one of Plat's (unpublished) alchemical works for publication. He was either convinced of its serious content or, as he had been known to do on other occasions, he did not examine the text very carefully.[5]

---

4   Nicholl, *The Chemical Theatre*, pp. 1–2; L.M. Principe & W.R. Newman, 'Some problems with the Historiography of Alchemy', in *Secrets of Nature*, ed. W.R. Newman & A. Grafton, Cambridge, MA, 2001, pp. 385–420.

5   Nicholl, *The Chemical Theatre*, pp. 8–9; James Shapiro, *1599*, 2005, p. 134.

ALCHEMY

Plat wrote much on alchemy but published virtually nothing. His caution may have been due to the fact that alchemy had been technically illegal since 1404, a felony punishable by death. Although the law was not used against gentlemen scientists (it was aimed at charlatans who peddled false transmutation recipes and the like), the very existence of such severe penalties made caution advisable.[6] Plat devoted most of one of his longest published passages on alchemy to distancing himself from what Charles Nicholl called the 'magical underworld' of false alchemy. He was not the first to warn of these dangers, which he acknowledges with references to Petrarch and Chaucer. His tale of 'an old smokie Alchemist' and the subsequent summary of sleights of hand would have been familiar fare to many of his readers. They would also have recognized the writer as someone with much knowledge of both the philosophy and practice of alchemy. Significantly, he begins by ranking the study with the subjects at the core of the university syllabus:

> It is a worlde to see, how everie Arte hath gotton his Counterfeite in these daies. Howe Logike is turned into Sophistrie, Rhetorique into flatterie, Astronomie into vaine and presumptuous Astrologie, that ancient and divine science of Alchimie into Cementations, Blaunchers, and Citronations, ending commonlie either in coosenage, quionage, or in *Capistro*, which made *Petrarke* to give a Caveat in these wordes. *Cave Alchimiam, semper rebus aliquid desuerit, dolis nihil.* And againe. *Chimista qui tibi autrum suum spondet, cum tuo auro improvisus aufugiet.* This made Chawser in his time to play so plesantlie upon the Alchimists hollow cole, & this hath made me to touch or glance at a few other sleights of later date, therby to admonish al yong gentlemen and others to take heed of al these mercenerie hirelinges. *Qui cum aliis mille aureos promittant, ipsi drachmam petunt. Locus posit fabulam.* A subtile marchant sorting himself of late, with an old smokie Alchimist for his better credit, as they became fellow travellers in the higher parte of Germanie together lighting by chance upon a young crewe of marchants that were wel monied, and ready for any rich prize that should be offered unto them, especially for *Iasons* barke

6   Nicholl, *The Chemical Theatre*, p. 13.

that was laden with the golden fleece, after some salutation had, and a few words of course enterchangeablie passing betwixt them, this cunning companion of the alchimists began to parlie with them in this manner. My maister and friends, you seeme to be men of honest parentage and condition, and most happily to be here met both for your owne good and ours. So it is, that if you will parforme that secresie which is requisite in so weightie a matter as I am in purpose to commende unto you, *I will make you the moste royall Marchantes of the whole Worlde*. Neither shall any of you make the hazzarde or adventure of one Deniere, untill with your owne eyes and handes you shall have seene and made a sensible proofe of this my friendlie offer. It is but in vaine to use manie wordes amongst friendes. You shal make a perfect proiection your selves upon Mercurie, *ad omne examen*; and because my selfe and my partener will bee free from all suspicion of deceipt, you shall bringe the Crucible, the Coales, and also the Quick-silver with you, and wee will but onelie deliver you one graine of the medecine which shall extend itselfe uppon a full ounce of Mercurie, which you your selves shall likewise let fall onto the crucible. A man would thinke that this were plaine dealing, and that unlesse these men were wilfullie bent to cosen themselves, that it were impossible, to deceive so many young eies, that watched so carefully for them selves. But now to the practise. The fire being kindeled, one of them setteth on the Crusible by direction of the Alchimist, under the nose of a paire of gold-smithes bellowes, who told him that for the better fixation of the Mercurie, there muste a reverberatorie blast be made now and then with the bellowes, after the Mercurie was once warme in the melting pot. Now this Imposter had before conveied into the nose of the bellowes, an ounce or somewhat more (to supplie that which the Mercurie with his fume should carrie awaie with him) of Sol sublimated by often reiteration of *Aqua Regis* upon it, as that it became almost an impalpable powder, which when the Marchant by the appointment of the Chimist, had blowne amongst the Mercurie, he was willed to drop in the medicine, being wrapped uppe in a smal paper, and then to leave the crusible in the fire, untill the medicine and the Mercurie were both incorporated together, and that the Mercurie were sufficientlie tincted into Sol, and within one halfe hour (after he had first caused them to melt downe an ounce of fine golde in an other crusible, and put the same to the first worke, for the better fusion of the powder) he willed to be taken out of the fire and conveyed into an ingot, and the same being colde became twoe ounces of perfect Sol, abiding both the hammer, the test, and the horne of Antimonie. It is not to bee doubted, but that these gallants were right ioyful of this good successe, desiring nothing more then to become Lullistes, offering to exchange their freedome both of the old Haunce and of the newe, for this multiplying Art. Now this geere worketh like wax, and the Alchimist demaundeth 2000 dollars, for the provision of coles, furnaces, saltes and Minerals, but especiallie to engrose all the Mercurie, that could be gotten, least either it should rise to an excessive price, or be transported into Spaine, for the refining of the Indian oare. The money is foorthwith delivered by weight, because there must bee no time lost in the telling, with a charge to use all expedition that could be for the gathering of the Mercurie together that was to be gotten far and neere. The substance of this historie is already delivered, I will not stande long upon the circumstance. The Alchimist having fingered the monie, beginneth to erect furnaces, and enterteineth them with a few distillations, calcinations and sublimations, teaching them howe to make *Lutum Sapientiae, Aquam seperationis, aquam regis, oleum vitrioli, salis, & sulphuris,* to congeale Mercurie with the spirit of Saturne, to make saccarum Saturni, to whiten their teeth withall, to blanch copper with Arsenick, to melt one part of Luna with 3.parts of Venus together, and then to forge plate thereof, and by a certain ebulation, to make the same deverse times to touch equall with our best startling, or higher according to the finenesse of the silver that was mingled with Venus, all this (with an infinite number of spagiricall experiments) was performed, both to passe away the time without tediousnesse, whilest the Philosophers egge (which required 10. monethes digestion, was hatching in *Coelo philosophicum*) as also to gain ye more credit with the marchants, wherby a man of these single gifts, might not be feared or mistrusted of his flight, which both he & his companions were dailie practising, and in the ende finding good opportunitie they put the same in execution, leaving them that had most need to blow at the cole. I doo verelie beleeve that if the old D. of Florence were alive againe, he would have out bidden the marchants for this secret, whose distillatorie vessels, furnaces, & other chimicall instrumentes, were all of silver as I have heard it often reported. But now to give a few Items more against these Impostors, before I conclude, Let every man that is besotted in this Art, and dependeth whollie upon other mens practices (himself not beeing sufficientlie acquainted with those great and hidden Maximes of nature) take heed also of all false and double bottomes in Crusibles, of all hollowe wandes or roddes of yron, wherewith some of these varlettes doe use to stirre the metall and the medicine together, of all Amalgames or Powders, wherein any Golde or Silver shallbee craftilie conveyed, of Sol or Luna, first rubified and then proiection made upon it, as if it were uppon Venus hir selfe: but speciallie of a false backe to the Chimney or furnace, having a loose bricke or stone closely ioyned, that may bee taken awaie in an other Roome by a false Sinon that attendeth onely the Alchimistes hemme, or some other such like watch-worde, who after the medicine and the Mercurie put together in the Crucible, enterteineth *Balbinus* with a walk and with the volubilitie of his toung, until his confedereate may have leysure enough to convey some Gold or Silver into the melting potte, which were able to decieve the best sighted Argus in the world. By these fewe legerdemaynes, I hope many thousandes will be sufficientlie warned, of these wandring & roguing Alchimists.[7]

In so publicly condemning and exposing false alchemists Plat was emphasizing his own probity as a true alchemist, following an ancient and noble art.

Plat succinctly summed up his studies in natural science as a whole as a mixture of reading, conference, practice and contemplation: we may conveniently examine his alchemy under these heads.[8] To begin with conference: much could (and was, by Plat) be obtained from books but, especially in the arcane world of alchemy, personal tuition was essential. Innermost secrets of alchemy were communicated only by word of mouth. One could try to cope with books alone but, as Plat wrote in his alchemical poem:

……when wearied are thie witts,
Thow must be firste for Maisters help to call,

7 Plat, *The Jewell House*, pt. 1, pp. 86–91.
8 Plat, *The Jewell House*, Preface, B4v.

a sentiment shared by the fifteenth-century alchemist Thomas Norton:

> ......no man cowde yet this science reche
> But if god sende him a master hym to teche.[9]

Moreover, the 'wonderful science, secrete philosophie', as Thomas Norton termed it, was,

> A singular grace & gyfte of almyghtie
> which neuir was fownde bi labour of man,
> But it bi teching or reuelacion bigan.

In other words, these were divine secrets to be handed down, not found by experiment. Moreover, Norton's master made sure that his pupil was truly worthy of the secrets to be bestowed on him, then taught him orally for,

> If I shuld write I shulde my foialte [oath] breke
> therefore mowthe to mowthe I must nedis speke.[10]

Stripping these opinions of their piety, we find alchemy little different from any other technology of Elizabethan England (skilled trades were all known as 'mysteries'). Eric Ash has summed up the position:

> The notion of expertise in the early modern period not only involved considerable practical experience with respect to a specialized set of skills, but also increasingly implied some sort of theoretical knowledge on the part of the expert, a deeper understanding of the practical, empirical skills at his disposal.... The rare and valuable knowledge and skill of the early modern expert could not be acquired through casual observation or transmitted through literary means alone, but could be truly learned only through long and painstaking personal experience. Expertise, in short, travelled with the experts: knowledge and experience were not easily separated.[11]

9 Thomas Norton, *The ordinal of Alchemy*, ed. John Reidy, Oxford, 1975, ll. 181–186, 215–6.
10 Thomas Norton, *The ordinal of Alchemy*, ed. Reidy, ll. 851–2.
11 Eric H. Ash, 'Queen v Northumberland, and the control of technical expertise', *History of Science*, xxxix, 2001, p. 215.

Plat was taught the innermost secrets of alchemy in a relatively short time, remarking that '3 moneths alone is a tyme sufficient for a true & faithfull teacher to unfold all those great & hidden groundes of Nature wrapped upp in ye threefold Philosophies whereof Hermes professeth himselfe to be Mr.' He added that 'Norton (yt English Paragon) acknowledgeth to have learned in 40 dayes.' Moreover, if teaching the secrets of the mineral stone were excluded 'the animall & vegetable philospohie may in 40 houres be easely had & obtained'.[12]

Plat was initiated into the secrets of alchemy by a master, who he is careful not to name, probably around 1580. We can eliminate some likely masters by noting the presence of their names in the same documents as references to the unnamed master. This excludes such likely candidates as John Dee, Plat's father-in-law Richard Young, and his long-time friend Thomas Gascoigne.[13] A remark by Plat in his letter to the Landgrave of Hesse in 1608 leads to the tentative conclusion that his master was the doctor Godfrey Mosan (usually known as Mosanus). In this letter Plat offered to train Mosanus' son in some secrets of alchemy, 'whose ffather did first tech me to close a helme and a bodye together'.[14] Indeed, Plat lists several recipes from Mosanus of 'lutements' – glue to stick distilling vessels together – and in the same manuscript, in a recipe for oil of coral, remarks of the apparatus 'the nethermost of theise two glasses is abowte 13 of my inches in higthe, & the uppermost glasse wch is called the helme'. The latter remark implies Plat is a novice at this point, reminding himself of the use of a helm. Further links to Mosanus are Plat's acknowledgement of medical recipes copied from a book owned by the doctor and another medicine that Plat has seen Mosanus make.[15]

Godfrey Mosanus, probably born in France, was practising as a doctor in England by 1571, qualifying as a surgeon in 1593, the year he died. He

12 BL, SL 2203, f. 26v.
13 BL, SL 2210, ff. 26–27v. Plat's earliest notes on alchemy are in SL 2210, a manuscript used by Plat for notes in the 1580s.
14 Mosanus senior was first prosecuted for unlicensed practising in 1581, which fits the time when I think Plat became interested in alchemy. Margaret Pelling, Frances White, *Physicians and Irregular Medical Practitioners in London 1550–1640* – Online Database, 2004; BL, SL 2172, ff. 18–21.
15 BL, SL 2203, ff. 205v, 207v.

had several brushes with the medical authorities in London for alleged malpractice and deficiencies in knowledge – almost certainly because his Paracelsian views upset the regulatory bodies. His son, a trusted acquaintance of Plat, practised medicine in England until he moved to Hesse in 1603, having obtained his MD from Cologne University. He too was criticized by the Royal College of Physicians for his 'continental' approach to medicine, being told in 1593 to go and study Galen for four years before re-applying to be licensed in England.[16]

If Godfrey Mosanus was Plat's master, he supervised Plat's approach to one of the foremost natural philosophers of his time, John Dee. Dee records in his diary a visit from Plat at Mortlake on 1 November 1582, accompanied by his first father-in-law Richard Young and a Prussian named Martin Faber. Dee saw other people that day, so Plat's visit may not have lasted long or, more likely, he saw Dee for a short time but spent the day in his library or talking to visitors and students resident at Dee's house. This resembled the museums established by several Continental rulers: the foremost library of natural philosophy and topography in England, together with collections of charters, seals, globes, maps and other curiosities and three laboratories for alchemical experiments. Visitors were frequent and Plat may have made other, unrecorded trips to Mortlake to see Dee and met like-minded gentlemen there. One historian has remarked of Dee's library: 'it proved such an attractive and accessible site that Dee was inundated with both wanted and unwanted visitors – so many that in 1592, when he found it necessary to advertise its status, he called his house the "Mortlacensi Hospitali Philosophorum peregrinantium" (Mortlake Hospital for wandering Philosophers).'[17]

Dee records nothing else of Plat but the two had more contact on alchemical matters. Plat recorded three recipes for medicines obtained

16 Margaret Pelling, Frances White, *Physicians and Irregular Medical Practitioners in London 1550–1640* – Online Database, 2004. An order to study Galen often implied a condemnation of Paracelsian ideas.

17 William H. Sherman, *John Dee: the politics of reading and writing in the English Renaissance*, Amherst, MA, 1995, pp. 38–45; *The Diaries of John Dee*, ed. Edward Fenton, Charlbury, 1998, p. 47; Deborah Harkness identifies Faber as a Lithuanian-born alchemist who was another of Plat's alchemy teachers, Harkness, *The Jewel House*, p. 222.

from Dee, one noted as 'this I learned of Mr John dee./ the xth day of may.ao 1580./', eighteen months before Dee's diary entry. Of more interest are three pages of questions and answers on alchemy found in Plat's manuscripts. They concern the deepest secret of alchemy, preparing the philosophers' stone. The questions were put to Dee and some were also posed to Plat's master. In other cases the master comments on Dee's response. Dee gave straight answers to some questions: the stone makes a noise like 'Sensible thunder' when being created, 'but like unto the frying of grease or tallow' when projected. To the question, 'How many miles off an Homunculus will serve'? Dee replied fully:

> 12. miles off. per J. dee. this Homunculus is a litle Creature engendered in a glas, & may bee termed the right familiar of the man, whose devine chimicall arte by the help of Godd, hath reached to the performance of ye same, of him the chimiste may learne many woonderful and devine secretes, he answereth by speeche.

Other answers were incomplete or evasive, at least in the opinion of Plat's master. Plat posed a 'trick question' to try Dee's knowledge, with inconclusive results:

> What hower, tyme or tymes in the night, a man must wach the Crow from sleeping, leaste hee fall into the redd sea? Note by confession of the Author that there is an error of purpose sett downe in this Question to sownde a trew Artist.

The answer was:

> As concerninge the hower tyme or tymes. Mr Dee saith that for full answere unto them, he will answere nothing, & yet therin hee answereth sufficiently but the night signifieth the putrefaction of the woorck during the wch a gentle fier must bee kepte, lest the croes bill, being rubified to soone make the worke unprofittable.

This did not satisfy Plat's master who commented,

> this answere is a printed answere, & nothing apte to the Question, and that yt sootheth upp [glosses over] the error that is couched therin, & answereth nothing at all to the woord sleeping.

Plat's master was jealous of his own reputation, as the following inconclusive exchange illustrates:

> There is a tyme in the greate woorke, when the glas beinge opened, a devine influence descendeth downe in to the glas & maketh animation to the earth or worke. qre what tyme that is per J. Dee. To this question my Mr wolde not answer unles Mr Dee wold first answer this Question v[iz]. To tell him the beginninge of that foode wherof he wold then norish his earth or worke. yet he accomted this for a very good question.

Over some arcane matters Dee and Plat's master were in open disagreement:

> per J. Dee all the metalls are to bee reduced into mercury in .3. howres space, & wth a thing no bigger in quantity then a cromme of bread. But my Mr thinketh yt not possible in all the metalls in so shorte a tyme neither wth so small a substaunce, & though yt were possible, yet the secret helpeth nothing to ye great woorke, for that is not mercury philosopherus, and yt is as hard a matter to open mercury as any other mettall. But by my Mrs opinion the mercury of the philosophers is, when the mettals are so opened, that they are apte to make division of their parts wth preservation of kinde.[18]

This confused exchange confirms the depth of Plat's alchemical education and his links with Dee. A further possible link with Dee's household may be a handful of secrets attributed to one 'Taylbois', many of them concerning colouring and testing gold and silver, which Plat copied from a book owned by this man. 'Talbot' was a pseudonym adopted by Edward Kelley, Dee's scrier, when he first joined Dee's household and it may not be too fanciful to identify Kelley as Plat's informant, especially when we note that Kelley was himself a noted alchemist who purported to make gold by projection and gained considerable fame and fortune on the Continent in so doing.[19]

18   BL, SL 2210, ff. 26–28v; SL 2203, f. 218.
19   BL, SL 2203, ff. 221–222v; Kelley as 'Talbot': see Benjamin Woolley, *The Queen's Conjuror*, 2001, pp. 141–2, 278–82, (a scrier is one who can see images in a crystal ball); Nicholas H. Clulee, *John Dee's Natural Philosophy*, 1988, pp. 227–8.

Dee's place as a natural philosopher with an international reputation and as a consultant to the Crown on many scientific matters make it almost inevitable that some of the alchemists whom Plat knew were similarly acquainted with Dee. Plat may indeed have met them through him. Men such as Sir Thomas Chaloner, who put his alchemical studies to practical use by supervising the exploitation of alum shales on his estate at Guisborough. Also Auditor Hill, who appears briefly in Dee's diary as a party to a monetary transaction; Plat is indebted to him for a recipe to turn mercury into lead which 'proveth Transmutation of mettalls'.[20] Thomas Digges was an eminent mathematician, pupil and close confidant of Dee known to Plat for his alchemical activities. In the middle of the questions and answers with Dee outlined above, Plat inserted a short secret headed, 'This is a gradation of the great worke by Mr Digge'. On another occasion Plat records the opinion of 'Mr diggs that ancient & painfull chimist, on the nature of ingredients to begin the 'great work'.[21]

Plat talked alchemy with men associated not only with Dee but with other alchemical and natural philosophy groups, such as those within the ambit of Sir Walter Raleigh and Henry Percy, ninth Earl of Northumberland. The chemical experiments of Raleigh himself were known to him: he may have heard by repute of Raleigh's weapon salve or his 'sweete oyle of Amber whiche Sr Walter Raleighe will not willinglye disciver for 500li which cureth the dead palsie and all crampes and convulsions'. But he also visited Raleigh's house in London, where he saw 'A lanterne to Carrie a candle in lighte in the greatest winde that bloweth' in October 1591. This lantern was invented by Raleigh's chief steward and his main advisor on natural science, Thomas Harriot, who was certainly well known to Plat and was also an acquaintance of Dee.[22]

---

20 Chaloner informed Plat that the foil for a looking-glass was made of 'Regulus Antimonii, leade, & Colophonia' and was fixed to the glass during its manufacture and he also advised on a slow fire for chemical operations; Charles Singer, *The Earliest Chemical Industry*, 1938, pp. 182–188; BL, SL 2210, f. 170; SL 2189, f. 138; SL 2216, f. 106; *The Diaries of John Dee*, ed. Fenton, 1998, pp. 259, 256.
21 BL, SL, 2245, f. 9; SL 2210, f. 27v.
22 BL, SL, 2212, f. 4; 2210, ff. 45, 108, 154v; Plat, *Jewell House*, p. 27.

Plat knew Adrian Gilbert, half-brother of Raleigh, who was also part of Dee's circle mainly because of his collaboration with Dee on plans for exploration of the north-west passage. This may link Plat with the alchemical studies of Mary Herbert, Countess of Pembroke. According to John Aubrey, she kept at Wilton House, 'her Laborator in the house Adrian Gilbert (vulgarly called Dr Gilbert) ... who was a great Chymist in those dayes'. Plat also refers to him as 'Dr', so probably met him when he had retired from the sea to Wilton.[23] It seems that several of the gentry employed alchemists, who were provided with laboratories in which to experiment. In 1580 Plat learned of an effective lutement for sealing the joint between two glass vessels from one Phillips, worker for Armigal Waad, sometime a clerk of the Privy Council and he talked chemistry with Nathias, a foreigner who worked for Mr Pope at Queenborough Castle in Kent.[24]

One is struck by the number of nobility and gentry to whom Plat talked about alchemy. As well as those already mentioned, he took advice from Sir William Bowes, a Durham gentleman and courtier. Presumably at Bowes' London residence, Plat saw 'An excellent Balneo to sett diverse glasses in wth a small fire' in July 1590. Bowes described in detail 'A conceipte how to worcke philosophically uppon any plante, wherby Nature shall devide herself into her earth and oyle', a long experiment said by Bowes to take seven months. Thomas Gascoigne thought this last experiment took seven years to complete – he was either with him at the Bowes' house or Plat asked his opinion as a friend who was deeply involved in alchemy. Plat obtained a recipe for mercury 'attenuated into water' from the Bishop of Bristol, Dr John Thornborough. Thornborough, who died aged 91 in 1641, published highly theoretical works on the philosophers' stone but was also a practising alchemist.[25] Link with the Court is provided by Francis Segar,

---

23 John Aubrey, *Aubrey's Brief Lives*, ed. Oliver Lawson Dick, 1982, p. 220; BL, SL 2203, f. 221.
24 BL, SL 2170, f. 17v; SL 2203, f. 219. Sir Thomas Smith, diplomat and scholar, employed two servants to carry out distillations for him and produce alchemical medicines, Charles Webster, 'Alchemical and Paracelsian medicine', in Charles Webster ed., *Health, Medicine and Mortality in the Sixteenth Century*, 1979, pp. 315–6.
25 BL, SL 2210, f. 122; SL 2245, f. 30v; Allen Debus, *The English Paracelsians*, New York, 1965, pp. 103–4.

brother of the painter and herald William Segar, and himself a painter. Francis advised Plat on making the stone and other alchemical mysteries. He was, after 1605, an agent for the Landgrave of Hesse, a monarch deeply interested in alchemy.[26]

As one would expect, Plat found craftsmen working with distilling equipment and chemicals who shared his interest in alchemy. Russell the distiller claimed to have made mercury into lead by art. Russell suffered paralysis from ingesting mercury fumes and provided Plat with a remedy 'To help a palsy gotten by the fumes of [mercury] and to drive the fumes owt of the body'. In addition, Plat makes many references to chemical advice he has sought from gold refiners in the city, and from apothecaries such as 'Maister Kemish, that aucient and expert Chimist' and enthusiast for Paracelsus.[27]

Plat obtained secrets in alchemy from a number of foreigners. Recipes collected early in the 1580s from 'Martinus' may have come from the 'stranger of Prushen, born at Regius Mons... Martinus Faber' who went to see Dee with Plat in 1582. Nathias, the alchemical worker at Queenborough Castle Plat knew, was 'stranger born' as, of course, was Plat's possible master, Mosanus. Maybe foreigners, bringing new and exciting ideas, were attractive to English alchemists. They certainly were of interest to the Crown, who on more than one occasion in Plat's time entertained foreigners who claimed to have skills in transmutation of base metals into gold. The government official Armigal Waad was involved in negotiations with one who promised to make 50,000 marks of gold for Elizabeth in 1565 but was still confined to the Tower without providing any results two years later.[28]

---

26 BL SL 2245, f. 31; Sir Roy Strong, *The English Icon*, 1969, p. 17. See below for links with Hesse.

27 BL, SL 2245, f. 29. Plat, *Jewell House*, pt 3, p. 9. Mercury poisoning was a problem for alchemists. Russell's remedy was a mixture of dried herbs boiled in goat's milk. Another of Plat's alchemical confidants, Thomas Gascoigne, favoured making 'a fresh hole in neate earth hold your head therin half an howre, at a time, & this helpeth asuredlie'. The after-effects of ill-advised chemical experiments may be a possible cause of Plat's eventual decline and death.

28 *Alchemy in the English State Papers*, www.levity.com/alchemy/statpap.html.

Plat read a large number of alchemical texts, some printed books, but mostly manuscripts which he either acquired, borrowed from friends (such as 'Talbois Book' from which he extracted many recipes, 'my olde booke of Alchimie' and 'my Blake Book of Bacon'), or most probably read in Dee's extensive library at Mortlake. He took notes from texts supposedly from the mythical ancient authors of alchemy: Hermes Trismegistus or Hortulanus; ancient Arab writers – Calid, Almadir; many medieval Continental alchemists, Bernardus Trevisanus, Geber, Portanus, Jodovicus Greverius, Arnold of Villanova; and ancient anonymous texts – the tenth-century *Turba philosophorum* for instance.[29]

He was particularly well read in recent Continental authors – Paracelsus, Quercetanus (Joseph Duchesne), Cornelius Agrippa, Giambattista della Porta – and also in English alchemy, Roger Bacon, Thomas Norton and George Ripley. He paid close attention to Norton, summarizing accurately the first five books of his *Ordinal of Alchemy* and he was rigorous in his examination of Agrippa. He prepared 'an Epitome of those three bookes de occulta philosophia written by yt great Clerke Cornelius Agrippa' for publication, describing in the introduction how he had coped with this work written 'dispersedlie, & wthout all method of purpose & in raunging stile', teasing out, 'ye great & hidden misteries & magisteries of Nature both Concerning ye animall & ye vegetable stone.... Being a volume of great bulk & Conteyning about 200 leaves in a Close letter, I have by devidingye Chaff from ye Corne, & ye shells from ye kernelles, reduced into les then 20 small leaves, & Could (if it were lawfull to send nature abroad naked & wthout hir veyle) Contract them almost into 20 lynes.' He also read new books such as *The Mirror of Alchimy*, a collection of treatises published in 1597.[30]

29 Charles Webster has noted the many manuscript copies of major alchemical works in circulation in Plat's time, and the number of gentlemen who collected such manuscripts or took notes from them for their collections, Charles Webster, 'Alchemical and Paracelsian medicine', in Charles Webster, ed. *Health, Medicine and Mortality in the Sixteenth Century*, 1979, pp. 310–11; BL, SL 2170, ff. 23b, 24; SL 2195, f. IV; SL 2203, ff. 221–222v. 'Bacon' refers to Roger Bacon, the thirteenth-century philosopher and scientist.

30 BL, SL 2223, f. 27; SL 2195, ff. 82, 100v; SL, 2203, ff. 47–55. SL 2174 (not written by Plat) is a MS copy of Norton's *Ordinal* which ends after chapter 5 – this may have been Plat's source.

But there are many problems, Plat finds, with published works on alchemy. Continental writers write in Latin, Italian or French and are not accessible to non-linguists. They often make mistakes in describing more complicated recipes. Some are mere theorists: they 'have published whole volumes by imagination onely', which when tried 'in the glowing forge of Vulcan, they vanish into smoake'. Others write in such a complicated way that they cannot be followed to the letter, and if experiments fail the excuse is that not all steps have been carried out exactly. In general, Plat finds that alchemy writers,

> all wright in Ridles, Metaphors, darck and clowdie speeches … they wright not methodically, for somme of them leave owt preparations, some begin at the midle part, some at the last part of ye worke and they doble as a hare doth in their delivery sometimes going forward & sometimes backward, thirdly many of them confess, that this science must either bee had, by revelation, or by a Maister: fowrthly they doo of purpose conceale many braunches and circumstances of the worke wthout the which it can never be perfected, being forced therunto partly by reason of their oath, partly by feare of prophaning so great and devine a misterie, and partly least they shold worke a confusion in the world, if every man might make himself as rich as he wold.

Despite his misgivings, Plat summarized, paraphrased and noted published works extensively, and what he took from these authors is indicative of what he wanted from alchemy.[31]

Plat was a practical, hands-on alchemist describing his efforts in verse in the introduction to *Delightes for Ladies*:

> From painfull practice, from experience,
> I sound, though costly, mysteries derive,
> With fiery flames, in scorching Vulcan's Forge,
> To teach and fine each Secret, I do Strive

In 1608 he summed up his work as the discovery of 'suche particuler Secretts of Art as I have with great chardge in long time, not without

---

31 Plat, *Jewell House*, Preface; BL, SL 2195, f. 14v–15.

ALCHEMY

great danger of my life and health attained unto' and his notes and printed works abound with recipes for philosophically prepared medicines and alchemical operations.[32] In an unpublished pamphlet commenting on the works of Quercetanus and other 'modern' writers, Plat boasts that his 'poor labours in ye animall & vegetable skill' exceed those of the writers he is criticizing. Whilst we may be sceptical of such claims, Plat's notes also reveal a keen interest in the most mundane skill of a practising alchemist – laboratory techniques.

He records many recipes for 'lutements' – cement to seal distilling vessels together to avoid any leakage, such as the one from his tutor Mosanus: 'A stronge lutement to close pottes one uppon an other that no spirts may goe owte be the substaunce wthin never so stronge'.[33] A safer option when conducting alchemical operations which demanded the total exclusion of outside influences was to melt the glass at the neck of a vessel and seal the vessel completely. Plat describes in detail 'Howe to nippe or close a Glasse with a paire of hot tongues, which is commonly called Sigillum Hermetis'.[34]

Many alchemical operations were slow and demanded a gentle heat for many hours, days, or even years. Plat recorded a slow-burning mixture of juniper wood and berries which John Dee claimed would last for a year. Plat's notes on a candle or lamp which could be left for at least twelve hours reveal his practical interest in provision of a steady heat, his mechanical ingenuity and, again, the influence of Dee. The notes also show how he worried away at a subject, piling ideas, problems, and possible solutions one upon another:

> C Cawse a candle to bee made so longe, as yt may laiste 12 howres, then you muste have an Instrumen, to winde the same upp by litle and litle (like the racke in a Clocke) as yt wasteth that yt may still continew in one place./ Conserninge your Lampe these rules following are requisite./ Somme hold opinion that a weike of lawne is most excellent./ 24. single threedes of cotton are a good proposition for the weike./ your weak must

32 BL, SL 2172, f. 18; Plat, *Floraes Paradise*, Preface.
33 BL, SL 2203, f. 207v.
34 Plat, *Jewell House*, pt. 1, p. 91.

bee dipte in the oyle before you light yt/ the weake must bee a good ynch above the oyle yf you will have yt to laist longe/..... Sallet oyle is the best oyle for sweetnes, & fowles your fuumers least, & if you use the samme you shall not neede to wipe your fuumers wth a feather more then once in 3. dayes./ let the lampe burne ½ hower before you set yt in the fuumers./ qre iff tallow, or tallow & oyle together will not make a good lampe/ In the midste of the cover of your Lampe, let a pipe rise at a funnell, thorough wch the flame may ascend, for so your fire will bee the whotter, & soote shall not fall downe into the oyle. you may place diverse wicks uppon severall wyers, every wier being platted like a trefoile in the bottom & the wyer thrust thorough a litle rownd flatt peese of leade to make yt stande stely./ your lampe wold bee made as large as the furnes will give yt heate, & therfore let the doore of your fuumers bee as great as yt may bee. for so the oyle will setle but a litle in a longe tyme & the heate will keepe much at a degree/ powre your oyle uppon water./ iff you wold have your weike to burne 12 howres at one heate, you must place the weike uppon a wier that declineth a litle in the topp, wch trewest sloping of the wier you shall finde owte by many trialls qre, of a weike wounde abowte a wyer./ Also iff your oyle becomme blacke either wth the cole of the weike, or by soote falling downe, keepe the same together till you have somme store therof, & then beate yt in change of scalding water till yt becomme white & fayre.... qre of a fier of brimstone simple or impure wth caphire./ qre of a weike dipt in brimstone whether the same will not burne in a clos glasse, the glasse being made rownde in the topp, and falling downward in foorme of a founnell as in this forme you may see [drawing of a glass bulb] or making the same in copper, uppon the top wherof you may place sand or ashes or make a balneo./ qre how the smoke will consume or whether the same will dropp downe in oyle or no./ per Mr Dee take oyle olive, common salt prepared, quicklime one pound, mingle them all together well, & distill the same gentlie by an Alembicke, then incorporate the oyle so distilled wth the feces agayne, & distill in that manner 3 or 4 tymes at everie tyme returninge the distilled oyle uppon the feces & yt will burne verie well in a lampe, & will becomme an incombustible oyle qd non credo/.[35]

35   BL, SL 2210, f. 95–95v. A 'fumer' is a perfumed lamp.

In 1595 Plat advertised that he had perfected an alchemical method of 'The Art Spagiricall which shall bee delivered for the drawing of all oiles out of gums, seedes, flowers, and aromaticall bodies, and of all waters, spirits and salts out of vegetables with all necessary circumstances belonging thereunto.'[36] At the end of his life Plat was sufficiently confident of his own skills that he could offer sophisticated alchemical operations, albeit with the usual obfuscations. In 1608, he described his favoured method of applying slow and gentle heat to animal or vegetable substances to produce a quintessence:

> Now, to give you som taste of that fire which the Philosophers call the Stomack of the Ostriche (without the which the Philosophers true and perfect Aqua vitae can never be made) you must understand, that it is an outward fire of Nature, which dooth not onelie keepe your Glasse, and the matter therein contained, in a true proportionable heate, fitte for workmanshippe without the helpe of any ordinary or materiall fire; but it is also an efficient and principall cause, by his powerfull nature and pearcing qualitie, to stirre up, alter and exalt, that inwarde fire that is inclosed within the Glasse in his own proper earth. And therfore heere, all the usuall Chymicall fires, with all their gradations, are utterlie secluded; so as neither any naked fire, nor the heate of fylings of Iron, of sand, of ashes, nor of Baln. Mar. [Bain Marie] though kept in a most exquisite manner, nor any of the fiers engendered by putrefaction, as of dunge and such like, no nor the heate of the Sunne, or of a Lampe, or an Athanor, the last refuge of our wandring & illiterate Alchymists, have here any place at all. So that by this fire and furnace only, a man may easily discerne a mercenery workman (if hee deale with vegetables onely) from a sound Philosopher: & if in anything (as no doubt in many things) then here especially *vulgaris oculus caligat plurimum*. This fire is by nature generally offered to all, & yet none but the children of Art have the power to apprehend it: for, being celestiall, it is not easilie understoode of an elemental braine; and beeing too subtile for the sense of the Eye, it is left onely to the search of divine wit: and there I leave it for this time.

36 Plat, *A Discoverie of Certaine English Wants*, 1595.

> The physicall use of this fire, is to divide a Coelum terrae, and then to stellifie the same with any animall or vegetable starre, wherby in the end it may become a quintessence.[37]

Alchemy for Plat was an intellectual challenge and an area where he could advance existing knowledge. Although it is often not possible to date his alchemical writings, we can detect through them the advance of his alchemical studies, through initiation by an unnamed master, much questioning of Dee and other practitioners on both practical and theoretical matters, extensive reading and summarizing of both ancient and modern alchemical works, extensive practical experimentation in his laboratory, and finally the confident exposition of his own thoughts, culminating by offering for sale both his arcane knowledge and alchemical preparations. The publication of his own thoughts was, for the most part, not to be. Works were 'made ready for the Presse', as Plat announced in 1608, and approved by the Stationers' Company censor, but never appeared in print. They may have been held up by a cooling of official interest in alchemy with the accession of James I or simply because Plat died before publication could be arranged, as so nearly happened with *Floraes Paradise*.[38]

The historian Charles Webster has described the contemporary appeal of alchemy: 'The esoteric aspect of alchemy blended well with the currents of religious thought which were gaining ground during the Reformation, and with the Neoplatonic philosophy which was beginning to attract the intelligentsia. The rather ramshackle edifice of hermetic and alchemical writings was stimulating, even exhilarating, to generations breaking away from the highly organized didactic regimen of scholasticism.'[39] In 1614, Sir Walter Raleigh described alchemy as a 'kinde of Magicke [which] containeth the whole of philosophie of nature; not the brablings of the Aristotelians, but that which bringeth to light the inmost vertues, and draweth them out of natures hidden bosom to humane use.'[40]

37 Plat, *Floraes Paradise*, 1608, p. 10.
38 Plat, *Floraes Paradise*, Preface, A5.
39 Charles Webster, 'Alchemical and Paracelsian medicine' in Charles Webster, ed., *Health, Medicine and Mortality in the Sixteenth Century*, 1979, p. 314.
40 Sir Walter Raleigh, *History of the World*, 1614, vol. I, xi, p. 2.

In Plat's clearest summary of his view of alchemy we see echoes of both of these commentaries:

> C having a long tyme vexed and wearied my selfe in the darke tedious and uncertaine quiddities of profane Philosophie, wthout either sound Comfort to my soule, true profit to my knowledge, or perfect delight unto my senses: at length uppon good deliberation I resolved wth my selfe yt I would no longer derive my future hopes from these Gretian and heathen Authors; but seeke for such learning as either I might draw from the welspring of life, (the only saviour of soules) or els from nature the true and faithfull handmaid of ye highest workeman. By ye one I find ye way open & a plaine path leading to salvation: by ye other a true & certain direction into all Philosophicall Courses: the one pointing to a Caelum Empyreum; the most glorious place of the highest heavens: ye other teaching how to make a Philosophicall heaven uppon the earth: yea first to make a Caelum terrae & after to stellifie the same wth any animall or vegetable star whatsoever. Here is ye true still of nature, not yt furius accedia so Comon wth Comon Alchimists, but yt stomack of the Ostridge working a sound & perfit digestion, wth a seperacon of purem ab impuro in any subiect whatsoever, and removing yt Infection wch for the sin of adam was together wth his posteritie laid uppon every plant & beast according to yt of Paul in the 8th to the Rom. Vers. 22 for we know yt every Creature growneth wth us also & travaileth in paine together unto this present: There is ye glorified bodie of Christ, who by his pretious blood hath washed & Clensed ye soules of many from ye guilt & punnishment both of originall & actuall sin. By this art we draw downe to ye earth those heavenly influences of the stars, wch by a simpatheticall nature (as being derived out of ye first Confused Chaos whereof each thing did precipitate) doe worke most aboundantly in their naturall recipients being truly prepared. But by yt heavenly knowledge Homo quasi ex humo doth rise & mount up to heaven, & is here united wth blessed societie of those heavenly spirits.[41]

---

[41] BL, SL 2223, ff. 25–26.

He joined his fellow-gentlemen alchemists in making more than one 'attack on Gretian and heathen Authors' and on the theoretical, Aristotelian methods of studying the natural world which predominated in the universities. In a more detailed critique he ridiculed the lack of practical knowledge of gardening in the writings of these 'Schoolemen, who have alreadie written many large and methodicall volumes of this subject (whose labours have greatly furnished our Libraries, but little or nothing altered or graced our gardens and Orchards)'. Plat's knowledge was derived from nature and Christianity – the study of nature leading to what probably sums up the goal of Plat as an alchemist: 'a Philosophicall heaven uppon the earth'. He hoped to achieve this by extracting the pure heavenly influences of the stars from earthly substances.[42]

Producing the philosophers' stone was a major goal of Plat's, as it was of most alchemists. He combed written works on alchemy for clues to its creation, composing elaborate tables setting out the stages of the 'great work' of making the stone. His notes were either short pithy formulas: 'Cofectio Lapidis. Bodie 3, Spirit 4, all purified 7.tymes, work uppon a doble base wth a doble heaven, rote and ferment thrice first wth the base & after wth the rich heaven, then let the Hen devowre her owne Chickens, then blont the medecine & prove it'; or they were large digests of published works arranged under the various stages of the work leading to the stone.[43] This is how he summarized this research:

> In the matter & manner of the working of the stone, I finde all the trew philosophers being truly understood fully to agree: yet somme of them to have only attayned the foundation of the stone which they cold not multiply, somme the white stone, and some both white and redd and many of them also not able to turne the philosophers wheele abowt, the fowrth tyme, becawse they knew not the rich ferment of the stone. many of them did worke wth severall owtward fiers and in severall kindes of furnesses, yet they all knew the trew and principall minerall fier. They varie much in their proportions according to the lettre, but not in sense and meaning. Somme of them had the longe, sure, and plaine way only, somme knew

[42] Plat, *Floraes Paradise*, Preface, A4.
[43] BL, SL 2195, ff. 2v, 4, et seq.

and practised as well that way, as the way of accutation which they found daungerous & verie casuall, and though in that course they gayned tyme, yet there fell owt also a diminution of perfection and yet very profitable. they do varie in the trew matter of the generation of mettalls, and according either to the sharpnes of their witts, or fulnes of their instrucyions they do much differ in the disclosing of the secrete & hidden operations of Nature and in the secrete cawses therof. yet he that knew least, did far excell any that was meerely an academicall, peripateticall or stoiacall scholler only. and they that knew most were worthie of the name of great Maisters, wrighting soundly though obscurely of the animal, vegetable & minerall skill, conteyned as well in their litle microcosmos, as in the great Macrocosmos of that great & mighty Archeas of the world.[44]

The fruits of Plat's own work, theoretical and practical, on the making of the stone, are embodied in his alchemical poem. This was composed after mid-1605, headed 'To the trew chimicall Reader', and signed 'H. Platt Miles. 264 verses'.[45] The poem is a description, in seven chapters, of the various stages of the 'great work', leading readers through the creation of the stone and its use to 'project' its properties onto other substances. In the introduction Plat modestly refers to himself as an alchemical novice, but quickly modifies this by placing himself amongst Nature's 'trew adopted Sons'. Distancing himself from the earlier English (and Popish) alchemical writers, he explains that he is,

A Knight, a Scholler, not mured in Cell,
Nor treyned up within a monkish mew.

The introduction begins a process continued throughout the work, a profession of openness and daring revelation of alchemical secrets followed be much obfuscation, for there is, 'No pen so plaine that ever penned all', and even those with some alchemical knowledge must, to know all, call on the help of 'Maisters' in the art.[46]

44 BL, SL 2195, f. 2v.
45 BL, SL 2195, ff. 119–123.
46 The poem is reproduced in full as Appendix II.

The six chapters which follow the introduction promise, in their titles, to take the reader through the process of creating the stone and multiplying its virtues. Chapter 1, 'of the matter of our Stone', begins with a comparison between the creation of the world and that of the stone, an analogy found in many earlier alchemical treatises.[47] Another frequent symbol, the chemical marriage, appears in this chapter. The importance of mercury as a foundation of the work is emphasized. Chapter 2 deals with the preparation of materials: they are to be ground on a 'worthy stone' of porphyry or serpentine. Then Chapter 3 tackles the proportions of substances to be used in the work, another theme common in alchemical writing. Although both sections promise much information, they are, like the rest, so veiled in allusion as to be meaningless to most, if not all, Plat's readers. In discussing proportions of ingredients, he mentions George Ripley's advice that the proportion of liquid to be used to achieve a solution after calcination should produce the same consistency as potters' loam. A few lines on, he returns to Ripley and his allusion to the dangers of 'Noaths flood' for those who do not have regard to proportions. This second comment on Ripley refers to his introduction to *The Compound of Alchemie* and Plat's drawing together of two items from different parts of Ripley's work demonstrates a close reading of it.[48]

Chapter 4, 'of the Philosophers' Vessell', is clearer, discussing various types of vessel suitable for the work, coming down in favour of at least three glass vessels, luted close with glass. Chapter 5, on heat used in the process, distinguishes between 'owtward' fires, on which he is relatively open, mentioning in passing the gentle heat of the 'bath of Baath where Charnocks Mr wrought', and the inward fires about which he is coy. Chapter 6, on colours, allows Plat poetic licence to describe the various exotic colours which the alchemist should observe during the work:

The spotted Panther wth the peacock tayle,
The rayneboe graced wth the Lion greene.

---

47 See for instance George Ripley, 'The Compound of Alchemie', in Elias Ashmole, *Theatrum Chemicum Brittanicum*, 1652, p. 122.
48 Ibidem, pp. 113, 132.

ALCHEMY

The process is long:

> Thie base must often tymes bee circulate,
> Before a perfect white thoe shalt obtayne
> And earth rotation in his circled coorse
> Bringes all the Coollors round abowt againe.

On the vital step from the white stone to the red Plat speculates from sources he has read that:

> Digestion will in tyme bring foorth the same.

But this is a secret,

> ......never yet disclosede
> By pen or press to any wordly wight
> Save where a secret trust hath ben reposed.

Plat tacitly admits that he had not himself reached this ultimate stage. He advises his readers,

> ......content the wth the white
> A perfect Stone most Pearl-like to the Eye
> And rich enough to him that knoweth how
> With kindly frement yt to multiplie.

Multiplication of the virtues of the stone, by passing the virtues to other elements, known as projection, is the subject of the last chapter. Again, there is much veiled allusion, the explanation being that the power to project, in a tyrant's hand, could be devastating and so must be kept secret.

What are we to make of this? As a piece of literature it reads well, with only occasional failures to scan. It bears comparison with some of the doggerel alchemical verse collected by Elias Ashmole in his compendium *Theatrum Chemicum Brittanicum* of 1652. References to Ripley and Charnock place the poem in the tradition of English alchemy. Ripley's *Compound*, a work from the fifteenth century, was printed in London in 1591, and Charnock's *Breviary* appeared in 1557. Several other alchemical works were published in the last quarter of the sixteenth century in London, reflecting an upsurge of interest in the subject. We know from

Plat's extensive notes that he had read the works of the most renowned writers, both English and Continental. Many early English alchemists were clerics, hence Plat's allusion to those trained up 'within a monkish mew'. Unlike the works of Charnock and Norton, which are heavily biographical, Plat's poem purports to deal with practicalities and his approach has much in common with George Ripley's, fitting the English 'native tradition' of didactic exposition sometimes combined with allegory, composed in couplets or rhyme-royal stanzas.[49]

In Ripley's poem, the process of creating and projecting the stone is dealt with in 12 chapters or 'gates'. Although the titles of the gates (Calcination, Solution, Putrefaction, etc., to the final Projection), promise a detailed guide to the work, meaning is heavily veiled and the poem is not capable of interpretation as a series of chemical procedures. Charles Nicholl concludes that Ripley's chapters, 'if approached on their own criteria – as successive stages of a journey inwards, a journey for which chemistry provides the vehicle but not the total meaning, … do form a pattern, a process no less real for being scientifically "invalid".'[50] Nicholl's words fit Plat's poem in equal measure. His chapters trace the steps to be taken to create the stone, dealing with practical matters in an opaque way. Although he places emphasis on the materials, proportions and apparatus, he covers the same ground as Ripley and attains the same end, a stone which can be projected and multiplied. Plat's is a much shorter poem and it may have been written as an introduction to the alchemical treatises which he had written for publication or private circulation among friends (or only among his fellow alchemists), a demonstration that he had achieved some mastery of the subject.[51] We may find the concept of transmitting scientific ideas by verse strange but there were advantages to practical alchemists. Despite the allegorical language, it was deemed important, as now, to conduct chemical operations with complete attention to detail. Verse form helped the reader to memorize important passages, and verses,

49 Robert M. Schuler, *Alchemical Poetry 1575–1700*, New York, NY, 1995, p. xxix.
50 Nicholl, *Chemical Theatre*, p. 35.
51 Many of Plat's fellow-alchemists wrote verse, including John Dee and his scrier Edward Kelley. Plat's contribution would surely have been regarded as a valuable addition to the genre by any who read it.

because of rhyme and scansion, were less likely to be corrupted when copied in manuscript or typeset by unlearned compositors.[52]

Probably more important to Plat were the social reasons for turning to verse, and its appreciation outside the narrow world of the alchemically adept. Poetry, love poems and other verse were a vital element in the royal Court to which Plat belonged for many years. Verse was used to attract attention, petition for patronage and, especially when dedicated to the Queen, to flatter and impress. The Queen was interested in alchemy throughout her reign and at various times encouraged foreign alchemists who turned up in London promising great revelations. She had a long association with John Dee who was deeply involved with alchemy. Her known interest in the subject led Sir John Davies to flatter her in an acrostic poem with the stanza:

> **R** udeness itselfe she doth refine
> **E** ven like an Alchymist divine
> **G** rosse times of iron turning
> **I** nto the purest forme of gold
> **N** ot to corrupt till heaven waxe old,
> **A** nd be refin'd with burning.[53]

Plat probably composed his verses in the reign of James I but Court life was much the same as before. Lastly, alchemical verse emphasized the 'age-old associations between poetry, magic, and the sacred'.[54]

We can enjoy the poem as a work of art, examine it for evidence of Plat's extensive reading in alchemy, and speculate to what extent he practised the experiments he wrote about so enigmatically. The rich symbolism and teasing puzzles of such works led Elias Ashmole to conclude that there was more to be extracted from alchemical writing than one would have discovered from conversation with the authors themselves:

---

52 Schuler, *Alchemical Poetry*, xxxiv–vi.
53 Nicholl, *Chemical Theatre*, p. 17.
54 Schuler, *Alchemical Poetry*, xxxvi, xxxviii.

There you may meet with the Genii of our Hermetique Philosophers, learne the Language in which they woo'd and courted Dame Nature, and enjoy them more freely, and at Greater Command ... then when they were in the Flesh; for they have Written more then they would Speake; and left their Lines so Rich, as if they had dissolved Gold in their Inke, and clad their Words with the Soveraign Moysture.[55]

Aside from the philosophical and practical quest for the stone, Plat sought to use alchemy for more immediate ends. We have seen how he hoped to harness beneficial celestial rays to increase soil fertility and how he claimed to have used a philosophical 'fire of nature' in food and drink preservation. The majority, however, of his recipes involving alchemical operations were aimed at making medicines. Plat was, as we will discuss, an unlicensed medical practitioner. With no formal qualifications, he was not trained in the orthodox, Galenist school of medicine of the universities, his interest in medical matters probably derived from his study of alchemy.

Plat was very much in the English tradition of alchemy for medicinal purposes. When a group of alchemists petitioned King Henry VI for a licence to practise alchemy, they noted that 'in former times' philosophers had made 'many glorious and notable medicines' from a range of animal, vegetable and mineral substances and 'chiefly that a most precious medicine which some philosophers have called the Mother and Empress of medicines, others have named the priceless Glory, but others have called it the quintessence, others the Philosophers' Stone and Elixir of Life; of which potion the efficacy is so certain and wonderful, that by it all infirmities whatsoever are easily curable, human life is prolonged to its natural limit, and man is wonderfully preserved in health and manly strength of body and mind'. These were sentiments echoed by Raleigh's succinct summing-up of the work of earlier alchemists: they 'understood the power of nature, and how to apply things that worke to things that suffer.'[56]

Publications on distilling in the early sixteenth century, which were followed in their turn by translations of Conrad Gesner's works towards

---

55 Elias Ashmole, *Theatrum Chemicum Brittanicum*, 1652, B4 v.
56 Webster, 'Alchemical and Paracelsian medicine', p. 302; Raleigh, *History of the World*, vol. I, xi, p. 2.

the middle of the century, fed an interest in some English medical quarters in chemical medicines. However, the principal impetus behind these new ideas came from the Swiss-German doctor and alchemist known as Paracelsus.[57]

Phillippus Aureolus Theophrastus Bombastus von Hohenheim, known by the pseudonym Paracelsus, was born at Einsiedeln, Switzerland, in 1493, into an impoverished noble family. His mother died when he was nine years old and he moved with his father to Villach in Carinthia. His father was a physician with an interest in alchemy. Paracelsus served as apprentice in the Fugger mines, learning at first hand about metallurgy. He is traditionally said to have been educated in part by a famous alchemist, Abbot Johannes Trithemius. When he was fourteen, he became a wandering journeyman and scholar, roaming all over Europe and the Middle East, working for some time as an army surgeon. He may have acquired a medical doctorate at Ferrara. He tried to settle down in his thirties but his bad temper and lack of tact dictated frequent shifts of residence (he was also condemned by later commentators for his drunkenness and coarseness). Through his prowess as a doctor he became city physician of Basle in 1527, allowing him to lecture at the university. His contempt for Galen and other traditional authorities, and insistence on lecturing in the vernacular, led to much opposition and his forced withdrawal from Basle. The remainder of his life was spent wandering from one town or city to the next, writing extensively and rapidly. He died in 1541, aged 47, leaving a large body of work, some published, some still in manuscript.[58]

Paracelsus' works (and those of his followers, some of which were erroneously attributed to him) convey a Neoplatonic view of the universe, full of mysticism and hermeticism. Natural magic was prominent: by Paracelsus interpreted as a need to observe closely and work with nature. In line with the microcosm-macrocosm theory, Paracelsians saw a correspondence between earthly objects and the cosmos: in medicine man should seek plants and minerals which correspond to the proper celestial bodies. Nature should be closely examined to observe the correspondences.

---

57 Allen G. Debus, *The English Paracelsians*, New York, 1966, p. 49.
58 Paracelsus, *Selected Writings*, ed. Jolande Jacobi, New York, 1951, pp. xxxvii–xliii; Debus, *English Paracelsians*, pp. 14–18.

For Paracelsus, alchemy was at the heart of any explanation of the universe. The very Creation was interpreted in chemical terms – it was in effect a series of chemical separations. Paracelsus believed God used one prime matter in the Creation, the Mysterium Magnum. Whilst acknowledging the Aristotelian elements, earth, air, fire and water, he introduced into chemical theory three other principles – sulphur, mercury, and salt. These concepts were used in confusing ways by Paracelsus and variously interpreted by his followers.

Following the Creation, Paracelsians thought almost everything else could be explained chemically – thunder and lightening, rain, the formation of the earth's crust, all could be replicated in chemical operations with heat and flasks. Man, too, operated chemically. Rather than the imbalance of humours posited by traditional medicine, Paracelsus taught that diseases often had external causes and were located in particular organs. Diseases were entities in themselves, not mere humoric imbalances. They were thought of in terms of chemical reactions. If normal chemical processes in the body malfunctioned, poisons could build up and produce disease. 'If the bodily functions reduce to chemical reactions it should be expected that chemically prepared medicines would cure diseases.' Alan Debus has pointed out that the use of distilled medicines, particularly those prepared using metals and minerals, despite earlier advocates, came to be associated firmly with Paracelsus within a few decades of his death. These were powerful and often poisonous remedies and Paracelsian doctors were attacked for reckless prescribing despite being careful to find the right dosage.[59]

In the years following his death, many of Paracelsus' works – including some mistakenly attributed to him, as well as commentaries by followers – appeared on the Continent and were read also in England. Seen as an attack on accepted medical traditions, they brought forth denunciations from the medical establishment which only highlighted their revolutionary nature. But with one exception, no English exposition of Paracelsus' theories was produced before 1600, whereas his medicines did appear in print in England and were, to some extent, taken up by the general run of doctors and physicians. Paracelsus was especially popular with

---

[59] Debus, *The English Paracelsians*, pp. 32–34.

unofficial medical practitioners, both for his medicines and his opposition to conventional medicine. Paracelsus' demand that it be replaced by a new theory founded on alchemy led many followers to see parallels between medicine and the Reformation – heathen, Galenist medicine replaced by a pure, ancient system derived from God-given primal wisdom.[60]

Plat had probably read much more of Paracelsus than most of his learned and literate contemporaries. He refers to three publications by name in his notes, the *Congeries* (collected works), the *Labyrinth*, and the *Archidoxis*. Thomas Hodges was a later owner of Plat's papers. If the booklist he produced includes Plat's books as well as his own, then Plat probably owned seven Latin texts of Paracelsus, of which the five dated ones were published on the Continent between 1568 and 1575.[61] That Plat valued Paracelsus' chemical medicines is attested by the scattering of recipes attributed to Paracelsus throughout Plat's medical manuscripts. Plat notes also some of his alchemical theory such as thoughts on the action of putrefaction on substances and a method of producing the white stone. Not least, he extracts a passage on the Creation.[62]

In the preface to *The Jewell House of Art and Nature* Plat justifies putting forward his work on distillation and chemical medicines by questioning the guidance on purifying such medicines in existing publications. The result, in his view, was that 'the most part of our chimical and paracelsian practisers' do not sufficiently refine medicines to remove 'their earthly and poisoned parts'. Plat's aim to improve the work of such practitioners places him firmly amongst them, as do the wealth of chemical medicines found in his papers.[63]

A brief but clear statement of Plat's knowledge of, and agreement with the philosophy of Paracelsus and its emphasis on the study of nature comes at the end of his pamphlet on 'Cole-balles' of 1603 :

> And therefore O happie and thrice happie are those wits, (but most infinitely happie are those kingdoms and countries which inioy them) who

60 Debus, *The English Paracelsians*, pp. 41–2, 49–80.
61 BL, SL 2242; SL 2210, f. 2; SL 2245, f. 48a–49.
62 BL, SL 2245, ff. 23, 25, 48v, 69v; SL 2203, ff. 203–204.
63 Plat, *Jewell House*, preface, B1–B4.

have drawne and derived their knowledge from the greate God of nature, from the firmament, from the foure elements, from the great Anatomie and from the little world, and the rest of those unwritten books, whereof *Paracelsus* in his Labyrinth maketh a large and learned discourse.[64]

In a manuscript pamphlet approved for publication Plat makes detailed criticisms of the work of Quercetanus, a follower of Paracelsus. Plat may be critical, but his standpoint is that of a firm believer in Paracelsian chemistry with its three 'principles' of salt, mercury and sulphur. He continues the theme outlined above of the importance of the correct refinement of chemical medicines, to strengthen their curative powers and to make them as digestible as possible. [65]

The many hours Plat said he spent in his laboratory distilling and refining medicines no doubt satisfied his intellectual curiosity and brought him some fame (or notoriety) when he published the fruits of his labours. They also furnished him with a way to make money. He dispensed his medicines to the sick, as we will discuss in detail, tried to sell them to other practitioners and made a concerted attempt at the end of his life to pass on his most treasured secrets for a considerable sum.[66] In August 1608, two or three months before his death, Plat wrote a revealing letter to Sir Francis Segar, an Englishman and courtier who was gentleman of the bedchamber to Moritz, Landgrave of Hesse. The letter tells us much about Plat's alchemy and also his strong desire to make money from all his chemical research. It was addressed to Segar but the letter's purpose was to sell alchemical secrets to Moritz using Segar's position as an intimate servant to provide a channel of communication direct to the Landgrave. Additionally, Segar was himself was an alchemist, once described by Plat as 'a trew Hermetist'.[67]

In general terms, he offers Moritz, 'whatsoever grounde and meanes that I have for the trewe procedinge in all philosophicall courses, concerninge minerall animal and vegitable bodyes which I doe here

64 Plat, *A new, cheape and delicate Fire of Cole-balles*, 1603.
65 BL, SL 2245, ff. 18–25.
66 See Chapter 10.
67 BL, SL 2172, f. 18; SL 2245, f. 30v.

ingeniouslye acknowledge that I have now done or finished', which amounts to 'that light into nature both in Chimicalls, Vegetables, and mineralls, as nether I dare valewe at anye price, nor yet can he [Moritz] make sufficient requitall of'. On a more mundane (and tangible) level, Plat offers,

> the oyle of mercury irreduceable yet not trewlye philosophicall, curinge the pocks in an easie and delicate manner as alsoe that sweete oyle of Amber whiche Sr Walter Raleighe will not willinglye discover for 500li which cureth the dead palsie and all crampes and convulsions. A spirit off wine made without anye fier at all times farr beyond all comon spiritt of wine though never so well rectified. The oyle of Talen that giveth any excellent and durable tincture to mercury but this I have not proved but have it warranted from the hands of a great scholler: mercury turned into perfect lead in halfe an howre/ Cinamon water made extempore a vessell of wood to boyle in without anye daunger of burninge./ Cinamon or a Rosemary sugar, and soe of all spices./ Howe to drawe out the Quinttessence of anye herbe with Spiritt of wine but not without some Mixture of the Elements but yet far beyond anye ordinarye extract or inhibition./ A profitable waye for the devidinge of the sweete parte of the water from the saltishe part by an attraction bodye

– and many more if his offer is accepted. Plat will also send, as a token of his serious intent, two pamphlets (these may be the treatises on alchemy which survive in BL, SL 2223). Further, Plat implies he has obtained an advanced knowledge of alchemy, writing knowingly of the glass vessels needed to prepare the red stone and telling Moritz he can lead him 'into the secretest corner of her howse yea unto the bedchamber of Nature her selfe'.[68]

Approaching Moritz was by far Plat's best hope of making a sizeable amount of money from his alchemical secrets. Hesse was a Protestant German principality with close diplomatic connections with England. Indeed, of Moritz it was said, 'His affection to Englishmen is so great, that no stranger of any part of Christendome can bee more welcome to him then an Englishman'. According to the historian Bruce T. Moran,

68 BL, SL 2172, f. 18.

'More than any other princely court in northern Europe, the Kassel court stands out as a centre of occult patronage in the years prior to the outbreak of the Thirty Year's War'. Moritz had 'an entire entourage of alchemists, Paracelsian physicians, Rosicrucians, and natural magicians'. Thomas Coryate, writing in 1611, declared that Moritz's learning was, 'confirmed by the testimonies of thousands of the learnedner sort,' and that his 'admirable wisdome' led to his designation as 'the Solomon of Germany'. In 1608, one of his leading advisors, Johannes Hartmann, presented Moritz with plans for a 'collegium chymicum' at Marburg and this commenced in 1609. Plat was hoping to interest Moritz in his secrets at a time when this, the first academic institution for the study of alchemy, was being created.[69]

Plat must have had high hopes – given his good connections with both Segar and one of the Landgrave's personal physicians, James Mosanus. I have already put forward James's father Godfrey, who died in about 1593, as the possible alchemical master of Plat. James, himself a Paracelsian doctor, was suggested by Plat as the person best suited, in default of Segar, to receive personally the secrets Plat wanted to sell to Moritz. James was in the service of the Landgrave from 1599. He received his medical degree at the University of Cologne in 1591 and gained most of his experience of medicine and alchemy outside England. He was an important part of the scientific entourage at Moritz's court, acting as physician to the royal family, preparing chemical medicines, reviewing new medicines recommended to Moritz and advising him on alchemical theory. He also acted as Moritz's personal envoy, in the course of which he visited leading Paracelsians and alchemists such as Joseph Duchesne and Andreas Libavius.[70]

James Mosanus advised Moritz against over-emphasis on attaining the philosophers' stone before a thorough exploration of new alchemical medicines and he may have viewed Plat's letter, if he ever saw it, as over-optimistic. Plat, however, was a skilled salesman. The approach to Moritz was a carefully worded attempt to sell secrets for a large sum. Early in the letter Plat seeks to persuade Segar that his own interests will be forwarded

69 Bruce T. Moran, 'The Alchemical World of the German Court', *Sudhoffs Archiv Zeitschrift Fur Wissenschaftgeschichte, Beiheft*, 29, Stuttgart, 1991, pp. 8–9, 11; Thomas Coryate, *Coryats Crudites*, 1611, pp. 577–8.

70 Moran, 'The Alchemical World of the German Court', pp. 68–85.

by putting Plat's proposals to Moritz, who will, as a result, 'reioyce… that ever a gentleman of your worth came under his cullors'. Plat soon mentions money, asking of Moritz 'what honorable requitall he will make for such particular Secretts of Arts as I have with great chardge in long time, not without great danger of my life and health attained unto'. He gives assurance that should too much be offered to him, he will repay any excess and he also assures Moritz that he is not a 'great Imposter' trying 'to sell you manye millions for a fewe thousands… my purpose is only to valewe things valewable'.

After detailing what he has to sell, Plat returns to his pitch telling Segar that what he has written are dangerous secrets and, if Moritz is not interested will he 'consecrate those fewe and idle lines to vulcan without reservinge anye memorye of them'. Moreover, a speedy answer is required 'for I assure you upon my creditte, I staye my conference with a great englishe Nobleman herein in respect of your lord whome I cannot long delaye'.

I have traced no record of an approach reaching Moritz. This carefully crafted letter either did not meet with Segar's (or Moritz's) approval or, more likely, events were overtaken by the death of Plat soon after it was written.[71]

---

71 BL, SL 2172, f. 18; personal communication from Bruce T. Moran.

CHAPTER SEVEN

Medicine and Medical Practice

In discussing Plat's alchemy we have noted its close links with medicine. English alchemy before Plat had the production of medicines as an essential goal and the absorption of Paracelsus' works into the English tradition in Plat's time reinforced this aim. Plat may have been initiated into alchemy by a physician and he must have been influenced by the apothecary-alchemists such as Kemish with whom he had frequent contact. No wonder then that he took a keen interest in medicine. Such an interest was by no means unusual amongst gentlemen of his standing or within the population of London as a whole. What was more unusual, but again not uncommon, was that he developed a medical practice, producing, dispensing and administering medicines and eventually attempting to sell both his secrets and practice.

From dates in Plat's published and unpublished medical writings we find his interest in alchemy predates by several years the first indication that he was producing and administering medicines, reinforcing the notion that the one subject led to the other. The earliest date in his surviving practice-notes is October 1593 though it is reasonable to suppose he had build up some skill prior to this. In 1588 he comments that a cold in his knee was cured by the application of an ointment he had made but the application was overseen by the surgeon Mathew Ker. At that date, he still felt the need of expert advice. In 1594 he published many recipes for oils extracted for medicinal purposes but he modestly directs readers who want more information to 'Master Bakers' translation of Gesner entitled the *Jewell of Health* or to the shop of his apothecary (and alchemist) friend Kemish, maybe an indication that he felt he had still much to learn himself. His practice-notes continue until the year of his death and he must have gained expertise over the years both in making and administering medicines. By 1600 we find him confidentially negotiating to sell medical secrets to a Northamptonshire gentleman. On a more public scale, he issued a broadside in about 1607 advertising medicines for seafarers and offered a medicine for sale in the preface of *Floraes Paradise* in 1608, the same year in which he offered medical and alchemical secrets to the Landgrave of Hesse. In sum, for about two decades prior to his death he had an interest in medicine serious enough to class him as an unlicensed practitioner. Before looking in detail at his medical work we need to consider medicine in the London of Plat's time, particularly 'irregular' practitioners.[1]

Despite the increase in 'alternative' medicine in recent years, medicine in modern Britain is predominantly a professional activity, practised by doctors who undertake a long and rigorous training at university and teaching hospital. They are closely regulated throughout their working lives by professional bodies. In the London of Plat's time it was a much more diverse activity. True, there was a professional body, the College of Physicians, which from early in the sixteenth century had attempted to regulate closely the medical profession in London and the suburbs up to seven miles distant. Its membership was small, between 20 and 40 doctors at any time in the period, and these men were theoretically the only ones authorized to administer medicine and oversee the 'internal' health of Londoners. Barber Surgeons, members of a London Company and numbering maybe 100, confined their activities to wounds, amputations and other 'external' medical operations, while about 100 apothecaries – members of the Grocers' Company – prepared and sold medicines to the professionals and, increasingly, to the general public. This neat tripartite model was but the visible, and countable, tip of an iceberg of medical activity. Twenty to forty doctors could not hope to provide a medical service to a population of, say, 200,000 in greater London in 1600, even if people could afford their fees. In their stead was a wide variety of practitioners, many of whom, ironically, can be identified only from the records of the College of Physicians when that body attempted prosecution

[1] BL, SL 2209, ff. 4–4v, 16; SL 2172, ff. 18, 28–29; Plat, *Jewell House*, pt. 3, p. 8; Plat, *Certaine philosophical preparations of foode and beverage for sea-men*, c. 1607.

to maintain its monopoly. Below the irregular practitioners in the 'pyramid' was a substructure of those who had some medical knowledge – enough to treat themselves and friends and family.[2]

Health – achieving it, maintaining it, and being on the alert to threats to it – was a preoccupation of early modern England, especially in London. The city had by far the largest concentration of population in the country and a high incidence of endemic and epidemic disease, as well as the accidents and occasional violence of living and working in a crowded setting. The historian Margaret Pelling has stated that 'early modern people were obsessed with health, its fragility, and with the means for preserving it'. Although the strong belief in God and His providence, especially when faced with epidemic diseases such as the plague, might be thought to have induced a passive acceptance of disease, such religious attitudes did not dissuade most people from actively seeking to maintain health. 'Case records and personal papers show that people constantly monitored their own state of health and were profoundly concerned about the significance of apparently trivial symptoms'. People worried about agues, colds, toothache, backache, skin blemishes and constipation, as well as more serious matters such as kidney and gall stones. The generative organs were closely monitored. There was much consultation on mental health. Ailments – especially eyesight problems – which might affect ability to work and therefore spell economic disaster for most people were another preoccupation.

This constant monitoring of health is reflected in the diary of the seventeenth-century Essex clergyman Ralph Josselin, who chronicled his own and his family's ailments. He was prone to colds and described symptoms in great detail. He listed his stomach complaints, agues, and travel sickness as well as long-term complaints such as inflammation of his navel and a swollen and ulcerated leg. He recorded minor symptoms, rheums in his eyes, a sore tongue, chest pains, sore bones, minor swellings and the odd sensation of 'coldness or moistness' in the crown of his head.

2   Margaret Pelling, *Medical Conflicts in Early Modern London*, 2003, p. 1; Pelling, *The Common Lot*, p. 240; Roger A.P Finlay & Beatrice Robina Shearer, 'Population growth and suburban expansion', in *London 1500–1700: the making of the metropolis*, ed. A.L. Beier & Roger A.P. Finlay, 1986, p. 45.

Lucinda McCray Beier concludes that, although no hypochondriac, 'Ralph Josselin was unwell, or at least uncomfortable, most of the time.'[3]

If this state of affairs was typical of most people, no wonder that medical matters were a frequent topic of conversation. Like the neutral topic of the weather today, Margaret Pelling found discussion of illness amongst Londoners went across social boundaries: 'When it came to illness, a Privy Councillor could learn from his laundress, a husband from his wife, a philosopher from an old woman, a gentleman from his servant, a bishop from a cunning man, an empiric from a collegiate physician, a Protestant from a Catholic, an Englishwoman from an Italian.' This willingness to discuss illness and cures was not an indication of relaxed social boundaries but a gauge of the importance all classes accorded good information on health.[4]

In many cases, the first line of defence against illness was the home – households dosed themselves against actual or anticipated infection and applied poultices, creams, plasters and bandages to wounds and external blemishes. We find frequent instances of aristocratic women involved in the making and administering of medical procedures to their households and friends. This was recognized by Gerard who with his *Herball* addressed the 'gentlewoman's skill, who helps their family and their poor neighbours that are far removed from physician and chirurgeon.' Lady Margaret Verney of Claydon in her will of 1639 wanted many of her papers destroyed but 'medsinable and cookery Boockes, such keepe'. The survival of many medical recipe books in family papers shows the care taken to preserve such knowledge in the household.[5]

Ralph Josselin, for all his own preoccupation with health, recognized that his wife was the main dispenser of medicine in the household. Women and men wrote down herbal remedies and prepared them from

3   Pelling, *The Common Lot*, pp. 5, 29–30; Lucinda McCray Beier, *Sufferers and Healers*, 1987, pp. 141–8, 154; Lucinda McCray Beier, 'In sickness and in health', in *Patients and practitioners*, ed. Roy Porter, 1985, pp. 101–128.
4   Pelling, *The Common Lot*, p. 1.
5   Layinka Swinburne, 'Of each a handful: Medicinal herbs in the country house', in *The Country House Kitchen Garden*, ed. C. Anne Wilson, 1998, pp. 177–93; Lynette Hunter, 'Women and domestic medicine', in *Women, Science & Medicine*, ed. Lynette Hunter & S. Hutton, 1997, pp. 89–107.

wild or cultivated herbs. Stillrooms were not uncommon in large houses, allowing the household to benefit from more powerful and complex medicines. Thomas Tusser in the 1570s advised wives in Maytime, 'Wife as you will, Now ply your still' for 'The knowledge of stilling is one pretty feat / The waters be wholesome, the charges not great'. Plat included a section on distilling in *Delightes for Ladies* as did Markham in the *English Housewife*. Markham's book also had a chapter on physic and surgery. Some may object that we are talking of London in discussing Plat's medicine, where the chance to grow or collect herbs was less and the availability of professional help greater, but herbs were available in the markets, 'suburban gentry' such as Plat had gardens in their out-of-town residences and gentry temporarily in London had herbal remedies sent from home when they were ill.[6]

Londoners did, however, have greater recourse to apothecaries for their self-administered medicines. In 1592, a satire on London life ridicules a courtier who frequents apothecaries because he 'cannot have a fart awry but he must have his purgatives, pills, and clysters, or evacuations by electuaries; he must, if the least spot of morphew come on his face, have his oil of tartar, his lac virginis, his camphor dissolved in verjuice.' The apothecaries, although supposed only to prepare and sell medicines, frequently acted as consultants (some were prosecuted by the College of Physicians) and were just one of the informal ways in which Londoners fulfilled their desire for medical advice.[7]

Informal medical aid was available in a variety of forms. Some of the gentlewomen who dispensed medicines to their families were more widely consulted for their expertise. Lynette Hunter instances Lady Lisle in the 1530s, Ann Dacre, Lady Arundel in the 1560s and '70s and Margaret Hoby in the early 1600s, who all had a local reputation for medicine. Margaret

---

[6] Swinburne, 'Of each a handful', pp. 177–93; C. Anne Wilson, 'Stillhouses & Stillrooms', in *The Country House Kitchen*, ed. P.A. Sambrook & Peter Brears, 1996, pp. 129–143.; Thomas Tusser, *Five hundred points of good husbandry*, ed. W. Mavor, 1812, pp. 147, 158.

[7] Pelling, 'Thoroughly Resented? Older Women and the Medical Role in Early Modern London', in Hunter & Hutton, *Women, Science & Medicine*, p. 70; Gervase Markham, *The English Housewife*, ed. Michael R. Best, Montreal, 1986, pp. xxviii–xxix.

Hoby, indeed, attempted surgery as well as internal medicine. One general practitioner at the end of the sixteenth century had no qualms about noting many recipes for medicines gleaned from local gentlewomen, and was even comfortable in jointly treating a patient with one.[8]

Literate Londoners read medical works, particularly those in the vernacular which were produced with increasing frequency in the later sixteenth century. Many were written by unofficial medical practitioners and some were translations of recent Continental works. 'Bills', prescriptions from doctors, were circulated for further use, particularly from dead doctors who could not object. And, as Margaret Pelling has indicated, the bustling city of London threw together casual acquaintances who traded medical advice with one another.[9]

There was a fine line between help to friends and commercial activity: 'men were extremely concerned about their own health and that of their friends and family, and … this concern readily slid into forms of practice, first within and then outside the circle of friendship and kinship.' More so in the case of women, who were most likely to be found practising medicine in a modest way, often purchasing their medicines from apothecaries.[10]

Established physicians and surgeons were often savage in publishing denunciations of unlicensed practitioners in the capital who they characterized as quacks:

> which doe forsake their honest trades, whereunto God hath called them, & do daily rush into Physicke & Chirurgerie. And some of them be Painters, some Glasiers, some Tailors, some Wevers, some Joiners, some Cutlers, some Cookes, some Bakers & some Chandlers &c. Yea, now a daies it is to apparent to see how Tinkers, Toothdrawers, Pedlars, Ostlers, Carters, Porters, Horse-gelders, horse-leeches, Ideots, Apple-squires, Broomemen, bawds, witches, cuniuers, South-saiers, & sow gelders, Roages, Rat-catchers, Ruagates, & Proctors of Spitlehouses, with such other lyke rotten & stincking weeds, which do in town & countrie, without order,

---

8 Lynette Hunter, 'Women and domestic medicine', in Hunter & Hutton, *Women, Science & Medicine*, pp. 100–101; McCray Beier, *Sufferers and Healers*, pp. 106–7.
9 Pelling, *Common Lot*, pp. 2, 30; McCray Beier, *Sufferers and Healers*, p. 21.
10 Pelling, 'Thoroughly Resented?', p. 70.

honestie, or skil, daily abuse both Phisick & chururgerie, having no more perserverance reason, or knowledge in this art, then hath a goose.

This diatribe of 1585 by William Clowes expresses the exasperation of the small medical élite when faced with a much larger number of irregulars (estimated to have been in the region of 250 in London in the late sixteenth century). Although he concentrates on those at the bottom of the social scale, his long list of tradesmen accords with other research into irregulars and with the attempts by the College of Physicians to prosecute these interlopers. One also has the feeling that the targets were not easy to hit, these people were not full-time physicians and surgeons, and there was a constant movement in and out of the medical business.[11]

The names of 714 irregulars occur in the annals of the College of Physicians between 1550 and 1640, many of them several times over as prosecution failed to deter their activities. For example Thomasina Scarlet appeared before the College of Physicians for the first time in 1588 and admitted administering purgatives and emetics and giving medical advice to up to 100 people. She was fined and imprisoned for continuing to practise in 1595, 1598, 1603 and 1610, making a final appearance in 1611 when an old lady. Some claimed to be little more than individuals who happened to get caught treating a friend but many were in business, proclaiming themselves with handbills and the like. Some had moved totally from other occupations into medicine. The shoemaker William Trigge in 1647 was described by his wife, 'he was by his breeding a shoomaker, butt now he made profession only of distilling waters, and that he did use to give certaine powders, and Cordials to such as were infected by the Plague'. Others, such as the travelling shoemaker in the mid-sixteenth century who turned up in Maidstone claiming to cure sore eyes, kept their dual status – he may have been one of many irregulars who were ignorant of most medicines but had expertise in just one or two areas.[12]

[11] T.R. Forbes, *Chronicle from Aldgate*, 1971, p. 89; Pelling, *Common Lot*, p. 240.
[12] Pelling, *Medical Conflicts In Early Modern London*, pp. 3–4, 149–50; Pelling, 'Thoroughly Resented?', pp. 70, 71–2.; Margaret Pelling & Frances White, *Physicians and Irregular Medical Practitioners in London, 1550–1640*, on-line database, 2004; McCray Beier, *Sufferers & Healers*, pp. 21, 38–9.

Many members of the College of Physicians did not help their attempts to control irregulars by withdrawing to the country in the face of plague epidemics, losing ground each time to apothecaries, surgeons, and irregulars who remained. They also lost the initiative by ignoring the new venereal diseases (it was difficult to fit 'new' diseases into Galenic theory) which much preoccupied Londoners and were treated by many of the irregulars, often with new chemical medicines.

The stubborn persistence of many of the practitioners prosecuted also taxed the patience of the College of Physicians which sometimes had to retreat when faced with the powerful allies of some irregulars. Aristocrats brought in their own unlicensed doctors from abroad and successfully defended them against the College. Such practitioners (like Plat's friend Mosanus junior, who fell foul of the College) may in any case have possessed foreign formal qualifications. Others had judicial support against the medical authorities. William Rich, despite being considered a quack, was released from prison by Dr Lewes, Judge of the Admiralty, to continue his practice in defiance of the College. Another unlicensed physician, John Clark, was imprisoned and fined £20 in 1603 but later released by the order of the Lord Chamberlain. He clearly reckoned his skills, as may have other members of the Queen's Court. Clark's case is of great interest because he had links with Plat: he claimed to have helped him write two medical treatises (which are now lost) and to have used mummia (see below) in a cure for fevers recommended by Sir Hugh. He also sold cinnamon water (a speciality of Plat).[13]

Many of the irregulars prosecuted were not, as portrayed by the College's members, poor and ignorant. One hundred and twenty eight out of the 774 identified between 1550 and 1640 had university degrees and the extent of literacy among them was high. That many were skilled physicians is reflected by the fact that 8.4 per cent subsequently gained admission to the College itself. Margaret Pelling has concluded that irregular practitioners were a diverse group: both rich and poor practised (clergy-practitioners often pleaded poverty as an excuse when prosecuted);

13 Pelling, 'Thoroughly Resented?', pp. 76, 80–81; Pelling, *Medical Conflicts In Early Modern London*, pp. 96, 103–4, 340, White & Pelling, *Physicians and Irregular Medical Practitioners in London, 1550–1640*, on-line database, 2004.

many had other occupations – the food and drink trades were frequently found combined with medicine; women were well represented in the group; some specialized in a narrow range of complaints; and the body of persons carrying out medicine was constantly changing as people started or left off practising.[14]

Medicine in the London of Plat's time was thus characterized by a wide range of practitioners of varying skills, social class, and ranges of activity, proffering their medicines and advice to a similarly diverse population of rich and poor who were, as a whole, very interested in maintaining health. Margaret Pelling describes the poor as consuming medical care in excess 'as a kind of consolation for insecurity and disadvantage'. Aptly, this has been described as a 'medical marketplace', where 'sick people chose freely from among the range of practitioners according to their own and friends' judgement and the nature and seriousness of their condition. Like Ralph Josselin, people might at various times treat themselves, consult friends, buy medicines from apothecaries, consult local irregular practitioners or consult physicians. Scales of fees varied according to the needs and expectations of the differing social classes, including a significant amount provided for nothing. The destitute sick might find solace through charity: a special collection was taken in the parish church at Aldgate in June 1587, yielding four shillings to pay to cure a young man 'trobled with the faling sickness'.

At the other end of the social scale, expensive drugs and treatments were consumed to enhance or reinforce high social status and, conversely, exotic drugs were advertised as having been taken by the gentry. In *The Gull's Hornbook* (1609) Thomas Dekker satirized the resort to high-status physicians for purgatives by advising, 'You may rise at dinner-time to ask for a close-stool, protesting to all the gentlemen that it costs you a hundred pound a year in physic besides the annual pension which your wife allows her doctor'. When the sick or injured were seriously worried about their condition, they consulted widely. Robert Curtis, a wealthy gentleman, was badly injured riding through New Romney in 1640. Taken to a nearby inn, a local woman was summoned as nurse and both a physician and a surgeon were brought from Tenterden. A man was then sent to Canterbury to ask

14   Pelling, *Medical Conflicts in Early Modern London*, pp. 141, 144, 188, 333; Pelling, 'Thoroughly Resented?', p. 70; Pelling, *Common Lot*, pp. 242, 245, 250.

another surgeon and two physicians to attend, and another local woman came as nurse. Despite all this activity, Curtis died.[15]

The case of Curtis may have been a panic response to a hopeless case but it emphasizes that patients could, and did, pick and choose their practitioner. People often self-diagnosed their illnesses and then went to an apothecary for the appropriate drugs or to the practitioner of their choice to ask for treatment which would lead to a cure. Payment was sometimes made on a contract basis, a payment on account with the balance on cure. Money-back might be demanded if the patient was not satisfied.[16]

This increased bargaining power on the part of the patient was helped not only by the diversity and numerical size of the body of practitioners available but also by developments in medical theory and practice in the late sixteenth century. The range of medicines expanded and there were competing philosophies behind their use. The Galenism of the College of Physicians was assailed by outsiders with an increasing emphasis on the therapeutic effectiveness of treatments aimed at the symptoms of specific diseases rather than correction of an imbalance of humours. Works by Paracelsus were imported from abroad and translations appeared in print of works by his followers, part of an increasing vernacular literature on medicine which was read by both medical men and laypeople. Emphasis was placed on chemical cures: distillations of plant extracts and chemical compounds. New plants for use in medicine were sought, particularly from the New World but also from England. The derogatory name 'empirics' given to many irregulars by College physicians, identified the emphasis of these practitioners on practice rather than theory. Single, strong cures were aimed at specific complaints, most noticeably in the use of mercury to cure syphilis. In alchemy, there was a renewed search for a universal medicine to cure all ills. All this was underpinned by a revival of Neoplatonism and also accorded well with the Protestant faith with its belief that man

15 Pelling, *Common Lot*, pp. 12, 20–21, 30; Forbes, *Chronicle from Aldgate*, p. 94; McCray Beier, 'In sickness & in health'; Gail Kern Paster, 'Purgation as the Allure of Mastery', in *Material London*, ed. Lena Cowen Orlin, Philadelphia, PA, 2000, pp. 196–90; Ian Mortimer, 'The Triumph of the Doctors', *Royal Historical Society Transactions*, XV, 6th Series, 2005, p. 97.
16 Pelling, *Common Lot*, pp. 30, 246.

could, by striving, regain both his spiritual and physical strength lost at the Fall.[17]

These developments resulted in a wider knowledge of, and interest in medicine amongst the educated laity, lessening the gulf between the knowledge held by the practitioner and that of the patient. In the words of Margaret Pelling, 'many aspects of medical culture, far from "emancipating" the educated medical practitioner from the layman, formed part of a common philosophy which partly explains the level of lay knowledge of, and interest in, medicine.' Thus both the status and bargaining power of practitioners were weakened by the decreasing knowledge gap.[18]

Sir Hugh Plat was an irregular practitioner and, as we shall see, he had many of the characteristics outlined above. But although he apparently operated openly, treating a wide range of patients, his name is not amongst those harassed by the College of Physicians. Some of his acquaintances in the medical trade were not so fortunate. The most likely explanation for his immunity is a combination of protection by some of his more exalted customers, senior government officials and courtiers, and his own rise in status which lifted him far enough up the ladder of the gentry to be too consequential a figure for the College to challenge. As I discuss below, however, the prosecution of his collaborator Clark may have been an attempt by the College to intimidate him.[19]

In his writings Plat is not forthcoming on how he developed as a practitioner. The link between his alchemy and the influence of Paracelsus on his medicine is clear: in 1594 he portrays himself as the 'philosophical vulcanist, or perfect paracelsian' who will only be content with distilled extracts of herbs as medicines, not mere raw juices. He was also indebted to the translated work 'written by doctor gesnerus, in a booke intitled the Iewell of health, and englished by Master Baker'. But most of his

17 McCray Beier, *Sufferers & Healers*, pp. 15, 20–21; Pelling, *Common Lot*, pp. 30, 33, 251; Margaret Pelling, 'Knowledge Common & Acquired', in *The History of Medical Education in Britain*, ed. V. Nutton & R. Porter, Atlanta, GA, 1995, pp. 250, 263, 269–70.
18 Pelling, *Common Lot*, pp. 30, 251.
19 Margaret Pelling & Frances White, *Physicians and Irregular Medical Practitioners in London, 1550–1640*, on-line database.

expertise in medicine, he wrote in 1594, 'I have founde out by mine owne experience, or learned of others' and this statement holds good throughout his career.[20]

He was, on occasion, contemptuous of qualified physicians, telling a correspondent in 1600 that, being possessed of one of his medicines, 'then may you as boldly challenge a velvet Capp, and a doctors gowne as the best graduate that ever ware the same in England'. He acknowledged however, learning medicines from them, including several remedies from a Dr Jordan, and an amber oil from 'tharchduckes Phisition of Stepney'. He hoped to sell some of his remedies to a number of physicians and surgeons, including, surprisingly, 'Mr Clowes', the surgeon with a great antipathy to 'empirics'.[21]

Given his style of medicine, it is not surprising that he had far more contact with irregular practitioners then licensed physicians. One such man, who was not a doctor but by his skill and eminent patronage probably had the same protection apparently enjoyed by Plat, was Mr Williams, the schoolmaster of Highgate, who 'purchased great credit by sundry cures' and was 'phisition to Popham L. chief Justice of England'. Plat records his debt to his mentor in alchemy Mosanus senior for a procedure to relieve both gout and scurvy by letting blood. Mosanus, a skilled physician, was condemned as irregular by the College of Physicians because of his foreign qualifications – a restriction aimed more at exclusivity than upholding standards. Many of the men and women who Plat records as giving him medicines were probably irregular practitioners.[22]

Plat had a close working relationship with one such practitioner, John Clark, who has been briefly mentioned above. The College of Physicians' records of proceedings against Clark give the impression of a slippery and insolent man, a surprising companion for a gentleman of Plat's rectitude.

20 Plat, *Jewell House*, pt. 3, pp. 8, 22–3.
21 Dr Jordan was possibly Edward Jorden, physician, born *c.* 1565, lived in Bishopsgate 1596–1600, licentiate Coll. of Phys. 1595, see Pelling & White, *Physicians and Irregular Medical Practitioners in London, 1550–1640* on-line database; BL, SL 2172, ff. 13, 29; SL 2189, f. 50.
22 BL, SL 2209, f. 5a; SL 2189, ff. 6–7; Pelling & White, *Physicians and Irregular Medical Practitioners in London, 1550–1640*, on-line database.

He practised from 1592 to 1609. Summoned to a hearing in August 1602, he excused himself from appearing for fear of 'the officials making arrests who were lying in wait for him everywhere'. In February 1603 he denied both practising following a warning and selling 'expensive waters'. He was told to disclose the composition of the waters. In the following month he appeared and refused to say why he had sold a pint of cinnamon water (possibly one that Plat had prepared) for £1. He admitted using medicines recommended by Plat and claimed to have collaborated with him on two medical works (of which there is now no trace), *Tuba Apollinis* (which Clark said he had written) and *Rei Dispositione*. Despite his insolence, he was given three days to cease his activities and, only when he did not, he was fined £20 and imprisoned. On 7 March he escaped and went to the Lord Chamberlain. He sent him at first to the Marshalsea but later freed him. The College drafted a letter to the Lord Chamberlain, agreeing that Clark could treat him but nobody else. The letter was not sent and instead verbal representations were made.[23]

Clark emerges from Plat's papers as an alchemist and physician using chemical remedies. In alchemical practice he recommended a mixture of 'fatt earth' and salad oil as a *bain marie*; claimed to have produced crude mercury from lead with a mixture of distilled waters and to have a recipe for silver from lead; had a recipe for oil of gold; and told Plat that spirit of wine would work upon powdered lead.[24] Plat was impressed by his 'aqua bezoutica', an extract from the 'Bezuar stone', a recipe for which he kept secret, although Plat did note the recipe for his medicine for gallstones. Clark gave him the names 'of diverse persons whome wthin these few yeeres hee hadd cured and eased greatly especially by openinge the urinary passages wherby diverse that hadd not made water in a longe time before, did presently mak greate store of water'.

The closeness of the relationship between these two men is shown by their collaboration on the two medical treatises and Clark's use of Plat's medicines. Plat records a success with his ague medicine in 1602: 'I cured Mr Raff Skipwith dwellinge neere the turnpike in hoborne

23 Pelling & White, *Physicians and Irregular Medical Practitioners in London, 1550–1640*, on-line database.
24 BL, SL 2245, f. 12–13; SL 2189, ff. 7–8, 10.

having ben undersheriff of Midd. the last yeere, of a tertian, wth my redd powder at the second takinge in the end of may. Mr Clerke gave the medecine.' Before the College, Clark admitted using mummia to cure fevers and a purge, both medicines recommended by Plat. Clark was accused of selling cinnamon water; a medicine Plat claimed to be able to make in some quantity. It is possible Clark was acting as his salesman. Equally, is it possible that the College, knowing of Plat's medical practice and production of medicines but also conscious of his protectors, was attacking Plat via Clark? It is significant that in the end Clark himself was protected by the Lord Chamberlain, the Privy Councillor in charge of the Queen's accommodation, wardrobe, attendants and entertainment, a major figure in the Court. He forced at least a partial climb-down by the College and probably ensured Plat was not bothered by them thereafter.[25]

As one might expect, Plat bought many ready-made medicines from apothecaries: Mr Kemish ('the auncient and expert chimist') supplied him with chemical remedies, Mr Russell sold him distilled waters, and recipes were supplied by Trowt (or Trout) and Garret. He remarks of an oil of roses: 'this I had of an Apothecarie for an approved and rare secrete'. Some of the apothecaries Plat knew flouted the College's prohibition on unlicensed practising. Richard Trout fell foul of the College in 1601 and James Garret pursued a long career in practice despite attempts by the College (five times between 1601 and 1626) to stop him. His medical practice is confirmed by Plat's remarks that Garret's cure for ague was 'a very safe and easie vomitt [and] he hath given it to his owne child being but .3. yeeres of age and hee hath had 12 yeeres experience therof.'[26]

Most recipes and advice came from people not identified as making all or part of a living from medicine. They were friends, relatives, courtiers and officials, foreigners resident in London, and the day-to-day encounters of London life. His cousin provides a recipe, as does his friend Mr Rich

25 BL, SL 2189, f. 8; SL 2223, f. 20v; SL 2209, ff. 18v–19; Pelling & White, *Physicians and Irregular Medical Practitioners in London, 1550–1640*, on-line database.
26 Pelling & White, *Physicians and Irregular Medical Practitioners in London, 1550–1640*, on-line database; BL, SL 2209, ff. 1v, 6, 9; SL 2189 f. 9, 44–47; Plat, *Jewell House*, pt. 3, p. 8.

of Lee. Another friend, the government official Auditor Hill, provides several medicines, and Señor Romero, the Spaniard from Naples, gives him details of an important anti-plague pill. From the royal Court 'Mr. Kelway sommetimes trompeter to the Queene' provides a recipe which 'hath ben often proved both in the psiatica and other aches, & benommed members with cold'. A homely bath to relieve gout of sage, pennyroyal and salt from 'Nurse Chetle' sits alongside 'the diet drinke made according to Sr Ulrick Huttons prescription' for the same affliction. An itinerant woad-man suggests curing sore limbs by thrusting them into warming heaps of fermenting woad for several hours. (Plat omits to mention the stink that woad gives off!) Occasionally, we glimpse mechanisms for the informal exchange of medical knowledge: 'this of Mrs Colm[an] a gentlewoman whoe learned it of Mr fleetwood the recorder of London whoe hadd often used the same' or, more exotically, 'this of Sr Romero whoe learned it of Cardinall Grannile sometimes viceroy of Naples'. London had its own 'traditional' remedies: 'yff the gowte comme of a cold cawse, anoynt the place greived wth a balsamme called Giles balsame whoe was an ancient man & sometimes a goldsmith of London, & the same is sold in somme parts of london…. This was highlie commended to mee by often experience from Mr Grene the Refiner as a generall remedy for aches'.[27]

Plat gathered together these recipes, plus those he himself created, in substantial compendia of medicines which are now lost.[28] By 1593 he was confident enough of his medical abilities to make and distribute a considerable number of anti-plague pills and in the same year his practice-notes commence, indicating that by that time he was seeing patients regularly. In 1600 he was negotiating to sell some of his secret recipes and also advising his prospective purchaser on how to deal with patients. In 1608, now a mature practitioner, he could look back on many years of experiments in chemical medicines and in their administration, although he was still modest enough to admit of quartain agues, 'I cannot speake much of Quartanes, because I have not had any great experience therin'.

27   BL, SL 2209, ff. 3v, 4, 4v, 7, 7v.
28   Many such references are scattered through some MSS, e. g. 'See in my booke of medecines, the making of an excellent Diet drinke. no. 277&731. &. 718', BL, SL 2210, f. 145.

MEDICINE AND MEDICAL PRACTICE

The prices he hoped to fetch for his secrets reveal his confidence in his abilities.[29]

To judge by the reference numbers of the medicines in his lost books, he had accumulated over a thousand recipes.[30] These must have been a mixture of his own and other remedies, probably the majority would have been provided by acquaintances or culled from books. Reviewing the recipes in his existing writings one is impressed by their eclectic nature. True, recording a medicine does not mean Plat approved of it or used it, but those listed at the start of each surviving section of his practice-books were the ones he relied on for each disease he tackled.

For example, the range of his medicines to cure 'Ache. Gowte, psiatica. &c.' was very wide. He might chose a homely mixture of strong ale, sugar and honey boiled to a syrup and applied as a salve, as recommended by Shattenden the woad-man, or a long-established orthodox remedy: 'by taking 8 or 10 oz of blood from the patient foorthwth'. Failing these, there was the London remedy mentioned above, Giles' balsam, or a powerful vegetable treatment such as 'a dose of Laudanum well prepared'. Plat might opt for something expensive, with foreign ingredients such as 'gomme copall' from the New World, or a blend of chemicals (litharge of gold and silver, oils of roses, lilies, and eggs) to produce an ointment. If all else failed, the magical remedy: ' yf the gowte tak you in your Toe, tye 3 great earth worms being alive by the heads together & so likewise by the tayles, & then winde them round abowte your Toe, lapp a cloth abowte them & lett them lye 12 howres', could be tried. This variety, in terms of both cost and ingredients might have been a reflection of Plat's range of patients: gentlemen were most likely offered an expensive salve with exotic or precious components matched to their status, while a humble Londoner might be sold a 'cheap and cheerful' cure containing things he or she could recognize.[31]

Just how wide-ranging Plat's medicines might be is no better illustrated than by his 'Charactericall Cure', one of the remedies he favoured for ague, which appears to have been a charm – magical characters probably written

29 BL, SL 2209, ff. 16, 21; SL 2172, ff. 20–21, 28–29v; Plat, *Floraes Paradise*, A8.
30 BL, SL 2189, f. 77v refers to numbers 1044, 1051, 1063.
31 BL, SL 2209, ff. 3–9v.; Gail Kern Paster, 'Purgation as the Allure of Mastery', pp. 196–90.

on a piece of paper. The first recorded cure with this remedy was in 1593 and he was still using it eight years later. Most of the cures effected mention 'taking' the medicine once or twice but seven entries in his practice-notes are marked by him '**' and in these cases the patients took no medicine. For instance, 'In february I cured Mr Auditor Hills man Edmond Sawer of a tertian wthowt taking any medecine', and 'the 3rd of march 1599. I cured Roger Alison a Cutlers boy in holborn wthout giving him any medecine'. In the case of Mr Brooke, the York herald, the medicine was 'lying by him which hee took not', and Plat's maid Joan took medicine for her first fit, 'thother 2 fitts I cured in her absence'. The clearest statement that magic was involved is in Plat's note that he cured John Wootton of his first fit of tertian ague with white mummia, 'but at his next 2 fitt dayes I made my characters, but sent them not.' This cure found its way into the heart of the royal Court for he prepared one at the request of a friend for 'Mr. Darcies weif of the privie chamber'.[32]

Whilst it is clear from these asterisk-marked entries that magic was involved, the other cures listed using this medicine are probably similar. In no case does he specify an actual medicine being ingested, as he does when he writes of cures involving powders or pills, the entries simply read 'I cured' the person concerned. And there is the curious advice about this cure that, while it 'seldome or never faileth at the second or third taking', yet he advised 'hee that giveth this medecine may not compound for or expect any reward least hee faile, but after his cure hee may lawfully take a free gratuity'. Was a cure – effected by magical symbols and sometimes in the patient's absence – suspect in the eyes of his patients because no medicinal substance was involved? Because they had no tangible proof that the practitioner had done anything towards a cure? Was it best to wait for a successful outcome before expecting any reward?[33]

Although Plat had a body of medicines to draw upon, he continued to search for more, concentrating in his laboratory on chemical compounds and distillations ('the long expected fruites of my chymicall labours') to produce pure, strong remedies. Paracelsus had proposed adding a third substance to the basic building blocks of nature put forward by alchemists.

32 BL, SL 2209, ff. 16–20.
33 BL, SL 2209, f. 15.

MEDICINE AND MEDICAL PRACTICE

To sulphur and mercury he added salt. These three substances were not elements, rather they were qualities, 'spiritual entities that, as it were, gave form to the classical elements'. In alchemical terms the salt of a substance was the residue of solid matter left after distillation and evaporation of its volatile part. Plat produced medicines based on such salts; here are his directions for extraction from vegetable matter:

How to make the Salt of hearbs.

Burne whole bundles of dried Rosemary, sage, Isop, &c. in a cleane oven, and when you have gathered good store of the ashes of the hearbe, infuse warme water upon them, and make a strong and sharpe Lee of those ashes: then evaporate that lee, and the residence or setling, which you find in the bottom therof, is the salt you seeke for. This salt, according to the nature of the herbe hath his operation or use in physicke, and in my conceit, doth worke greater effects in the stomach, than any of the aforesaid extractions... Some use to filter this Lee divers times, that their salt may be the cleerer, and more transparent.[34]

Plat hoped to extract quintessences from salts. A quintessence was, in alchemical terms, 'the very essence of the body of the metal or Stone, the incorruptible, pure and original substance of the world magically able to preserve all *sub lunar* things from destruction and corruption.' A quintessence was important in medical terms because its purity was thought to have strong healing powers, turning the impure and corruptible into pure and incorruptible.[35] One method of extraction, Plat noted, took three months, an alternative method was even longer: 'put the herb or flowers whose quintessence you seeke into a glas nip the glas, let it stand so one whole yeere, open the glas, & you shall finde an exceeding stronge smell of the herb, infuse spirit of wine upon it [*inserted*: thrice infused] & it will draw all the quintessenceof the herbe unto it in a very short tyme.'[36]

---

34 *Floraes Paradise*, A8; Philip Ball, *The Devil's Doctor*, 2006, pp. 261–2; Plat, *Jewell House*, pt. 3, p. 23.
35 BL, SL 2212, f. 19; Lyndy Abraham, *A Dictionary of Alchemical Imagery*, 1998, p. 75.
36 BL, SL 2212, f. 22.

Another way of extracting the powerful goodness or spirit of a substance for use in medicine was by distillation, and Plat's notes are full of distilled remedies, one part of the *Jewell House* of 1594 being devoted to these operations. His optimism that such spirits would be effective medicines is summed up in his speculation, 'Qre of fermenting the iuyce or decoction of all meates and plants as also of drawing the spirits from such fermented liquors, whether this bee a not a singular restoratif for old men, or weake patients. as also to multiplie the vertew of each herb.'[37]

Plat also had great faith in the powers of mummia – powdered mummified remains which apothecaries obtained from Africa. This should be taken ideally from the body of one who had died suddenly, the power of the medicine coming from the life-force trapped therein. He listed the 'Phisicall uses to be made of the parts of a dead man'.

> 1 The common mummia is good for an inward bruse
> 2 The powder of a mans skull that is sound & killed in the wars is most excellent of all other, taken in half a dram in a litle drincke for a tertian or quodidian.
> 3 The same powder in the same quantity made up into a suger past is excellent against paine in ye head.
> 4 The whole mummia of a man philosophically prepared worketh renovation in man.
> 5 A child having a wen uppon the eye lidd was cured by often stroking of the dead hand of a child uppon it. So I thincke of a stie, or any other superfluous excrescence of the flesh. qre if a man must not bee cured by a man, & child by a child vz a boy by a boy. &c.
> 6 of Scabbes & ich cured by the water which hath washed a dead man.
> 7 of the curing of the toothach by a deadmans tooth.
> 8 C The Quintessence drawne from the blood of a young lustie & healthfull man is excellent for renovation.
> 9. C A purgia hominis cured the sinews of Mr Harringtons hande being shronke so as hee cold not strech foorth his hand.[38]

37 BL, SL 2189, f. 10v.
38 BL, SL 2212, f. 24v–25; BL, SL 2209, f. 18v–19; Pelling & White, *Physicians and Irregular Medical Practitioners in London, 1550–1640*, on-line database.

The results of his experiments in the laboratory and his experience in practice are reflected in the selection of medicines he confidentially advertised for general sale towards end of his life. These 'Hermeticall medicines and Antidotes' were nature's 'most secret Iewels, which she hath long time so carefully kept, onely for the use of her dearest children'. In his undated broadside of about 1607 advertising food and medicines for seamen he offered: 'A safe, general & goode purging Powder, to be taken in white wine, working easily without any convulsion, or other offence to the stomacke. It is pleasant, and hath not any common or known purgative therin.' A similarity between the phraseology in this advertisement and his practice-notes on 'Obstructions' make it likely that the advertised powder was composed of substantially the same ingredients as the cure in the notes, including expensive guaicum, a tree gum from central America. Unusually, the practice-notes contain no less than 37 paragraphs of advice on the administration and effectiveness of this medicine, listing the many afflictions it could cure or ease. They conclude:

> this is an excellent medecine to prevent the gowte, psiatica, Jaunders, greene sicknes and all diseases arisinge of Rhewmes [*added*: or obstructions and so I have diverse times both used it my self & perswaded my deerest freindes to take it] superfluous humors or to cure them speedily beeing taken before they bee deeply rooted, and howsoever they bee rooted, it will greatly qualifie & ease them making them very tollerable to the patient.[39]

That Plat should strive for a good purgative was no surprise. Purgation was a medical procedure popular with practitioner and patients alike in early modern England. Humoral imbalance, and resulting disease, was thought by many to be threatened by local obstruction of the body, whether by constipation in the gut or a cold in the head. Purging restored 'solubility', an unimpeded flow throughout the body. And, as prevention was thought better than cure, a regular dose of purgatives was necessary to ensure continued solubility. Indeed, purging was thought as improving one's whole being , not just the body, and to undertake a regimen of such

39   Plat, *Certaine philosophical preparations of foode and beverage for sea-men*, c. 1607; BL, SL 2209, ff. 34, –36, 38.

physic, often at the house and under the close supervision of a physician, was a form of conspicuous consumption in early modern London.[40]

Another medicine which Plat developed late in life (it does not occur in his practice-notes prior to 1606 and was advertised in print in 1607 and 1608) was the 'spagiricall (i.e. alchemical) Antidote', advertised in the preface to *Floraes Paradise* in 1608. The detailed announcement is abruptly inserted into the preface and may well be the text of a printed flyer, now lost:

> It serveth specially to master, and extirpate the venome, of that most fearfull, and infectious disease of all the rest, which wee tearm the plague. It is also most excellent, in all violent and burning Feavers, and in all sorts and kindes of poison whatsoever; wherein, no Terra Lemnia, Sigillata, Bezoar stone, or Unicornes horne, that ever I could see, though taken in a double proportion, can match it, or shewe it selfe equivalent. And in Quotidians, Tertians, and double Tertians (I cannot speake much of Quartanes, because I have not had any great experience therin) I dare commend it, and will wage for it to any reasonable summe, against any animall, or vegetable medicine whatsoever.
>
> In the trembling or passion of the hart, it is singular: and till it please God to blesse me with the true Oleum Solis, that hath passed all his philosophicall Rotation in Caelo Philosophorum, I will hazard and set my rest upon it.[41]

This was not the plague medicine he distributed in 1593 because it provoked neither siege nor vomit. It was a strong, chemical cure, proof against all poisons, of which the plague was widely believed to be the strongest and, befitting a remedy aimed at 'the Nobility and Gentry of this Land most principally', it was expensive: 'the matter is exceeding scarce, and hard to come by, & the preparation long and tedious'.[42]

Plat's range of treatments expanded as he became more experienced. He responded also to the needs of his patients, offering cures for ailments which most troubled them as well as medicine to combat the major

---

40   Kern Paster, 'Purgation as the Allure of Mastery', pp. 193–205.
41   Plat, *Floraes Paradise*, A8–9; BL, SL 2203, f. 113; Plat, *Certaine philosophical preparations of foode and beverage for sea-men*, c. 1607.

epidemic disease of the day, plague. Although sore eyes might not seem an obviously pressing complaint, Plat had medicines for them because eyesight was of great concern to early modern Londoners who feared blindness would mean economic disaster.[43]

The index to his practice-notes (which covers some MSS now lost), indicates what he considered to be the most important diseases in his practice. Tidied up and arranged by page-number it reads as follows:

- 1 Ague etc
- 35 Plague
- 65 Sore eyes
- 99 Gonorrhoea, whites, running of the reynes
- 131 Falling sickness
- 163 Stone
- 179 Obstructions
- 180 Deafness, headache
- 181 Dropsie, jaunders, liver stopped, megraim, pockes, palsie, purge general, rhewm, surfett, spleen, stomack, sleep, wind.
- 183 Gowte, green sicknes, psiatica, greene wounds, ulcer in the stomack
- 184 Nodes helped
- 186 Ringworme, solublenes, serpigo, tetter, yche
- 195 Queenes evil [44]

The list of medicines he hoped to sell to a prospective practitioner in 1600 covers a similar range of illnesses:

> 1 a philosophicall extract for the gowte which comonly easeth the payne at one dressing, and giveth such ease as the patent shall take good rest the first night. ....
> 2 The like Remedie I have for the psiatica being the hipp gowte and this serveth most excellently for all cold gowtes. this is Coossen germain to the former in all respects.

42 Plat, *Floraes Paradise*, A11.
43 Pelling, *Common Lot*, p. 29.
44 BL, SL 2209, ff. 39.

3 An excellent extract for the running of the Reynes in Men, or Whites in woemen…..

4 A remedy for the piles.

5 Mophew

6 pilles to purge the head

7 Rhewme, greenesicknes and dropsie cured by an excellent & pleasing diet.

8 a most rare water to cure any skin, filme, knot, blister or inflamation in the ey

9 stinking breath.

10 sore mouth.

11. bruse swelling or sprayne

12 bloody flux or laske stayed

13 ringworm, sepigo, lett. &c.

14 bleeding of wounds or at the nose

15 heare lost by disease sudenly recovered.

16 an oyntnt for a high complexion or collor in thee face, used by the Earle of Essex.[45]

The list from 1600 enumerates more diseases than the practice-notes but many are 'social' ailments – bad breath, sudden hair loss, discoloured skin and bruising – major embarrassments for the courtiers in Plat's practice but of little consequence to more humble folk. Sensibly, Plat confined his practice to diseases he thought he could cure, concluding the 1600 list, 'I can add if neede bee but I hold it best for a gentleman to proffesse but a few and those to bee certen Cures'. To these lists we must add his cures for scurvy which he advertised to another section of society he hoped to profit from, seafarers.[46]

Very little in his writings tells us exactly how Plat undertook his medical work. Most of his practice-notes survive (everything up to folio 180 in the index above) but they are not informative on how he operated. He is more interested in recording the medicines used than how he effected

[45] BL, SL 2172, ff. 28–29v.
[46] BL, SL 2172, ff. 28–29v; *Certaine philosophical preparations of foode and beverage for sea-men*, c. 1607.

a cure, listing successes but with no indication of how closely, if at all, he supervised the patient. The note, 'Mrs Hill a shoemakers weif in Bredstreet having miscarried of diverse children at 13, 14, or 15 weikes, by receyving thaforesaid medecine went owt her full time, haveing the retentive virtew strengthend therby and brought foorth a Son', tells us nothing of Plat's attendance, if any, at her bedside. He must have supervised his own wife when administering Ulrich von Hutten's diet and course of bathing (the diet took 14 days), recording 'wth this diet I cured my wieff being extremely handled wth the greene sickness & yeeres together and having used the Counsell & advise of the best Phisitions that those tymes didd affoord & were familiar with her freinds, wthout any shew or feeling of amendement, insomuch as not long after this cure she did also conceyve wth child & from thensforth grew a fruitfull woeman.' Did he have patients staying with him whilst he administered diets and purges? Did he make regular house-calls to patients?[47]

In the case of his plague practice in 1593 it is clear he simply dispensed pills to the relatives of victims, or to heads of households and institutions affected, or to apothecaries for dispensing themselves but, unless he was exceptionally courageous, that is what we would expect. (We have noted above that he also cured some clients with magic without seeing them.) However, when he was treating other ailments, the notes sometimes indicate close observation of the patient. There was the cure effected on his coachman: 'Ruben my Coach man having had 2 or 3 fitts of a terti and having taken a large infusion of Tobacco at the third fit hee had cast extremely and greveously sick shaking very violently in the coldnes of his fit, I gave him of my Antidote, his shaking stayed presently, nether did he bring at all after contray to his former fitt, and at this fit his ague ended.' And in curing Robert Thorp, vintner at the Mitre of an ague, the precise observation that 'I began wth him the 17 of October 1593' certainly looks like a course of treatment involving attendance. The last entry in his practice-book, in spring 1608 when he treated a boy living near his London properties at Cornhill, is the clearest indication of personal observation (significantly, the patient was not of high status): 'Thommas

---

47   BL, SL 2209, ff. 8v, 9.

Lodge that wrought at Mr Aylesworths in Cornhill a boy of 12 yeres old being extremely cold and deadly sick uppon a flyght from countering wth one that had the plague, took 4 greines of my antidote and in half an hower, fell with an extreme burning, and presently there issued owt of his face many pores and blisters. the next day he was well, and the day after fell to a kind of shaking agew and took 4 greins more, & never after had any more fitts or payne of that disease'.[48]

When negotiating the sale of medical secrets in 1600 he undertakes to supply the secrets and 'each practize' associated with them. In reassuring the prospective purchaser that he had the 'Capacitie' to practice he advises that a practitioner might always appear knowledgeable when faced with a patient 'for a few written lines shall always reduce to memory the particular use of every particular medecine'. Thus we may conclude, on somewhat flimsy evidence, that Plat practised by dispensing and administering medicines and, to some extent, he attended his patients and observed the effects of his medicines upon them.[49]

Excluding for the moment his plague practice of 1593, household members figure prominently in the list of patients Plat cured. At various times he treated himself for ailments and he also helped his wife, his sons Richard and Robert, the nurse to another son William, his 'little girl', two of his maids, two menservants, and Ruben his coachman. Self-medication was often the first option in London households – his family and servants were probably grateful for his expert advice although they may have been subjected to experimental cures.[50]

Other than his household, he claimed to have cured other relatives, such as his 'shee cossen Walker' and his sister-in-law 'Gore'; friends, including close friends Thomas Gascoigne and Auditor Hill; and neighbours like 'Goodwiff Harsley my neighbor at Bishopps Haall', and young Thomas Lodge near his London office in Cornhill. Courtiers, Court officials, nobility and knights formed another distinct group but Plat, while he was apparently not considered skilful or eminent enough to treat major Court figures, was trusted with their relatives and employees. He

48 BL, SL 2203, ff. 113–114; BL, SL 2209, ff. 16, 22–25v.
49 BL, SL 2172, ff. 28–29v.
50 BL, SL 2209, ff. 16–31; SL 2203, ff. 113–114.

treated the secretary of Lord Sheffield, a lodger of Lord Arundel, the sons of Sir Vincent Skinner and Sir Michael Pallisser, one Fermer at the request of Sir Richard Hicks, Lady Cripps, and the wife of Edward Darcy, Groom of the Privy Chamber (for whom he prepared a medicine which was probably administered by her personal physician). Of Court and government officials, he claimed to have cured Brooke the York herald, the Lord Treasurer's barber, the Queen's ale brewer, the daughter of a royal portrait painter, and a yeoman warder.[51]

The remaining non-plague patients for whom he claimed success were equally divided (17 each) between those designated Mr or Mrs and those not. The 'Mr's were usually city merchants and tradesmen: a goldsmith, scrivener, mercer, linen draper, shoemaker, and barber surgeon. The others were probably the poor with whom he came into daily contact near his home or collecting his rents in the city: a painter, a cutler's boy, a midwife's son, a goldsmith's maid, Joane Gwin 'a poore widow'. We must bear in mind that Plat most certainly had many more patients than he recorded in his practice-notes either because, as in the case of eye complaints, 'I have helped diverse having filmes and skinnes uppon their eyes… or haveing white speckes in their eyes whose names I have not observed' or because he only recorded cures – we know nothing of those whose symptoms persisted or who died despite his ministrations.[52]

I have singled out for separate consideration Plat's response to the plague epidemic of 1593, not only because his notes of practice are detailed and interesting but also because plague was the most terrifying cause of mortality to the population as a whole. It was a disease of special concern to Londoners, a mass of people in close daily contact with one another: in the words of Paul Slack, plague 'had its own special, pernicious horrors because it was a calamity which passed invisibly but rapidly from person to person … everyone knew that plague was infectious.' Plague was present in London in most years of Plat's adult life, causing up to 10 per cent of recorded mortalities, but occasionally a virulent strain produced an epidemic. In 1593, in the city and liberties (an area which excludes some populous suburbs), at least 10,675 people died of the plague, and in

51 BL, SL 2209, ff. 16–31; SL 2203, ff. 113–114.
52 BL, SL 2209, ff. 16–31, 27; SL 2203, ff. 13–14.

1603 the total was at least 26,350. Life in the city was seriously disrupted during an epidemic, large numbers of people died in a short space of time, placing enormous strain on the civic authorities trying to collect and bury the dead, keep order, and impose quarantine and other public health measures on a terrified population. Markets, and economic life generally were disrupted as Londoners sought to protect themselves by avoiding contact with others. It seems that in Plat's time the rich were steadily less at risk of death from plague than the poor – either because they had better housing and sanitation and/or because they were more likely to flee the city when plague became epidemic.[53]

Plague is caused by a bacillus which is primarily an internal parasite of rodents, particularly rats. Fleas feed on blood of infected rodents and those that leave dead rats to infest people infect them by regurgitating bacilli. In sixteenth century London the lack of adequate sanitation and buildings largely made of wood, wattle and straw meant that rats lived very close to man. Plague is a very effective killer of people: 60 to 80 per cent of those infected die. It has a short incubation period, usually less than six days, and most people die within a month of infection. The symptoms are a high fever, headaches, vomiting, pain, delirium and, finally, coma and death. A blue-black carbuncle forms where the flea has bitten, the lymph nodes swell and form buboes. The disease thrives in a humid and warm climate making it mainly a summer disease in the England of Plat's time.[54]

Such a major disease brought forth books and pamphlets offering medical advice, cures, and religious consolation. Many medical practitioners thought plague was just the most extreme form of generalized disease, a dreadful fever, sometimes simply referred to as 'the sickness'. Most writers treated it as a virulent poison to be drawn or driven out of the body. Many cures claimed to expel poison: an ancient remedy, a 'treacle' with many ingredients including viper's flesh, a strong defence against poison, was popular but expensive. The most highly regarded, 'Venice treacle', came from Italy. There were a great many remedies put forward (expensive cures

53 Paul Slack, *The Impact of Plague in Tudor and Stuart London*, Oxford, 1985, pp. 17, 150–1, 164–9, 189.
54 Slack, *The Impact of Plague in Tudor and Stuart London*, pp. 7–8.

for the rich, homely remedies for the poor), including chemical-based Paracelsian cures.[55]

In 1603, Thomas Lodge, playwright, physician and acquaintance of Plat, described the feverish activity of plague-pamphleteers advertising by fly-posting, and complained of persons,

> that have bestowed a new Printed livery on every olde post, and promised such miracles, as if they held the raine of destiny in their owne hands.... Amongst these, one by fortune is become my neighbour, who because at the first he underwrit not his billes, every one that red them came flocking to me, coniuring me by great profers and perswasions to store them with my promised preservatives, and relieue their sicke with my Cordiall waters.

He was 'amazed, to see the ignorance and error of the multitude, who dare trust their lives to their hands who build their experience on hazard of mens lives'.[56] (There are no references to the large 1603 plague outbreak in Plat's writing. It is possible he and his family adjourned to family property in Hertfordshire and he took no part in trying to combat the epidemic.)

Plat was fully engaged in combating the plague of 1593. His plague practice-notes for that year are not contemporaneous, but written up carefully sometime after the event. Nor are they complete. Either he failed to remember all the cases he treated or the episode was so full of moment that he dispensed 'unto persons whose names are not heere reported'. The notes begin with a formal introduction: 'A notable defensative medecine against the plague beinge first practized in Millain upon a generall Infection there abowte the yere 1579. by Mr D. Sirnige, and after brought over & practized in England Ao. 1593. in the greate yeere of Visitation. the resceit where of is heere sett downe together wth the good success, thereof in diverse particular howses and persons as followeth.' This is followed by the recipe for the medicine, then, 'The practize of thafforsaid defensative in the plague yeere ao. 1593.' The list of those to whom he gave his medicine is arranged with the major recipients (both in terms of numbers of pills

55 Slack, *The Impact of Plague in Tudor and Stuart London*, pp. 25–32.
56 Thomas Lodge, *A treatise of the plague*, 1603, A3–4; BL, SL 2189, f. 135v.

and social status) listed first. This document is Plat's proud record of achievement, designed to impress either a prospective buyer of his practice or, more likely, someone at Court with the power to dispense patronage or honours.

The medicine, 'A notable defensative against the plague beinge taken before infection, and so of burning feavers, beinge also a safe and a generall purge' was made up into cakes of four drams each (or two drams with less sugar) which would last seven or eight months kept between papers in a box in a warm place. Unlike many plague remedies it was not composed of a great many ingredients (only eleven). The more recognizable were included to make the pills palatable: rose honey, sugar, quince conserve, and ginger. The rest were purgatives, a gum to bind the mixture, and Bezoar stone – a concretion found in certain ruminants' stomachs – long thought to be a good antidote to poison. The pill was presumably supposed to work by countering the poison of the plague and expelling it from the body. Bezoar stone and the purgatives were expensive ingredients, brought from overseas. For instance, the recipe called for half an ounce of 'scamonnie', the gum of an Asian convolvulus which grew originally in Syria. True, John Gerard the herbalist may have been able to supply Plat with some of this plant, as at one time it flourished in his Holborn garden until 'an ignorant weeder plucked mine up'. The plague-pill recipe was given to Plat by his acquaintance Señor Romero, the Spanish gentlemen from the kingdom of Naples, who himself 'gave greate nobles of them ao. 93. being a yeere of great infection'.[57]

Plat notes at the start of his list of recipients that 'There have ben dispersed abroade wthin London and middlesex great nombers of these defensative Cakes': 'many hundreths of them' were distributed throughout Middlesex by his father-in-law, the leading justice of the peace Richard Young. Plat provided 45 pills to the Lord Admiral and 60 to other members of the Privy Council. At this time Plat was trying to make a name for himself at Court and so he had self-publicity, as well as help to his fellow-men, as a motive. No other members of the nobility are listed as recipients and only one Court official (a sword bearer) was given pills,

---

57   John Gerard, *Herball*, 1597, pp. 414–5, 866–7, 1169; BL, SL 2209, ff. 21–6.

probably because the Court had moved away from London to avoid the epidemic. The Council took Plat's pills seriously, for Richard Young 'made one especiall triall in the parish of St Marie Abchurch where hee himself dwelled, uppon 9 visited howses in which there abode 33 persons which all were preserved from the plague, to the great contentment of your LLs of the Counsell whoe sent for the trew note therof unto him to bee fully informed of the report.' The local preacher (one of three given pills) did the dangerous work of distributing them to the sick: 'Mr Jarfeild preacher at st Mary Abchurch going amongst thinfected, and having taken the medecine himself, his weif and his mayd by the grace of God were preserved.' He may have been assisted by the constable of the parish who was also given a pill.[58]

At least two apothecaries sold Plat's pills: 'Mr Pichforck the apothecary sold diverse of it in his shopp', and, as might be expected, pills were provided for many friends and relatives, including four to Sr Romero who provided the recipe. Despite the mixture of strong purgatives, he boldly 'took the same 4 times in 15 dayes'. Of Plat's immediate family, only he and his servants Harry Jenks and Rowland took one, maybe because his wife and the rest of the household had departed for the safety of the Hertfordshire countryside. Most names on the list are of London merchants and tradesmen but a significant number (noticeably more than in his general practice) are below this class – servants, cloth workers, victuallers, and jewellery workers. Whilst householders were given pills for their whole household, ten individual servants, both men and women, were provided with them. Were these people who had been left in charge of houses when their masters and family had fled? Two Dutch workers on the list, a diamond cutter and a goldsmith, are representative of the large influx of Low Countries Protestants working in London. Gem cutters and gold workers are scattered through the list, a reminder that many Londoners relied for their livelihood on luxury trades.

A concentration of journeymen (and women) were provided with pills in Gutter Lane: a purse-maker, a gold lace-maker, a gold wire-drawer, victuallers, a ruby-cutter, and a tailor. If we look at the location of people

58 Mary Hill Cole, *The Portable Queen*, Amherst, MA, 1999, p. 21; BL, SL 2209, ff. 21–6.

on the list as a whole, many are scattered along, or just off a series of connecting streets from outside Aldgate in the north-east, into the centre of the city via Aldgate, Cornhill, Poultry, and Cheapside. This would have been Plat's route through the city from his suburban house at Bethnal Green, passing on the way his properties at Cornhill. Might we suppose that Plat, with his two immunized servants, travelled this route, darting up the side streets, Milk Street, Wood Street, Gutter Lane and the like, giving out pills to needy supplicants? Might Plat not have encountered in this way 'Goodwif Russell an herb weif in cheapside having buried her husband of the plague'? Plat records 'shee and her mayd were both preserved by this medecine'.[59] The pills were also used outside London and Middlesex. 'D. Flecher Bishopp of Worcester bought to the nomber of 50 of these cakes and dispersed them in the visited howses of the said Citie, and it pleased God to preserve them all from thinfection'. Plat also 'Sold to Mr Belfeild for thinfected howses of St Albans 10 cakes'. In all, the detailed list of pills distributed totals 461, as well as the 'hundreds' more not specified by recipient. As with his general practice-notes, only successful cures are recorded and we may be especially sceptical of the claim to have saved 33 people in St Mary Abchurch in already-infected households. Nevertheless, as usually at least 20 per cent of infected people survived the plague, and the majority of Londoners escaped infection, one would expect many of those taking Plat's medicine to survive despite his pills.[60]

Plat's acquaintance, the physician and playwright Thomas Lodge, complained during the 1603 epidemic 'of the loathsome imposition which was laide uppon me, to make my selfe vendible, which is unworthy a liberall & gentle minde, much more ill beseeming a Phisitian and Philosopher, who ought not to prostitute so sacred a profession so abiectly, but be a contemner of base and servile desire of money'.[61] In Plat's plague practice most of the pills were given away, as a charitable act to the poor or as a demonstration of public spiritedness when given in bulk to the Lords of the Council. There is no doubt, however that Plat generally attempted to gain, by money or goods, from his medicine although, as he was not

59 BL, SL 2209, ff. 21–6.
60 BL, SL 2209, ff. 21–6.
61 Lodge, *A treatise of the plague*, 1603, A1.

licensed as a physician, he could not legally charge a fee. In his more general practice he makes no reference to reward but he implies that he did usually negotiate a fee by specifying that, in the peculiar case of his ague cure by 'characters', 'hee that giveth this medecine may not compound for or expect any reward least hee faile, but after his cure hee may lawfully take a free gratuity'. He lists some of these: after he cured Mr Brooke, York herald, his 'weif sent mee a Ruff band'; lower on the social scale, a charcoal-maker of Bromley sent him '2 pulletts for a gratuity' and a poor joiner's wife gave him a dozen sugar cakes. Lack of a gratuity to which he thought himself justified made him angry: 'the 11 of June I cured Katherin Wright Mr Barnes the mercars weif in cheapeside of a tertian. shee presently fell into it again, and I refused to recure her, becawse shee did not pay so much as for my botehire to battersay'. Such gifts were small in comparison with the income he boasted he could receive in 1600. One medicine could 'gaine you great favour and Crownes enough (I speake of experience)', while another medicine 'hath alredy yealded cheines of gold to the practizer'.[62]

Plat's principal concern, however, was to earn money from the sale of his own medicines rather than acting as a physician ministering to patients. This was a wise strategy when disease was rife and most medicines so obviously inadequate, despite the optimism of physicians, to perform a cure. In such circumstances, if one waited for a cure to succeed before claiming the balance of a fee or accepting a gratuity, disappointment was likely. Far better to obtain the money from straightforward sale of medicine.

Plat tried many strategies to promote sales. In about 1607 he issued a broadside: 'Certaine Philosophical Preparations of Foode and Beverage for Sea-men, in their long voyages: with some necessary, approved, and Hermeticall medicines and Antidotes, fit to be had in readinesse at sea, for prevention or cure of divers diseases' which is probably linked to a manuscript list (in his best hand) of items for sale headed 'Waters, Oyles, Electuaries, & other Compositions necessary for the Sea'. The broadside shares phrases with an advertisement for medicine inserted in the preface to *Floraes Paradise* of 1608; they may both have origins in another advertisement, now lost. The 1608 advertisement largely concerns his 'spagiricall Antidote'. It demonstrates Plat's skill as a copywriter. He

62   BL, SL 2172, ff. 28–29a; SL 2209, ff. 15–22.

flatters his prospective buyers, assuming they are 'Nobility and Gentry' and scares them by 'advising every man of sort or abilitie, not to bee at any time without a dose thereof in his bosom; because no man knoweth, what suddaine and unexpected cause he may have to use it'.[63] The excellence of the medicine in curing many diseases is emphasized: it is the best medicine he has come across 'and till it please God to blesse me with the true Oleum Solis, that hath passed all his philosophicall Rotation in Caelo Philosophorum, I will hazard and set my rest upon it'. After a reassuring passage on how to take the medicine and its lack of troublesome side-effects, he assures his readers that, after extensive trials, he has 'cured many therewith, … neither have I seene it faile, but in one particular person to this day', a nice touch, to claim complete success might appear incredible.

Then comes the final deft twist: buy now while stocks last! 'Concerning the matter of this medicine, I wold there were such store as I could wish, and I would the provision & preparation thereof, could be effected in any reasonable time: or rather I could wish, that seeing the matter is exceeding scarce, and hard to come by, & the preparation long and tedious, that it wold please God to free this Land of such heavie visitations, as our late sinnes have iustly brought upon us: and that he would be pleased, rather to give us cause to praise him still for his mercies, then for his medicines.'

All is not lost, however, if the customer is slow to respond. Plat has another, albeit inferior remedy: 'Yet, to give some supply to this defect, if my store shoulde happen to faile me, I have also another extract for intermitting Agues, whereof I have had more experience in this kind; and which seldome faileth at the second or third taking, & often helpeth at the first; which also worketh no evacuation, except a little by way of transpiration in sweat. And it is pleasing enough to take: but a greater dose thereof is requisite then of the other.' The advertising copywriter style is also evident in the 1607 broadside. In a concluding paragraph he entices his readers with further benefits 'if it shall receive entertainment according to the worth therof and my iust expectation, I may happily be encouraged

---

63   Plat, *Floraes Paradise*, A8–A11; BL, SL 2172, f. 30.

to prie a little further into Natures Cabinet, and to dispense some of her most secret Iewels, which she hath long time so carefully kept, onely for the use of her dearest children'.[64]

Plat enlisted (or planned so to do) the help of others in selling medicines. Two apothecaries sold 50 of his plague pills in their shops and no doubt he charged them a wholesale rate. His fellow irregular practitioner John Clark was selling cinnamon water, a speciality of Plat, in 1603 and in a list of ways of making money Plat outlines an ambitious plan, 'Cinamon water spirits of wyne and usqa bath to sell them both to Apothecaries and to furnish London therewth by spectiall meanes as also to gett the Custome of all the Irish men abowte London & herin I may spend all my stronge wares'. (Irish whiskey is not strictly a medicine but some would contend otherwise.) In the same document are further notes on selling the following 'Remedies in phisicke & surgery':

> 1 Vulveria for women & ruptions inward/2 my pills for the Rheum/3 my balisoms for wounds/4 A water for running eyes/5 a water for a fiever in the dy/6 a remedie for the dropsie/7& a water for a stinking breath/ 8. a remedy for the whites & running of the rheyness/ sanguinacia./ my water for ulcers/remedie for woormes . perservative for health/ a plaister for ye gowte./

It was Plat's hope to 'vente these by barbers, Surgeons and Phisicians' and he listed those he thought might be interested, including Clowes the surgeon and Kemish the apothecary.[65]

Although selling medicines might produce a steady income, Plat was more attracted by the capital sums which he might get by selling the secret recipes for them outright. Medicines feature in several lists of secrets he hoped to make money from – some were long, others apparently brief 'what if' notes to himself.[66] We find him in serious negotiations to sell medical recipes for the first time in June 1600, in an exchange of letters with 'his

64 Plat, *Floraes Paradise*, A 8–A11; *Certaine philosophical preparations of foode and beverage for sea-men*, c. 1607.
65 BL, SL 2209, f. 23; BL, SL 2172, f. 13, Pelling & White, *Physicians and Irregular Medical Practitioners in London, 1550–1640*, on-line database.
66 BL, SL 2189, f. 1; SL 2197, ff. 7–9; SL 2172, f. 13.

most loving ffreind Mr Thommas Elkinton at his howse at Whellsborrowe in Lecestersheere'. Elkinton had offered to buy some of Plat's secrets and he is offered seventeen. No prices are mentioned, but Plat implies that the secrets are very valuable. His 'hard sell' technique is evident towards the end of the letter. He emphasizes the value of his medicines by urging Elkinton's secrecy, telling him to seal all letters carefully, returning Plat's if the negotiations break down and 'I pray you of all loves and as you esteame my love and ffrendshipp, to bee very secrete even from all men saving your self in all these particulars'. As further encouragement, or more likely a negotiating tactic, Plat chides Elkinton for lack of resolution, warning him not to delay acceptance of his offer, 'ffor I have at this present somme great Schollers that are very earnest wth mee abowte these secretes in Phiscke which I meane to suspend till I heere your answere, so as it bee speedy and cary also a speedy expedition with it'. No conclusion of the negotiations is recorded and most likely they were unsuccessful.[67]

In August 1608 Plat attempted to sell a mixture of medical and alchemical secrets to the Landgrave of Hesse via an intermediary, Francis Segar. He drafted a letter with a priced list of secrets. No record of the letter is believed to exist in the Hesse archives and it may never have been delivered to Germany.[68] The documents, with due deference to a foreign prince, set out the wares for sale with great claims for their strength and effectiveness. Closely following the model of the letter to Elkinton eight years before, Plat seeks to excite the appetite of the Landgrave by urging security, secrecy, and speed to forestall a possible sale to other interested parties requesting:

> that you would consecrate those fewe and Idle lines to vulcan without reservinge anye note or memorye of them for I knowe not the man except your lord, for whome I would by writinge have adventured this farre to wade in the deepe streames of philosophie wherein manye witts better than mine doe dayle plunge and sinke themselves.

67 BL, SL 2172, 28–20v.
68 Both the letter, in a fair hand, and the list, in a more every-day script, have fold marks, with the endorsement 'To Mr Francis Segar for the Landsegrave of Hessen', Maybe they were given to Segar who later returned them to Plat. BL, SL 2171, ff 18–21.

MEDICINE AND MEDICAL PRACTICE

I must allsoe yf I proceede with your lord crave the like secresie at his hands as I was forced to yeelde unto before I could drinke of this Ambrosia, and above all things I must crave to have a spedie answere with a full and perfect conclusion for I assure you upon my creditte, I staye my conference with a great englishe Nobleman herein in respect of your lord whome I cannot long delaye.

The novelty in the Hesse offer is the presence of prices in a side note: 'Or els if yr Lord shall like better to deale wth a few particulars at the first & upon the liking of those to proceede farther I will propound these 4 secretes wth their valewations'. Three of the secrets are medical: 'a sure and cheape way to make store of rosewater, voilet water, honysuckle water and gelliflower water &c. ... My Antidote of Antidotes ... The trew way to make Quintessence of all vegetables'. The prices proposed are, in contemporary terms, staggering, £500, £400, and 1000 marks (£333) respectively. This may well have been a tongue-in-cheek opening gambit but there is no reason to doubt that Plat sincerely thought his secrets worth a great deal. The cheap way of making flower waters differs from the other two secrets in being a cost-saving method of production rather than a new cure. Plat claims the method will make 200 gallons in two days and will take only as much fuel as a man can carry to make 10 tons. If true, a commercially minded purchaser might well consider this secret worth several hundreds of pounds. To Plat, at this time possibly ill or dying of a chronic complaint, sale of secrets to add to the value of his estate was a prudent move and tangible proof that his extensive medical research was a success.[69]

---

69   BL, SL 2171, ff 18–21.

CHAPTER EIGHT

Scientific Thought and Technique

Sir Hugh Plat has been described as an early example of an English 'virtuoso' – a gentlemen with a general interest in many subjects including science and technology, a type summed up by one author as 'fundamentally a man for whom learning is the means to dispose of wealth and leisure in the happiest fashion – and with the comforting assurance that he may also be serving the desiderants of philosophy, history, or art.' The word was first coined in the 1630s, sometime after Plat's death, and by the end of the seventeenth century was as much a term of ridicule as commendation. In its broadest sense, the description included an interest in many aspects of knowledge, including the arts and history, especially collecting rare and ancient coins and other artefacts, but narrowed down it was applied to gentlemen, most notably those of the early Royal Society, with a wide interest in natural history and scientific research.[1]

In 1605, Francis Bacon criticized gentlemen who studied in what he considered a frivolous manner: 'For men have entered into a desire of learning and knowledge, sometimes upon a natural curiosity and inquisitive appetite; sometimes to entertain their minds with variety and delight; sometimes for ornament and reputation; and sometimes to enable them to victory of wit and contradiction; and most times for lucre and profession; and seldom sincerely to give a true account of their gift of reason, to the benefit and use of men.'[2] Note the penultimate phrase 'and

---

1   Walter E. Houghton, 'The English Virtuoso in the Seventeenth Century', *Journal of the History of Ideas*, vol. 3, 1942, p. 58.
2   Francis, Bacon, *Advancement of learning*, in *Works of Francis Bacon*, ed. Ellis, Spedding & Heath, Boston, 1860–64, vol. VI, p. 134, Quoted in Moody E. Prior, 'Bacon's man of science', in *Roots of scientific thought*, ed. P.P. Wiener and A. Noland, 1957, p. 387.

most times for lucre and profession', a reason for study which does not sit easily with many of the virtuosi but could happily be applied to Plat, whose *Jewell House* was described by one historian as 'too utilitarian to be the work of a pure virtuoso'.[3] For Plat, although undoubtedly a gentleman with leisure to pursue such interests as he chose, was very interested in practical matters, the minutiae of technology, not the 'pure science' beloved of virtuosi. And he wanted to make money from his work, which was more about improving existing technology and new inventions than simply investigation of the natural world. He spent a great deal of time collecting information from fellow-Londoners, including many skilled artisans and tradesmen. The enthusiasm with which he approached gathering information from artisans contrasts with the diffidence of the virtuoso John Evelyn, a couple of generations later, in compiling his 'histories of trades'. Despite firmly advocating the project, Evelyn confined himself to 'the history of "aristocratic" arts such as engraving, oil painting, miniature painting, annealing, enamelling, and marbling paper', finally abandoning it because of his distaste at the need to 'converse with mechanical and capricious persons'.[4] Admittedly Evelyn was a snob, but he also reflected the sensibility of gentry to their status in post-Restoration London, as well as a long history of aristocratic disdain for the ingenuity of artisans. Evelyn would have agreed with Xenophon, 'The arts which we call mechanical, are generally held in bad repute'.

Robert Boyle, Evelyn's contemporary and a dominant figure in the Royal Society, sought to encourage extracting knowledge from artisans, but as a necessity, not a pleasure: 'Knowledge of many retired truths cannot be attained without familiarity with meaner persons, and such other condescensions, as fond opinion, in great men disapproves and makes disgraceful'.[5] Plat's willingness to engage with London artisans may have been a reflection of his status as the son of a tradesman. His brewer-father

[3] Houghton, 'The English Virtuoso in the Seventeenth Century', p. 70; Charles F. Mullet, 'Hugh Plat: Elizabethan Virtuoso', *University of Missouri Studies*, vol. 21, 1946, p. 95–6.
[4] William Eamon, *Science and the Secrets of Nature*, 1994, p. 331.
[5] Quoted in Clifford D. Conner, *A People's History of Science*, New York, 2005, pp. 8, 332.

did not die until 1600 and Plat may have considered it unseemly to be too concerned with his gentle status while his father was alive. Indeed, he was proud of his brewing heritage. Using legal puns to imply he had as much knowledge of brewing as one who had served an apprenticeship, and reminding readers he was a brewer by patrimony, he declared, 'I am in by discent, and have continued five yeares in possession at the least, and therefore am not easily to be removed without a philosophical action commenced against me.'[6]

Having been brought up in the world of commerce, Plat was ever conscious of the value of practical knowledge, both in monetary terms and in its social value as a 'secret'. His alchemy, we have seen, was rooted in the idea of revealed knowledge imparted by adepts and masters, jealously guarded from those not worthy of it.[7] He often manifests the same attitude outside his alchemy, showing an instinctive bias towards secrecy in his publications, even when he is in fact providing information. In the middle of a detailed section in the *Jewell House* on ways to preserve fresh fruit, he imagines that ideas he has so far put forward will excite housewives for more, especially the secret of preserving fresh cherries. 'To give these Ladies some more content, I will unfolde a scroule which I had long since as carefully wrapped up as ever any of the Sybels did their fatall prophesies, wherein I wil make them as cunning as my selfe.' He then adds a rider, 'saving onelie that I wil reserve one strange venue to foile a scholler withal if need be'. He then teasingly reveals the secret *philosophically*: 'The secret in short, let one element be included within another'. Finally he reveals all – lead or glass bells with hooks inside are to be placed in vessels of water (treated to avoid putrefaction), with cherries suspended inside from the hooks, clear of the water.

Later in the same work he describes in detail the many advantages of 'a perspective glasse' in drawing scenes from life but gives no detail of this instrument 'because it is consecrated unto Arte, [and] I dare not profane the same too much by delivering it unto unhallowed hands'. Here are echoes

6 Plat, *Jewell House*, pt. 1, p. 18.
7 In the middle of notes on alchemy he inserted the following: 'He that publisheth any of the Sacred grounds of Nature to ye world, he breaketh up so many Seales of the mightie God. R. Bacon.' BL, SL 2245, f. 33a.

of the adept alchemist's concern that recipients are worthy of secrets. Such concern is made explicit when, in discussing methods of distilling a spirit of roses, he acknowledges that any apothecary can divide the spirit from a liquor he has philosophically prepared but 'they will stagger at the firste digestion, and though they should either reele, or fall, I may not lende them my helping hand, otherwise then I have donne already, unlesse I were assured that they were of the nomber of Hermes sonnes, and not begotten by some base Alchimist'. He was also, in some cases, careful of the trade secrets of artisans when he thought their livelihood might be damaged by disclosure. His attitude to disclosing his own discoveries reflected as much his father's successful life as a London tradesman as his own status as courtier and gentleman. Many of his improvements on existing technology, his inventions, and his new medicines were to remain secret unless either sold to an individual or revealed in return for reward from the state.[8]

Plat was in many ways a man of his time in his outlook on natural science: traditional and backward-looking in his alchemy but at the same time pressing for change and advance in the way the natural world was investigated. (Paradoxically, the resurgence of interest in alchemy, of which he was a part, was new in the sense that it challenged the university-taught Aristotelian view of the natural world.) I do not, however, contend that he was a major thinker in the philosophy of science – his philosophical musings were few and scattered, and some strands of his thinking have to be inferred from what he did. Perhaps it is better to see Plat's approach to natural philosophy within the context of the evolving concept of a 'scholar and a gentleman' discussed by Steven Shapin. Shapin explores the changing attitude towards learning in the late sixteenth and seventeenth century. The period *c.* 1580–*c.* 1640 saw a significant increase in the higher education of gentlemen, measured by attendance at the universities and Inns of Court. But professional (i.e. university-based) scholars were viewed by the gentry as contemplative and isolated figures, immersed in deep study of impractical subjects, insecure and rude in polite company. Predominantly from a lower class, they were able but often poor men

8   Plat, *Jewell House*, pt 1, pp. 1–6, 40; pt. 3, p. 43.

trained up as clerics and academics, lacking the gentle attributes of active engagement with society: decorum, courtesy, and self control. Shapin traces how higher education was reformed to make it fit for a young gentleman and, more importantly for our present discussion, more fitting for the study of natural philosophy.[9]

The reason for the behaviour of scholars, gentlemen reasoned, was the nature of their curriculum and teaching methods. University-based scholars were pedants, relying on books and precedent for all their knowledge; reclusive, they lectured rather than conversed. They proceeded largely by disputations, fostering a combative nature. 'While the identification of the scholar with the pedant was quite general in polite Tudor and Stuart society, particular *forms* of scholarly life and endeavour were marked out for special odium. Literally trivial pursuits – grammar, rhetoric and logic, and, by imputation, the whole form and substance of Scholasticism, were increasingly condemned as nothing but pedantry. In the view of those advocating a reform of learning, what was wrong with Scholasticism was that it proceeded from, and fostered, a form of life that was in no way suitable for a civic gentleman.' The playwright John Webster summed up the vices of pedantry: 'a civil war of words, a verbal context, a combat of cunning craftiness, violence and altercation, wherein all verbal force, by impudence, insolence, opposition, contradiction, derision, diversion, trifling, jeering, humming, hissing, brawing, quarreling, scolding, scandalizing, and the like, are equally allowed, and accounted just'. Francis Bacon, in his *Advancement of Learning* (1605) recognized the unsuitability of the scholastic tradition and proposed that the 'distempers' of such learning be removed to make learning again fit for the gentry and rulers.[10]

Plat joined this growing criticism. In 1593 he identified himself with the philosophy of Raymond Lull, with its emphasis on acquiring and organizing knowledge, and the next year claimed that the arts of traditional learning were being debased, 'everie Arte hath gotten his Counterfeite in these daies…. Logike is turned into Sophistrie, Rhetorique into flatterie,

9   Steven Shapin, '"A scholar and a gentleman": The problematic identity of the scientific practitioner in early modern England', *History of Science*, vol. 29, 1991, pp. 279–327.
10   Ibidem, pp. 292–5.

Astronomie into vaine and presumptuous Astrologie'. The lowest alchemist, he maintained, 'did far excell any that was meerely an academicall, peripateticall or stoiacall scholler only'. When arguing with a Galenist the merits of chemical medicine, he would 'not promis as a Logitian to argew pro & contra wth him, but as a naturall & working Philosopher I will draw him to ye forge & fiers of nature'. At the end of his life he ridiculed theoretical scholastic writing on gardening, 'I leave method at this time to Schoolemen, who have alreadie written many large and methodicall volumes of this subject (whose labours have greatly furnished our Libraries, but little or nothing altered or graced our gardens and Orchards)'.[11]

Plat was, as we have seen in relation to medicine, heavily influenced by the writings of Paracelsus whose outlook has been succinctly summed up as, 'knowledge of nature is gained through direct observation of particular objects and that nature is known through the hands and the senses rather than through texts and the mind.'[12] Those who looked to classical learning were, Plat thought, not only hindering progress by their sterile philosophy, they were also denying the possibility of progress. He was wholeheartedly behind the idea that technology and natural science were capable of sustained progress, stating this most forcefully in his early publications – perhaps to establish a rationale for his stream of printed works, especially those detailing his 'new and profitable inventions'. In 1593 he decried those who refute the possibility of progress, citing clocks, maritime charts and cannon as three technologies not known to the ancient philosophers. He bolstered his case by citing an example of progress in London, 'the excellent water-force at the bridge' (a large water-wheel raising water to feed a piped supply to part of the city) which had succeeded despite the scepticism of many leaders in Court and city, as well as examples elsewhere such as a new slitting mill near Dartford and a 'new found meanes of charcoaling the peate and other turfe in Yorkshire'.[13]

11   Plat, *A Briefe Apologie of Certaine New Inventions*, 1593; Plat, *Jewell House*, p. 86; Plat, *Floraes Paradise*, 1608, A4; BL, SL 2195, f. 2a; SL 2223, f. 24.
12   Pamela H. Smith, *The Body of the Artisan*, Chicago, 2004, p. 239; Conner, *A People's History of Science*, p. 282. (In 1594 Plat allied himself with the 'trew english Paracelsians', *Jewell House*, pt. 2, p. 58.)
13   Plat, *A Briefe Apologie of Certaine New Inventions*.

A year later, in the preface to the *Jewell House* he expanded on his idea of progress:

> Why then should we think so baselie of our selves and our times? Are the pathes of the aunciant Philosophers so worne out or overgrown with weeds, that no tract or touch therof remaineth in our daies wherby to trace or follow them? or be their labyrinths so intricate, that no Ariadnes thread wil wind him out that is once entred? I cannot think, much lesse believe, that Nature hath dealt so niggardly with al the world besides, as first to make her staples and storehouses of skill and learning onely within Aegypt or Greece, and then also to cut off al trade and traffique with them from all other Nations. Nay rather, why should not little England ... seeke to raunge her selfe in the foremost rankes and troupes of all Minervaes crew, and not onelie reach with a victorious arme at the golden fleece, in despight of all the fierie Buls of Thessaly, but also wrest and wring the victorie even out of the victors hands.

It was, he wrote in 1595, 'time, and high time to let the world, and all posterity to understand, that if our english Artists ... were sufficiantly emploied in the fulnes, and height of their spirits, that they would bring foorth so many, so rich, and so inestimable buds, and blossomes of skill'.[14]

This idea of progress in technology was being expressed with increasing confidence in Plat's time. Citing three examples of progress was becoming a commonplace: printing, the compass and the cannon being the conventional trio. Many of these optimistic writers were Continental and Plat would have read and approved of such sentiments in works by writers such as Cardano but he may have come across the Englishman Richard Eden's book on navigation in which are ideas very similar to Plat's, 'this our age maye seeme not only to contend with the Ancients but also in many goodly inventions of Art and wit far to exceede them.'[15]

14 Plat, *Jewell House*, Preface to reader; Plat, *Discoverie of Certaine English wants*, 1595.
15 R. Eden, *A very Necessarie and Profitable Booke Concerning Navigation*, London, 1579; Alex Keller, 'Mathematical Technologies and the Growth of the Idea of Technical Progress in the Sixteenth Century', in *Science, Medicine and Society in the Renaissance*, ed. A. Debus, 1972.

Plat may have embraced the possibility of progress but he did so within the language and ideas of his time: classical mythology was commonly employed to convey his message, especially in his verbose prefaces. His concern to search 'into the workes of Nature' was couched in language which identified Nature as a goddess 'with her Cornucopia in her hande', secrets kept in her 'bosome', 'whom no man as yet ever durst send naked into the worlde without her veyle'. Nature 'hath not made anything in vaine' and it was difficult, if not impossible to go against her laws 'to turne nature out of her bias'. Given the chance, 'nature would play so infinitely, and at her owne pleasure'. He was careful not to 'offer .. disgrace to nature' for she was 'not bounde to cast up the treasures of her full gorge amongst us' but it was possible 'to provoke Nature to play, and to shew some of her pleasing varieties, when shee hath met with a stirring workman.' Plat does not employ the savagely violent and sexual imagery used by many contemporary writers in describing man's handling of nature but the conventional sexual nuance is present in his offer to 'leade my ffriend by the hand into the secretest corner of her howse yea into the bedchamber of Nature her selfe'.[16]

Plat's scattered observations on what may be broadly termed the philosophy of science can be fleshed out by looking at how he collected and arranged information, how and why he observed others involved in technology, how he himself worked on projects in his laboratory, and the progress he claimed to have made. We will examine in particular his collection of information from artisans ('histories of trades') which may determine whether his work was merely 'a weak "anticipation" of Bacon' or 'a very strong stimulus to Bacon's philosophy of science.'[17]

Plat's enthusiasm for examining artisans' working methods was not without precedent. 'In the late fifteenth and early sixteenth centuries, artisans were experts on the processes and transformation of nature, and individuals who wished to know (and take possession of) nature looked

---

16   Plat, *Jewell House*, pt. 1, pp. 3, 36; pt. 3, p. 37: pt. 2, p. 9; Plat, *Floraes Paradise*, A4; Plat, *The new and admirable Arte Of setting of Corne*; BL, SL 2172, f. 18; Conner, *A People's History of Science*, p. 364.

17   Conner, *A People's History of Science*, pp. 300–301. See footnote 202, where Conner relates a conversation with Deborah Harkness to this effect.

to *ars* as the medium through which to accomplish this. This is true not only of physicians and scholars, like Paracelsus, but also of princes and city governors who came to regard artisans as holding a key capable of unlocking the productive powers of nature. Out of the dynamic that formed between artisans and these others, new attitudes towards nature and a new discourse about it emerged.' Occasionally, learned men had gone outside their studies to seek information from artisans in earlier times, Roger Bacon in England and Nicholas of Poland both recognized the value of artisans' knowledge and experience in the thirteenth century. The German philosopher Nicholas of Cusa wrote in about 1450, 'wisdom can be found in the streets and market place, where ordinary weighing and measuring occur'. Closer to Plat's time, Juan Luis Vives wrote in 1531 that scholars should not 'be ashamed to enter into shops and factories, and to ask questions from craftsmen, and to get to know about details of their work'.[18]

Two writers read extensively by Plat may have stimulated his interest in the industries and crafts he came across in his daily visits to London. Paracelsus, the outspoken and revolutionary Swiss-German physician, advocated close observation of the natural world: 'Be not ashamed to study diligently the astronomy and terrestrial philosophy of the peasantry ... purchase coal, build furnaces, watch and operate with the fire without wearying. In this way and no other, you will arrive at a knowledge of things and their properties.' Significantly, Paracelsus wished scholars not only to observe artisans and peasants but to carry out experiments using the same materials and processes they used in their daily work.

Plat also relied heavily, and with much approval, on the geological knowledge of Bernard Palissy in works on salts and marls in *Diverse new sorts of Soyle*, published as part of the *Jewell House* in 1593. Palissy was born a peasant in France in 1510. After training as a glazier and a stint as a surveyor, he taught himself pottery, and won fame for his ceramics. He was a tireless experimenter in pottery and his books, published from 1563 onwards, display a wide knowledge of natural science. He was a firm advocate of practice, not theory, and unlike most of those who collected information from artisans for publication, he was himself an

18   Pamela H. Smith, *The Body of the Artisan*, Chicago, 2004, p. 151; Conner, *A People's History of Science*, p. 284, and Vives, quoted in Conner, p. 283.

artisan, writing from his own experience. It may be significant that Plat, introducing *Diverse new sorts of Soyle* (with its heavy reliance on Palissy), begins: 'having found by sundrie observations, drawn from experience herselfe the undoubted mother of all true and certain knowledge.'[19]

Plat probably also read William Gilbert's work *De magnete* of 1600. Gilbert described detailed experiments to investigate the mysterious properties of magnetism and he has been hailed as a pioneer of scientific experimentation: 'the first printed book, written by an academically trained scholar and dealing with a topic of natural science, which is based almost entirely on actual observation and experiment'. Close reading of *De magnete* leads to the conclusion that Gilbert's scientific method was derived from artisans – blacksmiths, miners, sailors and instrument makers – whom he consulted and observed. Sometimes he merely replicated as experiments the working processes of iron manufacturers, 'At least part of his laboratory must have looked like a smithy'.[20]

Finally, we must remember that a contemporary of Plat's (although there is no indication that they were acquainted), Francis Bacon, was writing a series of works which sought to establish a rigorous scientific method and thereby promote scientific and technological advance. In so doing he wanted to take over the methods of knowledge-accumulation and experimentation of the trades and industries of his day: 'Bacon's contribution was to codify the artisanal construction of knowledge that was already going on around him. In the process, he appropriated the empirical experimental knowledge and transformed and elevated it by declaring it to be philosophy, the task of high intellect, not lowly artisans.' According to Conner, 'Bacon, sensing the potential power of the artisanal knowledge collected by Plat and others like him, was concerned that it not be unleashed without regard to its political consequences but controlled to forward the interests of the governing class he represented'. Bacon, if he

19   Conner, *A People's History of Science*, pp. 296–299, 301–5; Pamela H. Smith, *The Body of the Artisan*, p. 151; Plat, *Jewell House*, pt. 2, p. 3.
20   Edgar Zilsel, 'The Origins of Gilbert's Scientific Method', in *The Roots of Scientific Thought*, ed. P. Weiner and A. Noland, 1957, New York, pp. 219, 233; in BL, SL 2189, f. 23a Plat describes 'An artifical compas from baked earth' which was described as 'according to .D. Gilbert'; Conner, *A People's History of Science*, p. 287.

saw any of Plat's works, presumably read only what appeared in print: if he had seen his extensive manuscript notes on, observations of and talks with artisans, and his own experiments, his concern would have been magnified.[21]

Bacon, in his proposed 'history of arts', concentrated on those 'that exhibit, alter and prepare natural bodies and the materials of things such as agriculture, cookery, chemistry, dyeing, the manufacture of glass, enamel, sugar, gunpowder, artificial fires, paper, and the like'. This list accords well with the major topics found in the many arts and trades in Plat's notes and publications. Some were scattered notes from books, friends and artisans, plus his own thoughts and observations; others quite large bodies of undigested material; and others again organized into what can justifiably be termed a 'history of trade' in the Baconian sense. He covered (amongst other things): marbling – on paper and leather; brewing; bread and baking; wines – making, counterfeiting and keeping; food technology and preparation; comfit-making and confectionery; dyeing cloth; saltpetre; paints, inks and colours; gilding, printing and engraving; precious stones and jewellery; fishing, fowling and hunting; aspects of gardening and agriculture; leather dressing; moulding and casting in various metals; papier mâché and other materials; a whole range of military technologies; cosmetics and perfumes; and distillation of all types. These are just the major topics, there are many more scattered observations and stray secrets, and of course we have the details of his own 'professions' as alchemist and medical practitioner.[22]

We can chart the development of trade histories in Plat's manuscripts. An early notebook, SL 2210 (containing dated entries from 1581 to 1593), on which Plat drew extensively for the *Jewell House*, reflects the haphazard nature of the first part of that book, a collection of secrets with little attempt at organization. However, the manuscript does contain 'Rules for the Arte off castinge off any flower, fruite, beaste, or birde according to the liff, as also for the casting off, of any patterne of wax, or mettall, into golde, silver, or other mettalls wth diverse profittable and excellent uses

21   Pamela H. Smith, *The Body of the Artisan*, p. 129; Conner, *A People's History of Science*, pp. 300–01.
22   Conner, *A People's History of Science*, p. 251.

therof', notes which formed much of the section of the *Jewell House* on casting and moulding and are undoubtedly a trade history. Other lengthy sections display some attempts at organization and comprehension such as, 'GENERALL & SPECIALL Rules for graftinge, plantinge, settinge sowinge &c. off trees, cions, buddes, seedes, rootes, slippes. &c.' (37 paragraphs over several pages), and 'Diverse Conceipts for the strange plantinge or forwardinge of all vegetables'. A section containing much on gardening, 'Approved & coniecturall experiments in trees, fruits, flowers, hearbes, seedes, Rootes, &c. Elements, Animalls, stones, & mineralls' is a mixture of gardening, agriculture, preserving, wine making etc. – as the rubric says, more a collection of loosely related secrets, but within it is the 'history' of hop growing and extensive notes on 'artificiall wynes', a topic developed later by Plat. When suggesting improvements in dressing leather he explicitly advocates a history, advising 'first yt shall not bee amisse, to learne all the order of the Tanner and Currier'.[23]

Notebooks used after the publication of the *Jewell House* display more organization. One has the feeling that Plat was moving towards more systematic expositions of craft techniques. Rough notes in one manuscript are refined and augmented in a later notebook. For instance, a relatively short paragraph of information from the painter Francis Segar on refreshing old oil paintings in SL 2189 becomes a much longer memorandum in SL 2216, incorporating parts of the notes made previously, along with material from other sources. Plat also refers to instructions already in print in the *Jewell House*. SL 2216 is a compilation of all he has discovered on the topic.[24]

The acquisition by Plat of the manuscript by 'T.T.' which forms the core of SL 2189 and which was probably acquired after the *Jewell House* was published, may have been one spur to Plat's attempts to produce trade histories. T.T., writing probably in the 1550s, produced material on many topics including food preserving and cookery, bread, beer, gardening, candle making, dyeing, leather, ink, paints, and organ pipes. Some sections were, in effect, short trades histories. The section headed (and augmented)

23 BL, SL 2210 ff. 23, 29, 35, 105, 115.
24 Plat, *Jewell House*, pt. 1, pp. 5, 51; BL, SL 2189, f. 33; SL 2216, ff. 23, 81 et seq.; SL 2210, f. 46.

by Plat, 'A rule to teach to make leather how and in what sorts yt ought to be done', methodically, if briefly, sets out how leather is to be made and coloured. 'To make tallow candles by dipping them into mouldes', includes a table of the ratios of tallow to length of candle and length of wick, followed by numbered steps in making candles. Plat add his own observations to these and other sections.[25]

Plat's extensive notes on making, keeping, and counterfeiting wines may also be have been stimulated by acquisition of an existing manuscript. In the *Jewell House* he announced he had the secrets of vintners – 'almost the whole art as it is this day in use amongst the Vintners, written in a pretty volume called *Secreta de pampinei*'. Years later, in the introduction to *Floraes Paradise*, he implies *he* has written this work: 'in my booke, entitled Secreta De pampinei, not yet published, I have set downe many particular practises'. That the work was to be published is confirmed by Gervase Markham who based much of his notes on wines in *The English Housewife* (1615) on *Secreta de pampinei*, a manuscript which was 'preferred to the stationer' and 'came to me to be polished'. An incomplete manuscript copy, SL 3692 ff. 26–30v, is not in Plat's hand and, more significantly, not in his writing style, so despite his implication, this seems to be another manuscript which he acquired and augmented with his own research.[26]

Another reason for a more systematic noting of trade histories may be a change in the way Plat regarded the information he collected. We have seen that he consulted John Dee on alchemy in the 1580s, a time when he was much absorbed with that subject and he may, in his early notes, have thought of the operations of, and possible improvements to trades in alchemical and 'philosophical' terms. Thus, in SL 2245 a good deal of general 'secrets' and information about trades is mixed in with extracts from alchemical treatises and speculations on alchemical matters. Practical matters, such as soil fertility, are discussed in philosophical terms – the operation of aqua coelestis on soils to produce 'a strange Earth for fructification'. Plat's philosophical preparation, the 'fire of nature', is listed as of use in 54 different situations, many to do with manufacturing, such

25 BL, SL 2189, ff. 75v, 87 et seq.
26 Plat, *Jewell House*, pt. 1, p. 65; Gervase Markham, *The English Housewife*, ed. Michael R. Best, xviii–xix.

as preserving or improving beer and wine, making aqua vita, preserving foods of all types, making an extract of logwood, strengthening metals, preserving ink, strengthening saltpetre, oil of sulphur and preserving oils in general, and extracting essence of herbs. Contrast this with a later notebook (SL 2247) in which there is no alchemy and many sections for proposed histories: 'English wines most excellent', 'The Art of molding and casting of beasts personage &c.', 'Advauncement of Silk to great profit', 'The art of marbling or clowding, uppon satten, taffata, paper &c.', and 'delicate conclusions concerning Gardeins and Orchards'.[27]

Plat's early notebooks, most notably SL 2210, are replete with extracts from Continental secrets books, popular printed works composed of all manner of recipes – medicines, household hints, domestic and industrial technology, alchemy, magic and much more – some copied from much earlier works. To take just a few of the more popular: in SL 2210 we find over 50 extracts from Girolamo Cardano, *De rerum varietate libri xvii*, 1557 Basel (or 1558, Avignon); 23 from Johann Jacob Wecker, *de secretis libri XVII*, Basel 1582; 11 from Alessio Piemontese, whose works in English translation appeared between 1558 and 1569; and 14 from Thomas Lupton's *Thousand notable things of sundrie sorts*, 1579. This latter is itself a compilation based on many Continental secrets books. Many of these extracts subsequently appear in the *Jewell House*, and while complaining of their shortcomings in his preface, Plat acknowledges his debts to 'Albertus Magnus, Alexis of Piedmont, Cardanus, Mizaldus, Baptista Porta, Firouanta' and 'Wickerus'. Plat's recourse to secrets books diminishes in later manuscripts and publications, especially when he is compiling trades histories. It is as if he found this recycled knowledge so unsatisfactory that he realized he had to go to practical men for his information.[28]

---

27   BL, SL 2245, pp. 95v–101; SL 2247. I deduce from a reference to 'my new booke of Gardening' (f. 3) this manuscript may have been written in the 1600s. It is incomplete and may have been written shortly before Plat's death.

28   BL, SL 2210; Eamon, *Science and the Secrets of Nature*, p. 258; 'Wickerus' is Johann Jacob Wecker, op. cit.; 'Mizaldus' is Antoine Mizauls, *Memorabilium aliquot natures arcanorum etc.*, Frankfurt, 1592; 'Firouanta' is Leonardo Fioravanti, *A compendium of the rational secrets etc.*, trans. John Hester, London, 1582.

It may seem disingenuous to include gentry in Plat's informants on trades but he gained much information on trades, household remedies, and general 'secrets' from his friends, relatives and the gentry he met in the course of his visits to the royal Court, the haunts of lawyers and other places of resort in London. As one might expect, the secrets divulged by gentlemen and women reflected their interests and status: foods, wine and beer are prominent, as are inks, colours and writing equipment, distilled waters and cosmetics, ways to colour precious stones and keep moths from clothes. Some informants were such casual acquaintances Plat had no name for them – a Cambridge scholar, or a French gentlewoman – others were close members of his own family: his son John, father-in-law Young, and many cousins. Famous names such as Sir Walter Raleigh, Sir Francis Drake, Thomas Harriot, and John Dee are acknowledged, but for the most part the ladies and gentlemen who Plat consulted have not merited entries in the *ODNB*.

Plat obtained a good deal of information on food and drink at sea from the great naval commanders of the day, most notably Sir Francis Drake. For instance: 'I have heard Sr Francis Drake constantly affirme uppon his owne and often practize, that iff you draw owte of a viniger caske 3 quarters of the viniger, and then fill upp the same againe wth stronge beere, that the same at the Sea in 10 dayes will become good viniger'; and 'Sir Francis Drake hath found by good experience that one vessell of viniger will make greate store of good beverage, and as I remember he spake of one parte of viniger to 20 partes of water which hee comended greatly for his cooling and refreshing Nature'. Whilst Sir Francis was undoubtedly a gentleman and courtier, he was peculiar in that being a sea commander he was much closer to his subordinates and their needs than most gentlemen. This is reflected in his understanding of the difficulties of keeping food and drink fresh at sea.[29]

It is rare to read of a gentleman demonstrating something to Plat – 'How to rase owte any letters clenly and neatly from paper or parchment – probatum per Mr Auditor Hill in my presence', being an exception. Instead, as one would expect, Plat reports conversations with gentlemen

29 BL, SL 2216, f. 50–50v.

at formal or social gatherings – 'I heard a gentleman report it upon his own experiment'; 'Sr Tho Challenor assured me'. Often he dates these conversations: 'This resceit I gott of Mr Arthur Gregorie the 16 of July 96', 'This of Mr Rich my .L. Riches brother for a most rare secrete the 31 of August 96'. The degree of intimacy in conversations with other gentlemen depended on relative status, so Plat may have obtained his information from Drake not in direct conversation but as part of an admiring crowd of courtiers clustered around the great naval hero listening to his reminiscences of life at sea. Similarly, his conversation with the Spanish Ambassador Mendoza about a lamp may not have been an intimate one. Other information came second or third hand, with modifications on the way. Of a 'quick way of making shell [liquid] gold', Plat notes 'this Resceite I hadd of Mr Gregorie in August 95. Whoe hadd yt of Mr Warde one of the Secretries of the Counsell, and it is thought to bee Chappells manner of liquid gold, but Mr Gregorie commendeth a resceit of his owne much better.'[30]

Strings of recipes and ideas from the same person were noted down by Plat over a short period of time. Mr Rich of Lee (brother of Lord Rich, and a descendant of the prominent courtier Richard Rich of Lees Abbey in Essex) in August and September 1596 (probably either during a visit to Court or a stay in London for some other reason) provided recipes for musk sugar, cinnamon water and other oils and waters, and how to preserve the fashionable eryngo roots. In this year and a year earlier Plat gathered a good deal of information from Arthur Gregorie and Mr Spencer at Court. A visitor to London who talked to Plat in 1597 was 'James Dakeson gentleman of Dorset neere Blandford'. This latter informant and Plat obviously got on well, for they swapped horticultural and agricultural secrets; the Dorset man told Plat how to sow 'white Italian or suger pease', and 'Mr Dakeson hath promised mee to send mee som of the pease to sow. & I have promised him somme extra ordinary English barlie in lieu of them'.[31]

Plat was impressed by the veracity of the advice he was offered by gentlemen, often adding phrases such as 'per verissimo', 'per experto', 'per

30   BL, SL 2216, ff. 74, 106, 111; BL, SL 2210, f. 91.
31   BL, SL 2216, ff. 74 et seq., 141–141v, 158.

verissime' to his notes. He was particularly trusting of advice from close friends such as Auditor Hill, Thomas Gascoigne, Garret the apothecary and Sr Romero, the Spaniard from Naples. Steven Shapin notices a similar phenomenon in the seventeenth century, concluding, 'Roughly speaking, the distribution of credibility followed the contours of English society'. Hill's green-tinged isinglass sheets to read with – the colour was restful on the eyes – was 'warranted me by 20 yeeres practize of Mr Auditor Hill'.[32]

His friends also compiled or acquired manuscript books of secrets and Plat copied recipes from such books in the possession of Hill, Garret, and Gascoigne. But it was not only gentlemen who compiled secrets books. Plat notes recipes from tradesmen's books – such as a green ink recipe 'E libro Cotford' (a goldsmith) and notes on the use of gold leaf in limning from the surgeon Watson's book. Plat makes references to 'my written book of secrets': was this another manuscript Plat had acquired or a general commonplace book? He also noted any secrets his servants possessed – a way of stopping leaks in wine barrels was garnered from 'John Tey my servante whoe stopped a wine vessel of myne therwth that leaked extremelie'.[33]

A number of Plat's informants were Court officials or servants. Some were quite high-ranking. Hill was an Auditor of the Exchequer; William Segar, who contributed recipes connected with limning, was a herald (his brother Francis also provided information on painting – he was a painter and representative of the ruler of Hesse). Arthur Gregorie (or Gregory), was described by John Fuller in his *Worthies* as a man from Lyme in Dorset who 'had the admirable art of forcing the Seal of a Letter; yet so invisibly, that it still appeared a Virgin to the exactest beholder. Secretary Walsingham made great use of him about the Pacquets which passed from Forraign parts to Mary Queen of Scotland. He had a pension paid unto him for his good service out of the Exchequer'. Such a man, a key technician in the government's espionage effort, might be expected to be a fount of ingenuity (Plat described him as an 'inginer' in January 1594).

32   Steven Shapin, 'The house of experiment', *Isis*, vol. 79, 1988, p. 376; BL, SL 2216, f. 110v.
33   BL, SL 2216, ff. 63, 68v, 69–70, 77v, 87v–92v; BL, SL 2189, ff. 98v, 104, 105.

SCIENTIFIC THOUGHT AND TECHNIQUE

Gregorie was particularly forthcoming on the subject of coloured inks, a speciality one might expect of an expert in tampering with diplomatic correspondence. But Plat also consulted the Queen's trumpeter, her rat catcher, purveyor of wine, ale brewer, and a member of her Privy Kitchen. This was Mr Webber, a confectioner skilled in sugar work but also Plat's father's tenant, so that Plat probably met him collecting his rent.[34] But did he also make a nuisance of himself badgering other servants at Court for secrets? More generally, how did he collect information from gentlemen? Did he engage them in casual conversation and hope to learn something of interest, or did he have a list of topics on which he wanted information and try, subtly or not, to make his wants known? A wants-list in his manuscripts, which one can imagine him slipping in a pocket before meeting new acquaintances reads as follows:

> A Note of Such Secretes as are worth the search and not as yet founde owte in any perfection by the Author
> 1 C The Secrete of Turnsole
> 2 C the Secrete of Vermillion
> 3 C The Secrete of Verdigreece
> 4 C The fixation of ye Coolor of Logwood
> 5 C The refininge of Aloes hepatique into aloes cicotrina
> 6 C The makinge of orpin/
> 7 C The makinge of indico owte of Anele, or Anele owte of Woad, or both owte of woad
> 8 C The hardninge of Suger in a short tyme
> 9 C The composition and colloringe of hard wax
> 10 C The making of all enamels
> 11 C Cherries and flowers to comme early
> 12 C To make worm seede to grow in England
> 13 C The refininge of Mirrlye
> 14 C A multiplyinge earth for peter./

[34] John Fuller, *The Worthies of England*, vol. I, 1811, p. 316; David Piper, 'The 1590 Lumley Inventory: Hilliard, Segar and the Earl of Essex, II', *Burlington Magazine*, vol. 99, no. 654, pp. 295–303; BL, SL 2189, ff. 71v, 146; SL 2210, f. 89v; , SL 2216, ff. 12, 24, 29, 74, 86, 111v.

15 C To make annys seedes to grow in England
16 C To sophisticate rennish wyne
17 C To cast pillers of marble, Jett, porphyre etc./
18 C Artificiall sandiner
19 C Artificiall Tinglasse./
20 C To marble well upon wood or leather
21 C To refresh pictures wrought in distemper. qre if it wilbe don wth a size or wth isinglass dissolved./
22 C a good lampe or Candle to bee made of Rosen, pich or tarr by taking away the fume therof.
23 C a safte or hard wax to bee made of pich to seale with./
24 C White copper and silveringe over of Copper
25 C To melt downe yron stone wth seacole wth strong potts in a winde furnace.[35]

It is not inconceivable that Sir John Harington, in satirizing Plat's enthusiasm for new inventions, might also have been reflecting exasperation in some Court circles at his single-minded pursuit of new secrets. He may, however, have been treated sympathetically by many gentry acquaintances: that a number of his friends also collected secrets is clear from the manuscript books he borrowed from them.

Over 50 different trades were cited by Plat amongst the named artisans he consulted. These included food trades, wines, beer and ale; distilling; medicines; agriculture and gardening; luxury trades such as goldsmiths and gem cutters; dyeing and other cloth trades; military supplies – saltpetre, and armour; the dissemination of information – inks, writing, printing; as well as oddities such as a royal rat catcher. Some names and trades occur but once or twice, others were a mine of information. They reflect the diversity of commercial activity in London and his eclectic interests. He obtained information from a number of foreigners, many Protestant refugees, such as the Dutch diamond cutter who recommended radish oil to clean cloth, the French clockmaker dwelling with Mr Kirwin who told him how to plate copper with silver, and a Dutch silk dyer who talked to

---

35   BL, SL 2216, f. 177.

him in July 1598. Maybe these refugees, highly skilled but not part of the London craft hierarchy, were more willing to converse with gentlemen than were native craftsmen.[36]

As with the gentry, there is much evidence that he had to obtain his information from tradesmen by talking to them rather than observing their activities. In the many cases when he is recording snippets of information overheard there is no doubt as to how he gathered his information: he 'heard Baylie the Dier affirme' an opinion on preparing the dye indigo and he 'heard a wise Carpenter affirme' that green timber should be used when piles were placed in water. One can imagine Plat recording the carpenter, probably in his cups, saying that 'hee did greatly disallow all the piles of London bridge and those at the new wharf at the Custome howse which were driven in drie.' Information gained from artisans whose names were unknown to Plat, such as '[blank] the glasse seller' who gave him a recipe for cement and another for porcelain powder to blot ink, must surely have been the result of chance meetings. Tradesmen well known to Plat, such as the apothecaries Garret and Kemmish, passed on recipes, probably in general conversation when Plat bought chemicals and drugs from them. Plat's medical activities brought him into contact with a wide cross-section of London society – no doubt he used these contacts to talk to traders.[37]

When noting more substantial information, Plat tended to conclude with, 'all this I had of', or some such phrase, often dating his encounter: 'All this I learned of Rowland [blank] the fine painter dwellinge neere broken wharf the 10 of aprill.1597', and 'All this of Mr Webber one of her Maties privie Kichin december.95'. (Note again that although Plat

---

36 BL, SL 2216, ff. 44, 113; BL, SL 2203, f. 87. The trades identified by Plat were: ale brewer, apothecary, armourer, attorney, barbary merchant, barber surgeon, brewer, carver, chemist, clothier, cook, coster, cutter in wood, diamond cutter, distiller, draper, dyer, engraver, fine writer, gardener, gem cutter, gold maker, goldsmith, grocer, haberdasher/cabinet seller, horseman, joiner, merchant, miller, musk melon gent., nurse, painter, peterman, pewterer, physician, printer, purveyor of wine, rat catcher, seal cutter, shipmaster, scrivener, soap boiler, surgeon, traveller, tree nurseryman, vintner, wine cooper, woad man, writing master.

37 BL, SL 2216, ff. 19v–20, 54, 111, 148.

gathered the painter's Christian name was Rowland – he did not recall a surname, which was almost certainly Lockie – now remembered mainly for his version of Holbein's painting of Thomas More and family now in the National Portrait Gallery, London.)[38]

It is conceivable that Plat observed some of the artisans' operations he records in his more detailed notes, but the fragmented phrases, many alternative substances or operations suggested, and the queries inserted by Plat (things after the encounter he realized he should have clarified with his informant), tend to argue that he was recording conversation. So, when noting a method of recovering 'faint wine' with 'powdered Argoll' he wonders 'qre if the caske must not first bee sented'.[39] Likewise, after taking copious instructions from a goldsmith on 'The Arte of Limminge upon Glasse', he realized he forgot to ask if the pigment 'sanguis dragonis' is dissolved in spirit of wine *before* being tempered with mastic, and if it was advisable to fast before breathing clove-scented breath on the glass at the start of the operation.[40]

As with some of his encounters with gentlemen, Plat tended to pump his artisan acquaintances for as much information as possible at a single meeting, often recording a string of recipes or secrets from the same person.[41] On some occasions he set out to obtain information, rather than taking advantage of a chance encounter. In a section headed 'How to make printers yncke both good and cheape' he lists several queries about ingredients, followed by a recipe from Cardano. Armed with these notes he then went to see Dawson the printer and added details of his techniques: he commonly boiled seven gallons at once, recommended a copper kettle to boil in and used balls of brown paper to soak up the fat in his ink mixture.[42]

Plat, by patrimony, was a member of the Brewer's Company and his father, lessee of a London brewery, lived until 1600. One would expect, therefore, that he had access to breweries and to brewers' secrets.

38   BL, SL 2189, ff. 71v, 101.
39   BL, SL 2244, f. 5.
40   BL, SL 2216, f. 32v.
41   For instance, notes taken from Hall the Cooper, BL, SL 2244, f. 55 et seq.
42   BL, SL 2216, f. 37v.

Moreover, he also claimed 'I have some reasonable skill in my trade'. But conclusive evidence that Plat has observed other craftsman at work is rare – I have found only four instances in his notes. In surely one of his last memoranda, dated 6 August 1607, he records in great detail the use of a lee of ashes mixed with red lead used by 'English the Travayler' to clean his oil paintings, informing us 'this I did see well performed upon diverse of my pictures'. He watched the process of engraving on metal by spreading a thin layer of wax, incising a pattern in it, then etching the pattern onto the metal with acid, 'this I saw donn by Hendrickes the goldsmiths boy'. Perhaps significantly, he observed an apprentice, not a master, and this may have been a chance observation not an authorized visit. He did, however, at the same time see 'Edward Partridge grave very well yron' using an alternative method.[43]

In September 1596 he recorded several pages of instructions under the heading: 'Mr Norris his arte in Candyinge Conserving and preserving of rootes flowers &c. wth tharte of casting massive & hollow in suger past.' One of the recipes, that for candying marigolds was subscribed 'probat me presente'. He may have observed only this one operation but the whole section from Norris reads as if Plat had observed him at work. The notes are clear and confident, with no asides, queries, alternatives, or hasty punctuation. The operation he says he observed is recorded thus:

> Boyle suger on rosewater a little upon a chafingdish and coales, then put the flowers being thorouglie dried either in the Son or by the fier into the suger, and boyle them a litle, then strew the powder of refined suger upon them, & turne them, and let them boyle a litle longer taking the dish from the fier, then strew more powdered suger upon the contrary side of the flower they will drie of themselves in 2 or 3 howers in a whott sonny day though they lie not in the son.[44]

We must be cautious about assigning all confidently written and detailed notes to direct observation. One such, on saltpetre production, was nonetheless the product of a conversation, concluding: 'this receipt I writt

43 Plat, *Jewell House*, pt. 1, p. 15; BL, SL 2189, f. 34; SL 2210, f. 72v.
44 BL, SL 2216, ff. 142–6.

from Wilkinson's mowth the Peterman 1597'. Similarly, a clear and detailed exposition of suburban rabbit production was transmitted verbally: 'Mr Parsons affirmed unto mee'. A subjective review of his notes, however, leads me to believe that most of his information on trades was collected without direct observation of artisans at work.[45]

We sometimes come across nascent histories of trades in Plat's writings as he attempts to draw together scattered memoranda into a history. The three folios SL 2216 ff. 64–67 contain an index of 50 references to paragraphs scattered throughout the manuscript on secrets of painting and limning. The index is headed 'Diverse Rules Circumstances and Secretes, concerning the art of painting staining, lynminge and ordinarie paintinge wth all the Appendicies belonging thereunto'. Had Plat drawn these together, they would have formed a new history. The topic had long interested him. In an early manuscript he records recipes from a clergymen called Bateman which were subsequently printed in the *Jewell House*. The manuscript Plat acquired of notes by T.T. (written in the 1550s) contained several pages on inks, paints and so forth superscribed by Plat, 'Secretes in wrighting painting, printinge, guildinge &c.' T.T.'s notes are followed by many more pages of 'Novae additions' drawn from artisans, gentlemen, friends' manuscripts and published books.[46]

While he also wrote extensively on wine-making and brewing, there are two other topics on which he attempted histories that are worth looking at in detail: moulding and casting, and dyeing. The nineteen-paged section of the *Jewell House* (1594) entitled 'The Art of moulding and casting' is Plat's earliest published piece of trade history. Beginning with no preamble, directions are given to make a loam 'coffin' on a board as the base in which to form a mould. Instructions then follow on how to make the 'papp' for a mould (various ingredients are put forward), making the mould itself in various forms, cleaning the moulds, and details of how to cast objects in various precious metals, papier mâché, and other materials. This is a didactic piece, paragraphs begin 'You Must…', 'You may…', 'When you meane to…' No artisans or gentry are named in the text as sources of information but clearly many persons were consulted,

45 BL, SL 2189, f. 109; BL, SL 2216, f. 50v–52.
46 BL, 2216, ff. 2–6, 64, 92–106v; SL 2210, f. 71; Plat, *Jewell House*, pt. 1, pp. 36, 44.

including several goldsmiths ('And som use to cast copper, and latten works in Highgate sande, some in lome only, some in cuttle bone, and divers other substances…'). The alternative methods and substances may reflect the degree of experimentation and technical improvements carried out by the goldsmiths and other artisans Plat consulted. We glimpse one experiment for stiffening leaves when making a mould of a branch to cast in precious metal: 'Qre if you may not keepe them stiffe as long as you please in a stove. This I have not proved but I had the same of an excellent woorkeman, who assured me upon his credit of the truth thereof'. Of this secret Plat writes, 'hitherto I have not disproved the same, and a small time, or charge would serve to make the proofe therof', clearly implying that he has carried out many of the operations he here describes, an implication backed up by his inclusion of 'myself' with 'such others as have spent their time, and thereby attained to any exquisite skill in this art of casting'.[47]

Plat must have spent a good deal of time talking to those working with moulds (goldsmiths in particular), learning how they cast precious metals. He was careful not to disclose all their secrets, breaking off discussion of one type of mould made from glue to observe, 'I holde it not convenient for the great hindrance, to all the Iewellers, and workemen in golde and silver, to discover all the secrets either of this composition, or of the rest that are contained in this discourse'. Later on he states, 'How to colorish your patternes in golde, and how to boile those that are cast in silver, I must refer you to the Goldsmithes, although I could easily set downe both the matter, and the manner thereof, but because therein, I should discover a secret, that concerneth their whole trade, I have thought good to suppress it for this time.' Not that he himself obtained all their secrets: he advises readers to 'Learne of the Goldsmiths' about the true heat for casting gold and silver, otherwise let the Goldsmiths heat their moulds 'in their apt furnaces'.[48]

This published history of moulding was compiled largely from notes in SL 2210. Comparing manuscript and publication, one is struck by the terse style of the manuscript. This reads like abbreviated notes taken during or

47   Plat, *Jewell House*, pt. 4, pp. 49–58.
48   Plat, *Jewell House*, pt. 4, pp. 57, 62.

immediately after a conversation, in comparison with the comprehensive instructions in the published version. For instance, the note:

> iff you take 1 parte of flaunders melting potts to 1 parte of bole armeniack both well powdred, put them in water, & shake the water upp & downe, still powring the same away till yt becomme clere, & then keping all the thicke pudled water together, & let yt setle, & then drie the powder that this powder is most excellent to caste in.[49]

In print, this is transformed into:

> Some do greatly commende the fine powder of Flaunders melting pots that be new, and bole Armoniak mingled together in equall partes, you must put this powder in water, and mak agitation of them together, and then power away the same water sodainely into some cleane vessel, and put in more water, reiterate your agitation as before, and so continew this worke until your water which you power away from the powders becom cleer; then let al this thick water so gathered together, settle wel, and then dreine away the water by declination, and after drie this powder and keep it to make pap thereof at your pleasure. And this for an excellent receit.[50]

As one might expect, the manuscript provides the names of some informants: Cooper the Goldsmith advises on 'papps' for moulds made from burnt alabaster and wax, and one Laude advised stiffening the wax with coal dust. Plat's Spanish friend Romero passes on a tip from the Irish gentleman Andrews. But generally there are few names in the manuscript.

Plat recognizes the many uses of moulding and casting, reflecting the growing employment of small metal objects in daily life and in scientific research:

> I thinck yt will bee profittable to cast of great latten son dialls, wheeles for Iacks or clocks in latten or copper, hilts of swoords and daggers in silver

49 BL, SL 2210, f. 102.
50 Plat, *Jewell House*, pt. 4, p. 60.

or gold, clusters of arrows as they grow, into [gold] or [silver], knotts, for Bookbinders, keyes, scocheons & flowerworck for locksmithes, Antiquities, walnuts in silver hollow to put muske in, al instruments of Astronomy or Geometry, leaden letters for Book binders, &c.[51]

Casting in papier mâché, sawdust, powdered stone, and similar substances opened up great possibilities in beautifying the parlours and halls of the gentry to reflect the latest taste in interiors at much less cost than similar items carved in wood and stone:

> yt is an excellent and proffitable Course, to mold of, ffaces, doggs, lions, borders, & Armes &c. from Tombes or noble mens galleries, first in wax and glew, and after in the papp of paper, wch being paynted you may garnish your howse or gallery wthall. or after you have them in glew, to mold them in sawduste, & fishglew dissolved, for so they will seeme to bee all of wood & if this way prove, then may you make pillars, balles, leaffes, ffontages &c. to garnish beeds, tables, Courtcobards &c. wth the sawdust, or filing of box, brasell, Ebony, Elephants tooth &c.[52]

In later documents Plat came back to moulding in papier mâché and similar substances referring to 'my paper papp' in a way which, with the absence of acknowledgements for ideas from others, leads to the conclusion the Plat developed these ideas himself, by experiment. He recognized that papier mâché was light but also strong, and could be used to make shields and plate armour capable of warding off blows, but his main hope was that the techniques could produce consumer goods to make money from. The cheap adornments to houses outlined above were fashion items, and Plat pondered what would sell well at home and abroad – figures from classical antiquity, Greek and Roman literature, emblems, the royal family, and current English heroes:

> Theise and such like may be chosen for patrons to carve your moldes by. vz. the 9 worthies C the 12 romane Emperors, C the 4 Elements C the 9 muses.

---

51   BL, SL 2189, f. 162; BL, SL 2210, ff. 103, 103v–104, 126.
52   BL, SL 2210, f. 104.

C the choysest of whitneys emblems. C The pictures of her Maiestie, her ffather Brother and sister and of such noblemen as are likeliest to bee of good sale C All the kings of England. C All the Cities cutt owt of Munster. C The stories in ovide naturphilos. C The Queenes armes cutt large and compleat./ my Ld of essex picture and Sr ffrancis drake will sell well in England but better in Barbary, Turkie, venice &c.[53]

The other aspect of moulding of interest to Plat was casting in sugar to produce fancy conceits for banquets – hollow sugar lemons stuffed with comfits and the like. Here he was indebted to the cook Mr Norris for information.[54]

In contrast to the art of moulding, which was dealt with at length in the *Jewell House* in 1594, Plat has nothing to say in that book on the subject of dyeing cloth. He did not subsequently produce a history of the trade in print, although he made extensive notes for such a work. In 1596 he knew very little about dyeing woollen cloth although he had started experiments in silk dyeing. Plat's earliest information on wool dyeing came not from dyers but from experts in woad growing and processing. Woad is a fleshy-leaved plant which is processed by quite complex 'seasoning' into a powder added to dyers' vats to produce a blue dye or the basis of various other colours. The powder, in barrels, had long been imported to England from various parts of Europe, and the dye was an important ingredient in the finishing of English textiles. Good grain harvests and subsequent low grain prices in the 1580s stimulated the interest of landlords and farmers in woad as an alternative crop. Because of the initial capital outlay required to produce woad, gentlemen and larger tenants at first took the lead. Then, as both growing woad and processing it were specialized skills, after a few years itinerant 'woadmen' took on these tasks, contracting with landlords to grow and manufacture the dyestuff. They erected temporary sheds to process the crop. Leaves were fermented in heaps, milled into a rotten paste which was formed into balls, then dried on wooden racks in airy sheds before being pounded into the powder used in dyeing. Woadmen also assembled the necessary labour to tend, harvest, and produce the finished

53 BL, SL 2189, f. 162; SL 2197, f. 16.
54 BL, SL 2216, f. 142 et seq.

dye. This was a labour-intensive operation and the government became worried about the threat to public order of small armies of poor people moving from place to place tending the woad crop, as well as the effect on other industries of a draining of labour into the woad fields. The offensive smell of fermenting woad was also a problem and processing was banned within eight miles of a royal palace in 1585. A government commission in 1585 and 1586 found a total of 8519 acres under woad, almost certainly an underestimate. Government estimates of profits from woad of £10 to £14 an acre after all charges indicate a very lucrative crop.[55]

It is no wonder that woad was of interest to Plat, a man keen both on novelty and profit. We find him in January 1596 taking instructions from Mr Coggaine, probably a woadman, on the place of woad in rotation with barley, and on how to manufacture woad dye. Coggaine also tells how to test the quality of the manufactured dye 'so as the dier shall not uniustly dispraise it', and he is also knowledgeable about the costs involved in growing another capital-intensive dyestuff, madder: plants for three acres of ground cost the large sum of £100 (and the crop took three years to mature). If one could stand the smell arising from close contact with fermenting woad, the heat generated by the process could be used medicinally: 'a great woadmaster' (probably Coggaine) advised putting aching or gouty limbs in fermenting woad to gain relief, and the same treatment was said by Coggaine to cure the pox, scabs, and ulcers. A kinsman, Plat's cousin Shattenden, told him how to process 'stubborn woad' which would not work, by re-fermenting it with a liquor of bran and ferdinando bark. Shattenden, either a woadman or wholesale dealer in dyestuffs, claimed to have bought such woad for 20 shillings the ton and resold it after processing for £20 the ton.[56]

In these early notes on dyeing, Plat absorbs the information on woad growing and processing and, characteristically, suggests improvements to the processes. Will the addition of lime-water help woad to rot quickly? Is the discarded juice of raw woad of use as a dye? Can the process of

55 Thirsk, *Alternative Agriculture*, pp. 79–96; Richard Hoyle, 'Woad in the 1580s: alternative agriculture in England and Ireland', in *People, Landscape and Alternative Agriculture*, 2004, pp. 56–73.
56 BL, SL 2189, ff. 124, 132, 134; Thirsk, *Alternative Agriculture*, p. 105.

drying woad be speeded up? As he becomes more involved, so he is drawn into the topic of dyeing. He notes the process of dyeing stockings, many of which were dyed blue with woad. At this time he has no first-hand knowledge of commercial woollen dyers' secrets, he merely reports that, 'Som thinke that the diers at this day use first to die theyre clothes wth woade, then they madder them, and then they logwood them and last of all they woade them wth a weake wood liquor & that therby they save a great charge.' Moving on from woading woollen cloth, aided by information from two dyers, 'Godfrey the dier' and 'Huchins the scochman a dier', he writes in some detail about dyeing silk in a range of colours, using brazil wood, turmeric, madder, cochineal, logwood, oak galls and the like. He speculates that he might be able to dye 'cloth or woll in the same manner as I dye my silk', disclosing that he has already experimented himself. He describes one possible experiment, and the cookery which inspired it:

> qre if kuchenell pounded and standing in a gentle balneo a sufficient time wth a true proportion of argoll water in it, I meane equall pt of argoll, to the kuchnells, being boiled awhile, and after the argoll devided, if so the kuchnells by this deepe tincture will not extend a great deale faster. also prove an infusion of logwood in water set it in a gentle balneo until the whole strength of logwood be extracted, if so the same will not hold his collor longer. this conceipt I have from peares baked in a clos coffin wch will become very redd, wch if they boyle open will have no collor but white.[57]

If Plat was busy gathering information on dyeing and starting experiments of his own on silk in 1596, by 1598 he had accumulated enough information for a detailed history of silk dyeing. He produced a treatise which begins with a table of contents, an indication of the scope of the work and the degree to which he has organized his material:

Discharging pag.9.   {of the gom
                            {of a coller
                            {of advancement.

---

57   BL, SL 2189, ff. 39v, 123 et seq., 132.

SCIENTIFIC THOUGHT AND TECHNIQUE

Advancing.   {of raw silke
                              {of discharged silke.

Dryinge off all manner of collors.
Alteringe of one collor into an other.

                              {by the fier
Dryinge   {in a stove
                              {in ye son
                              {in ye winde or Aer

Stringinge or pinninge of Silke.
Paperinge upp of Silke
Water to discharge, to drie, & to clense wthall
Vessels useall for this arte.
Tooles or instruments.
Skewinge how procured & how avoyded.
Glosse of Silke, preserved or destroyed.
Crispinge [58]

The operations begin with discharging – washing raw silk with soap to rid it of adhering fat. This Plat calls 'gomme'. He explains, 'All kinde of throwne silke doth usually hold 4 oz of gomme or a litle more or lesse uppon every pounde, so as the same beinge discharged by sope in boylinge liquor, the silke for the most parte doth hold but 12 oz uppon the pounde before it bee died.'[59]

Next follows advancing – using a mixture of water and dyestuffs to colour the silk. As well as colouring, this process added weight to the silk, although the amount varied considerably in Plat's time according to the methods employed by dyers. Unscrupulous dyers could add considerable weight (adding value to the silk when sold by the pound for weaving) but the glossy finish which accompanied the weight soon wore off when the finished silk garments were worn. Plat begins his section on advancing by condemning the most flagrant use of advancing by London dyers:

58 BL, SL 2247, f. 5v.
59 BL, SL 2247, f. 6.

first there is both light and a doble weight in theyr London die as they terme it. Yt is a black Collor & performed wth the barck of Elder, filings of yron, and the slipp of a grindstone ye color & glosse is so badd, and the silck so rotten by longe lyinge in thafforsaid mixtures before it have gained his doble weight, that I hold the secret therof rather meete to bee touched or glanced at then to bee publiquely discovered & made common by the hande of any honest wrighter.[60]

Under the heading 'Colours' Plat details the many substances used by dyers to colour silk and the distinct operations required to obtain a precise shade. He records a simple recipe: 'make a high pale yellow with Turmerique, turne it [i.e. the silk] in a tubb full of wast turmericque liquor having abowt a potle of wast spanish black well incorporated wth it, it is don very neere in 20 tornes this can not be to high of the yellow at the first'.[61]

Several hints on drying are given, such as 'The Smoake of seacole will eate of an ash collor or russett in the dryinge', and 'No greene may endure any whott son, nether may they bee longe in driyinge'. In 'Paperinge upp of Silke' (wrapping it in paper), 'the papers wherin brimstoned whites have lien' will leach colours out of other silks. Plat tackles other problems which occur during the process of dyeing, such as skewing – the uneven take-up of colour – which is avoided by turning the silk many times in the dye vat or dyeing for a long time in a relatively weak solution. 'Crisping' – silk yarn curling under the influence of the dyeing liquor is avoided, Plat opines, by turning it 'either in a maister Lee of pott ashes bee it never so weake, or uppon the fatt'. Some of the sections outlined in the table of contents, stringing or pinning, and vessels and instruments, get cursory mentions only but this is nevertheless an attempt at a comprehensive 'history'.[62]

Only two artisan dyers, Godfrey and Tilton, are named as contributors to this account although Plat is clearly conversant with, and critical of, many of the techniques of 'the silckmen of our time' and the 'ordinary dier'. Nowhere, however, does he tell us that he has seen the dyers at work – his information has come from conversations with those who will talk to him.

60 BL, SL 2247, f. 8.
61 BL, SL 2247, f. 17v.
62 BL, SL 2247, ff. 34, 36.

Gaining knowledge in this way has its limitations – Godfrey tells him that 'the overgalinge of Silck for Spanish black, doth make the color much worse then otherwise it wold bee', but Plat is left wondering 'how much it may bee galled if you wold have an excellent black that shold keepe color' and 'also if somme yndico mixed in the die will not bewtifie the collor'. [63]

Much of Plat's text is didactic. There are straightforward instructions such as: 'To have a light and faire straw collor, turne your silck in your turmericke liquor before it bee blood warme, or els put a litle of a stronge turmerick liquor to a good quantitie of faire water & turn therin'. And a good deal of these instructions carry the added assurance that Plat has tried and tested the techniques. The text abounds with phrases such as 'I have proved', 'I hold it best', 'I hold it necessarie', 'I finde', and 'probat'. No doubt much of his laboratory work simply replicated what he learned of the dyer's craft in conversation but he also went further and carried out experiments to refine techniques. Take for instance the following, where Plat describes in detail an experiment in dyeing silk using oak galls, together with further thoughts on the results:

> I did advance organzi undischarged to 4 oz ½ uppon evry pounde vz. first I did beate 6 oz of gaalls, I putt therto a potle of water and boyled it an hower & more, then I streyned it, and put therto asmuch cold water as that which was lefte, and beinge cold I did putt in my silck, & left it so for howres, then I kept a reasonable fier under it 2 howres, more, then I wrought it owte by hand, then I washed it owte and wrought it & dried it but the collor was not very faire, if you keepe it warme but one hower it gayneth the lesse qre if this weight will dishcharge if the silck were putt into London black. C I did also take throane silck, & to evry half pound therof, I took ¼ of gaalls, beatinge them & boylinge them one hower in a potel of water, then I did put as much cold water therto as that which remayned and so the liquor beinge blood warme I did spread my skynes therin, & in 3 howres I gayned 2 oz ½ uppon ye pound. qre if turninge this silck, so advanced, in a weake lee of pott ashes will not make yt feele more gentle & safter under the hande.[64]

63   BL, SL 2247, ff. 8, 20, 30, 33.
64   BL, SL 2247, ff. 9–9v, 68, 11, 13, 16, 17, 17v, 18v, 20v, 23.

These linked experiments show clearly the way in which Plat worked in the laboratory to perfect his dyeing techniques (and his mastery of the technical terms used by dyers):

> 15. The second of August 98 I died a Tawney, the galled silck beinge first steped in a stronge allome liquor of a l. of allom to evry gallon of water, for 2 howres the liquor beinge kepte good & whot but not scaldinge whott all that tyme, I maistered it in exceeding strong maister of pott ashes after it was brasilled & blackt sufficiently & yet I lost not in 2 l. of silck a dramme of my advancement which was 2 oz and ¾ uppon the pound. … peradventure allom in a whott liquor advaunceth as well as gaalls, peradventure it only bindeth down the advancement so strongly that it can not dischange in the dyinge.
>
> 16 I did not like the tawney suprs no.15. and therfore I layd it 7 or 8 howres in a stronge allomme water, & then I died it againe as before in fresh brassill liquor, and I gayned half an oz & better uppon evry pound of silck, over & besides my first advauncement. & I made a reasonable good tawney .3. August.98. heere the allom wthout all question advanceth & yt greatly for that the tawney as appereth ante no 15. was first doble rensed after a doble sopinge wherby at ye least ¼ of an oz is discharged of the first advancement, and then the silck yet gaining ½ more, then at the first, argueth an advancement of ¾ of an oz uppon the pound.
>
> 17 and yet layinge white silcke discharged but unadvanced 7 howres in such a stronge cold allom liquid as is mentioned ante no.15 and after rensinge owte the silck in cold water, I found that I gained not a greine uppon the oz. but the allom liquor beinge kepte whott, vz never a scalding heate for 2 howrs spare I found that I gained ⅔ of an oz upon the pound of white silck. qre if the silck had ben kept 3.4. or 5. howres in such whott liquor, but then it is to bee feared leaste the glosse of the silck shold bee impayred./ nihil valet.
>
> 18 In a skillet I didd make an advancing liquor wth 4 oz galls to a potle of water ½ oz of allom, I kept my silck therin 4 howres scalding whott, & I gained after 4 oz uppon 18 oz of silck, the silck gott a yellowish ground but the glosse therof did not greatly please mee I suppose ¼ of an oz of allom hadd ben a better proportion. this discharged very much in the

SCIENTIFIC THOUGHT AND TECHNIQUE

dyinge and the gllosse utterly decayed.

19 Qre what collors my deepe advaunced silk beinge after strongly allommed will beare either wthout the fatt or with the fatt. I suppose both good yellows & greens, I have found most excellent yellows this way of silck advanced but to 3 oz uppon ye pound.[65]

So, being dissatisfied with the tawny colour he first achieved, Plat varied the experiment and tried again the following day, making observations from the results about the role of alum-water in 'advancing' his silk which led to further experiments with that liquid. Finally, he considered the results of his labours and suggested further investigations he might make. Here Plat is using the techniques and equipment of London dyers to reach his own conclusions by trial and error. Was he trying to push forward the techniques of dyeing rather than conduct experiments prompted by his inability to learn first-hand the innermost secrets of the artisan dyers? For Plat was an inventor, an innovator, an ideas-man, and the mere recording of the techniques of tradesmen was not enough. He is keen to improve on the methods he records. Ever present is the thought that he might make money from any improvements but he also had a mind that constantly ran ahead of the information he obtained, suggesting alternatives and modifications to the processes, methods, and ingredients he described.

We have seen that his notes of conversations with artisans and gentlemen include many supplementary questions about the information obtained but many more of his 'qre' are suggestions for improvements or variations on existing techniques. On many occasions he simply throws out a thought in the course of writing notes – 'Qre if the blackberie wyne will bee of any strength', he wonders whilst writing an 'artificial wine' recipe. Thinking more deeply on the same topic he asks, 'qre iff all sorts of Gascoigne wyne & other wynes were boyled away to a certain proportion, when they are in theyre muste, iff so, they wold not prove a farr mightier & more laisting wyne'. After learning that radish oil will clean cloth he wonders if he can improve the method of extraction, if the oil may be

---

65   BL, SL 2247, ff. 13v–14.

obtained by decoction rather than expression, and can a similar oil be found in other plants, such as poplar leaves and buds, or beach mast?

Sometimes he merely 'tweaks' a recipe with his own ideas

> You may make an excellent claret or white gascoigne wyne in this manner vz: first wash yr malagey fruit (qre yf rotten will not serve in a greater proportion, & perhaps reasons of the son will prove most excellent) in a chaunge of water devidinge the stalcks and breaking the clusters, infuse 3 l. of this fruit so washt in each gallon of conduit or pumpe water (qre of thames water) let them infuse until you see the liquor to worke and hizze, and that the uppermost parte of the fruit which is risen have lost all his sweetnes abowte one ynch in thicknes.

With some confidence (he considers himself expert both in wine-making and beer production), he introduces 'How to make a most excellent Beere or ale in a new manner' by asking :

> qre of stoving of a good stronge woorte accordinge to my manner of reason wyne. qre also if it bee not the better way to putt the worte to stove being warme, and also the stove beinge warme when it is put in. qre of stoving of malt hopes and water together ... if this latter course take effect, wee shall neede neither milles, copper kettles, nor yeast & yet the drinke will last much longer and drinke more cleane and full of spirite then our ordinary beere or ale doo.

On many occasions Plat's 'qre' crowd round his original notes like buzzing bees and frequently he writes a whole speculative paragraph: 'A ghesse at a better making and fining of ypocras then is usually knowne', or 'A Conceipt for the Sophisticatinge off Rennish wyne good cheape.' Simple digressions or notes for possible variations in method shade into proposals for future experiments. [66]

Plat could take an idea and elaborate it with many queries and digressions until the resulting notes read as a programme for a series of

---

66 BL, SL 2216, ff. 44, 113v, 155v; SL 2210, ff. 47v, 120v.

SCIENTIFIC THOUGHT AND TECHNIQUE

experiments. An instance of this is 'A conceipt for making of verdegrece' dating from October 1595.[67] Plat was interested in verdigris as a pigment to make the blue-green colour verditer. Almost all sentences in his 'conceipt' begin with 'qre' beginning with his basic idea: 'Qre of dipping plates of latten [a brass-like metal] or copper in vineger, and then hanging them only in the aer of the vineger .v. in a vessell not half ful and the plates not touching the vineger, and the vessell also placed in a seller or somme dampish place.' Following this exposition is a mass of supplementary queries, extending to the side margins, modifying the composition and form of the metal used, the ingredients of the liquor and the method of operation:

> qre of making a lay of grape huskes, and a lay of copper plates, and so burying them either in firrs [furze?] or earth in a dampish place in a pott, or in the bare earth/ qre of sprinkling the plates continually wth vineger as they drie....qre of dissolving salt in the vineger / qre if the vineger were exceeding sharp, and acrated [sharpened?] wth calcined salt & so the peeces, shreds or portions of copper or latten were layd therein/... qre if latten powder may not bee hadd good cheape of those that make instruments or vessells of Latten./ qre if this must not bee don in glasse. [in the margins: 'qre of allom put into the vineger insteede of salt/ qre of copper dissolved in the vineger./ qre what proportion of all these salts is best/ qre of salarmon. [Sal ammoniac] / qre of tartar/qre of all or somme of these put together.]
> qre if [?] or small shredds of copper will make verdegres, for these may bee had for vi d. a pound/. qre if urine or strong brine will not serve insteede of viniger and sal armoneack.

Surely Plat is here giving his thoughts on solving a problem – the good and cheap production of verdigris – by setting out a programme of experiments to find the best method and combination of substances?

On the folio following this entry is another paragraph on verdigris-making, advocating filings of latten as the base material because 'the stirrup

67   Verdigris, 'A green or greenish blue substance obtained artificially by the action of dilute acetic acid on thin plates of copper (or a green rust naturally forming on copper and brass), and much used as a pigment, in dyeing, the arts, and medicine; basic acetate of copper', *OED*–online edition.

makers make store therof'. This method, a combination of the queries listed above, is – although itself surrounded with further queries – a confident exposition of a method, beginning 'Take the filings of latton', instructing that the soaking liquor should be drained after half an hour, the powder spread on copper plates, verdigris gathered as soon as it forms and washed 'in a little pannacle wth fayre water'. Either Plat has himself conducted the experiments he set out and arrived at this method, or he has taken his queries to an artisan or gentleman with knowledge of verdigris production and recorded the result.[68]

In contrast to the mass of queries, Plat's outline of an improved method of making saltpeter is calm and confident. Nevertheless it reads like a plan for a full-scale experiment using his new techniques:

> A proffitable Conceipt in peterwork.
> C first I wold have a large barn wherin I wold eather lay up an artificiall heape of earth to engender peter, or els I wold bring from all places neere my barn sufficient store of rich earth to work uppon a long time all this I wold do before I set up any furnes. C I do thinke that thordure of man mixed wth earth, but specially wth such earth as is fit to yeald peter, or wth such earth as hath once been wrought uppon, will in time engender a most rich earth for this purpose, so of sheepes donge, wch may be had from sheepe folded in great quantity, gotes doung harts doung old lime slaked or unslaked may be placed amongst som of it to calcine the dounge the sooner. Then having store of rich earth, I wold erect a furnace in bignes answerable to the quantity of my earth then wold I infuse my liquor uppon my earth, and devide the first third part by it self wch will be both of great strength and deepe in collor, this liquor only wold I boile when I have sufficient therof, then I take the last two third parts wch will be clearer and weaker then the first, and infuse that uppon fresh earth, and of this liquor I will draw only so much as is equall in tast, strength and collor wth the first, the rest, either I will put of it self uppon more fresh earth, if there be sufficient for one or more water wth it, & then devide as before. but becawse I will loose none of the strength of my liquors remayning in the earth I powre fresh liquor uppon that earth wch hath had any stronge

68 BL, SL 2216, ff. 109–109v.

liquor in it before, and then I powre that liquor uppon fresh mold, and so gather all my liquors that I will boyle of one strength and coller as neere as I can ghess. and thus in a ketel of 40 gallons I will make as much peter as they do comonly in a vessell of 60. gall wherby ⅔ of the fire is saved and ⅔ of the labor. & cariage of liquors only there is somme more laboring in often infusions in the beginning, wch is doble requighted in the ende. qre if liquors remayning uppon the earth 48. howres will not be stronger then at 24 howres.[69]

This experiment may have been successfully carried out, being the 'new course in the making of Salt Peter farre exceedinge the common practize at this day used as well for expedition as proffite unto those which shall execute the same' for which he attempted to gain a privilege and then sold to the monopolists of saltpetre, complaining in 1603 that 'bringeth in yearly & freely many 100 pounds to the Patentees (my self not having received one half yeeres profit for the invention)'.

There is strong evidence that Plat, as well as proposing experiments to improve technology, put some of his proposals to the test. We have seen as much when examining his notes on dyeing and there is no doubt that for many years he set aside part of his house for practical scientific and technological experiments. Although some Continental rulers were creating the first spaces set aside specifically for scientific research nothing like that existed in English palaces or universities in Plat's time. Alix Cooper has concluded that much innovative research took place 'within the seemingly humble and prosaic spaces of natural inquirers' own homes and households', and into the seventeenth century, 'by far the most significant venues were the private residences of gentlemen or, at any rate, sites where places of scientific work were coextensive with places of residence, whether owned or rented.'[70]

By 1593 Plat had assembled a number of inventions in his London house for 'a publicke viewe ... by some of her maiesties privie Counsell, and diverse other Gentlemen and Citizens of good worship and account'.

69 BL, SL 2245, ff. 74v–75.
70 Pamela H. Smith, Ch. 13, 'Laboratories'; Alix Cooper, Ch. 9, 'Homes and Households' in *Cambridge History of Science*, vol. III, Cambridge, 2006; Steven Shapin, 'House of experiment in seventeenth-century England', *Isis*, vol. 79, 1988, p. 378.

This was probably the visit of 'the Lord high Treasurer of England, my good Lord and master, who vouchsafed me his honourable presence at my house, when both my fire, tub, portable pumpe, and boulting hutch were made readie for his coming'. In 1603 he also mentioned a string of military inventions, some of which he had 'already shewed to divers of his honourable and private friends'. The exhibition in 1593 probably took place over a number of days, involving visits from both the nobility and gentry, city aldermen and London bakers (who expressed scepticism about his new bolting hutch). We may assume that these machines were made to his design under his supervision in his own workshop. Plat devotes much of his early pamphlet of 1593 to this exhibition, and refers to it again in 1594 and 1603. Shapin, in his article 'The house of experiment', comments on the private and secluded nature of experiments carried out in a gentleman's home. By dwelling on his exhibition of inventions in his own home to men of quality, Plat may have been trying to underline the importance of the occasion and to emphasize the veracity of what he exhibited by associating the inventions with the high-born names he quotes.[71]

We know he employed workmen in his workshop and laboratory: in 1608 he described how he had 'caused my workman (...) to make a cock of gold...' to his new, hollow, design. His collaboration with the alchemist and medical practitioner Clark appears to have been a master and servant relationship, with Clark possibly involved in making cinnamon water and other waters in bulk at Plat's workshop. Pamphlets and the advertisements appended to his books abound with offers of machines, medicines, foods, and other things – like the boot-blacking liquid and 'oily composition defending iron workes' of 1595, or the forty 'Waters, Oyles, Electuaries, & other Compositions necessary for the Sea' offered in a manuscript flyer. Some of these goods may have come from apothecaries and other tradesmen, being merely packaged and offered for sale, but surely others were created at Plat's house? He may have supervised workmen in producing the goods but he was responsible for the development and initial experimentation.[72]

71 Plat, *A Briefe Apologie of Certaine New inventions*, 1593; Plat, *Jewell House*, pt. 4. 'An offer of certaine new inventions', p. 72; Plat, *A new, cheape and delicate Fire of Cole-balles*, 1603, D3; Shapin, 'House of experiment', pp. 373–404.
72 BL, SL 2172, f. 30.

We have noticed evidence of his practical work in alchemy, his interest in how apparatus was set up and how to maintain a slow fire, as well as his claims to have perfected alchemical and distillation operations. In his writings on moulding and casting and dyeing there is ample evidence he was carrying out his own experiments. At a more mundane level, he used his wife's kitchen for many essays in cooking and preserving, made pasta with his machine to supply naval expeditions, and wines on a large scale. And his garden at Bishop's Hall served as a horticultural trial ground.[73]

There is little indication how large his laboratory was but he needed space to house several vats, together with hearths to heat the vessels for his dyeing experiments. At the end of one such episode he lists equipment he must acquire, 'I wante a back, a greate furnes to discharge in and to boyle my gaals and a tubb to breake my gaalls in.' Alchemical experiments often required long, slow heat sources, together with many glass, metal, and clay vessels. Others, like the operation to whiten blue sapphires (which Plat may well have tried himself), required a number of crucibles and 'a wide furnace, wth great charcole under & above' with which to 'continew a greate fire for 6 howers'.[74]

Plat is at pains to inform us that he spent much time and effort in experimentation. In 1603 he poetically claimed:

> From painful practice, from experience
> I sound, though costly, mysteries derive
> With fiery flames, in scorching Vulcan's Forge
> To teach and fine each Secret, I do strive.[75]

At the end of his life he looked back to the time when Dr Mosanus senior first taught him to distil and wrote that he had since discovered many secrets 'with great chardge in long time, not without great danger of my life and health'. Indeed, as he was conscious some months before his death that he was ailing, he might have succumbed to an illness brought on by many years in a fume-filled laboratory, such as mercury poisoning.[76]

73 BL, SL 2245, ff. 36–36v; SL 2216, f. 114. And see Chapters 3, 5, 6.
74 BF, SL 2203, f. 89v; SL 2245, f. 87v.
75 Plat, *Delightes for Ladies*, 1628, A3.
76 BL, SL 2172, f. 18.

The many and varied notes of experiments bear out Plat's contention that he spent a good deal of time in practical work. Instances include, 'A most trew rescete for making of Hardwax which I found owte in .July.1595' and the ink recipes he tried out by writing the notes on them with the ink in question. One of these came from Arthur Gregorie and another was founded on a by-product of silk dyeing: 'Take of the liquor prepared by selckdiers for Spanish blacke & put therto so much copperes, gals, and gomme all cold as will make it upp to an excellent black ynck which I have don wth very small charge and these lynes are writt therwth. filter yt and so use it'.

Further experiments included making dry litmus produce a substance 'in collor farr beyond all the ordinary drie Litmus wch the grocers doo usually make'. Auditor Hill tells Plat how to make a 'Sette and delicate mowth glew' but by experiment 'I find the best way of all somewhat differing from that which I learned of Mr Hill'. Both Plat and Hill tried to make clothing water-proof to wear in the rain. Hill wore a rainproof scarf out riding but Plat thought his method of rain-proofing better, 'I doo finde by myne owne experience that if you mix 8 spoonfulls of oyle of turpentine with 4 spoonfulls of oyle of spike, that the same is a good proportion... [and] I thinke this to exceede the scarf of Mr Auditor Hills'. Plat tested his rain-proofing by pouring 'a sponfull of cleane water' on his prepared cloth which stood 'for ¼ of an hower before it dropp through' He repeated the experiment with Hill's scarf which held the water for only two minutes.[77]

A stray note at the end of a manuscript about the use of seed-steeps reveals a closely controlled agricultural experiment:

> 1603
> 15 of november at night, I sowed wheat at the north end of a peece of ground that had ben stiped in a liquor wch had run through 3 earths before (wch earth had lien fallow & had not ben dunged in many yeeres before) for 48 howres, and after lay 12 howres to drie. the 16 of november in the morning I sowed wheat that had ben steped 3 dayes & three nights

---

77 BL, SL 2189, f. 102; SL 2116, ff. 20v, 49, 96v, 128v et seq.

in an infusion uppon soote: this I sowed nigh to the former wheat and next to this I sowed more wheat prepared as in the first manner, save only the wheat lay.3. dayes & 3 nights in infusion./ as this was turned under furrow after it was sowed in Taylors ground the ground wherin I sowed was utterly spent and barren./ but the rest of the field was dunged.[78]

An ambitious programme of experiments headed 'Plattes Desires in probable Secretes' lists 21 projects, the major part of which are aimed at improving industrial processes:

1 C To make artificiall pillers and baalls of marble.
2 C To refine Camphire and barrace.
3 C to refine Aloes & to make it transparent.
4 C To draw owte the strength of the Elements of ye Earth in march & Aprill, by distillation, by digestion, by infusion of water & evaporation, woorcke uppon naturall Earth, that hath not ben enriched. prove also uppon earth proceeding of mens bodies in their graves.
5 C To make an oyle or spirit of Suger.
6 C To make viniger speedely. & in great quantity.
7 C To devide spirit of wine from his phlegme wthout distillation.
8 C To make peeter shoote wthout ashes.
9 C To multiplie the vertew of hoppes.
10 C a lasting lampe to make an equall fier withall.
11 C To make yron barrs to last longer in seacole furnesses.
12 C To make a more excellent gonpowder then is yet knowne, either for lasting or use.
13 C to make a still powder.
14 C A more speedy way for making of gonpowder.
15 C To make a speedy distillation by reason of a large headd.
16 C To bring great store of bread, wine, beere, ale &c into a litle roome.
17 C To make use of the waste flamme, in peterworcks & brewhowses.
18 C Trie the fixation of Logwood & brasill by somme stronge & proper salt fermenting them longe together. brasill & urine agree well. last longe in a cold infusion./

[78] BL, SL 2245, f. 114.

19 C Aqua fortis made easely & cheape
20 C oyles aromaticall made in doble quantity
21 C Tabacco well strengthened were a rich secret.[79]

Some of this plan we will encounter when discussing his inventions and his attempts to make money. For the rest, although not taken forward, they underline his interest in the practicalities of industry and technology. The list is another instance of the methodical way Plat went about his research. He was not a dabbling 'virtuoso'. The planning, identification of worthwhile projects, careful questioning of informants, practical experimentation, questioning of results and queries about further experiments add up to at least the bare bones of a scientific method. Unlike Bacon, Plat was not sufficiently interested in the theory of natural scientific enquiry to stand back from what he was doing and produce a treatise on his methods. Later historians, myself included, have done this for him.

---

[79] BL, SL 2189, pp. 24–25.

CHAPTER NINE

Inventions, Technology and Practical Applications of Technology and Natural Science

Given Plat's interest in technological improvements and novelty what were the fruits of his labours – his inventions – and how did he try to exploit them? And what, in Plat's terms, was an invention? No doubt Plat would have regarded all inventions as 'secrets', knowledge known to few, or only to the inventor. But 'secrets', as used by Plat, cover a wide range of subjects and ideas, from major 'philosophical' secrets like the attainment of the philosophers' stone, through technological advances such as the better making of saltpetre, to the 'household hints' and party tricks he culled from secrets books, many of which he set out in the first section of the *Jewell House*. So Plat's many lists of secrets are too broadly drawn to define his inventions. But if we restrict inventions to new technology invented by Plat, our definition is too narrow. While some things – oil to preserve armour or a new shoe-blacking – appear entirely or substantially his own, the majority of the novelties which Plat wrote about and considered important were improvements on existing ideas, often originating abroad, like the pasta machine, coalballs, or sandbags to fill breaches in sea walls and fortifications. Some of Plat's many lists of secrets are of schemes from which he hoped to make money ('All the necessary and helpinge means that I can imagine for the purfect gaininge of money either by artificiall means, services or otherwise'). Select lists of this type appear several times in print in pamphlets and appendices as offers to the public of secrets he will sell or disclose in return for some remuneration. Such ideas were at the time generally called 'projects', defined by Joan Thirsk as 'a practical scheme for exploiting material things; it was capable of being realized through industry and ingenuity'. This description fits many of his ideas, especially those he sought actively to exploit by advertising for buyers or patronage. It is a good working definition of his inventions.[1]

Looking over various lists and publications, the inventions Plat was most anxious to promote fall broadly into four categories. There are those pertaining to military matters such as guns, gunpowder, bullets, armour and its maintenance, and preserved food and drink for campaigns. There are devices and processes relating to industry like a wooden vessel to boil in, a bolting hutch, mortar from soap-ashes, the use of a range of industrial wastes, improvements in wine, spirits, oils and sweet waters, saltpetre and starch-making. Then there are the schemes that have broadly domestic application such as a cheap candle, coalballs, a portable pump, shoe-blacking, rain-proof garments, or papier mâché mouldings to adorn houses. And finally there are his ideas that have an agricultural use, most notably an artificial compost. Of course, he also promoted a wide range of farming and gardening practices, and recipes for domestic food preservation, perfumes and cosmetics, etc., but many of these were recipes and advice freely published, not ideas held back in hope of financial reward.[2]

[1] BL, SL 2172, f. 13; J. Thirsk, *Economic Policy and Projects*, Oxford, 1978, p. 1.
[2] There are many secrets lists in Plat's manuscripts and the summary here is not exhaustive. Some lists have no heading but several which are titled occur at the start of the early manuscript BL, SL 2210, including: 'The principall Secretes off this book', with 32 items, including some familiar items from other lists or advertisements: '25 To boyle in wood'; '26 The secret of wynes' etc. This list is clearly of items with moneymaking potential. It is followed by another headed: 'The Secretes wch I meane firste off all to putt in practise are these'. As the title implies these are practical operations, again including 'to boyle in wood' as well as other profitable secrets like 'to grave a seale wthout a toole' and 'Beere wthout hoppes'. Yet another list follows: 'The choice & moste principall secretes off all my longe Travell & experience'. A long list with 175 items; some are potentially profitable innovations such as: 'How to kepe armor & iron woorke cleane'; ' cheape Candles good for poore men', but also included are many minor household secrets culled from secrets books: 'How to grave any fancy or letters uppon an egge'; 'How to keepe Gooseberries for sawce all the yeere'; 'How to make any dogge or Mastiffe

It would be tedious to plough through all Plat's inventions but many of his ingenious proposals deserve examination. In assessing his work as a promoter of inventions I have outlined some of those which he advertised most assiduously in print or spent some time writing about in his notes.[3]

In a sizeable body of manuscripts (and a few publications) by Plat there are inventions of use to the army and navy. That he had an interest in military matters is not surprising. Although he was a Londoner with, as far as we know, no military training, as a gentleman he knew the military obligations of his class. He was aware of the military activity in which his countrymen were engaged for much of his adult life: countering the threat of Spanish invasion; long-term campaigns abroad in the Low Countries and Ireland; and the naval adventures against Spain, Spanish bases and shipping. As a courtier he rubbed shoulders with many military men: professional soldiers such as the Continental adventurer and renowned fighter against the Turks, Thomas Arundell, and Captain Josias Bodley, soldier and military engineer, brother of the founder of Oxford University's library, who campaigned in the Low Countries in the 1580s and later spent

---

too follow one, as to still any Curr in the night from barkinge'. Another long list in SL 2245, headed 'A Monopolie of Profitable Observations' includes many of the topics on which Plat had spent some time researching as well as much more speculative ideas, such as 'To make use of petrifying waters, by putting in carved or square peeces of wood, mosses leaves of trees &c to grave any worke withall' or 'To make use of the Clay at Cheme which serveth to make stone vessells to kepe the glas mettall in their furnaces now that those stone potts stande a long time in a flaming fire wthout either melting or cracking'. A list in SL 2216 is clearly a 'to do' list of potentially profitable secrets – 'A Note of Such Secretes as are woorth the Search and not as yet founde owte in any perfection by the Author'. A list of 66 items at the start of SL 2247 is, as are many of the others cited, not followed by any details of the topics listed, it is just inserted into the document by Plat, and the frequency of these lists implies that Plat was preoccupied with potentially profitable secrets, reassuring himself with lists that these projects were viable. BL, SL 2210, ff. 5–7; SL 2245, f. 106; SL 2216, f. 177; SL 2247, ff. 2–3v.

3   Some of the things he advertised as inventions, such as pasta and other preserved foodstuffs and drink for the army and seafarers, his ideas on wine-making, and his secret formula for fertilizer, are discussed in other chapters and will not be re-examined here.

many years as a soldier in Ireland. His Spanish informant from Naples, Sr Romero, was knowledgeable on military matters. Was he related to the Spanish soldier Julian Romero, whose career until his death in 1577 included service for Henry VIII against the French and the Scots as well as in the Low Countries on the side of Spain? Plat claimed acquaintance with the naval heroes Drake and Raleigh, and Captain Plat, a probable relative, sailed with Drake. In seeking patronage for the *Jewell House* from the Earl of Essex, Plat identified that hapless soldier with Achilles.[4]

Plat lived at a time when guns and gunpowder were playing an greater role in warfare on land and at sea. Such developments provoked improvements in personal protection and fortification against the new weaponry, as well as changes in battlefield tactics. Armies became larger and this, as well as tactical improvements, led to greater emphasis on discipline and organization and threw up the necessity of efficient supply. As Henry J. Webb has shown, an impressive array of books and pamphlets on warfare were produced in England at this time, covering all aspects of military life. Plat's writings reflect primarily his own interests: nothing is said about strategy and very little on tactics, he ignores military discipline and organization. In short, he is not interested in the theory of war. Instead, he concentrates on technology – on invention – on problems facing contemporary armies and navies which coincided with his own research (and perhaps problems described to him by military men).

Frequently, an advance in performance threw up its own problems. Better gunpowder (and more of it, via improved saltpetre production) allowed soldiers to shoot further, with more power and accuracy. This meant in its turn that body-armour required enhancement. More effective use of infantry in the field called for technologies to counteract the new threat – fragmentation bullets, chain shot, poisoned smoke and the like. While it was thought desirable to make better use of cannon in battle – which led to improvements in gun-carriages and temporary fortifications – this also meant that trenches for the defence of the infantry needed to be better designed. More strongly built ships could carry more armaments,

4 See *ODNB* for Thomas Arundell and Captain Josias Bodley; see Chapter 4 for Plat's seafaring acquaintances; M.A.S. Hume, *The Year after the Armada*, 1896, pp. 75–121; Plat, *Jewell House*, A2–3.

but attackers required more penetrating shot to destroy them. As we have already discussed, longer sea voyages and larger armies kept longer in the field required food and drink which could both be kept palatable for a more extended period and could be transported with more ease. These large armies required such innovations as collapsible boats and temporary bridges to move quickly, especially over the watery terrain of the Low Countries. Finally, what might seem a minor matter, keeping plate-armour free from rust, was a subject on which Plat spent a good deal of time. This may have simply intrigued him, and have been a preoccupation of the military gentlemen he talked to, or more prosaically, it may have been something from which he hoped to turn a tidy profit.[5]

In 1603 Plat summarized what he then considered to be some of his most useful military inventions:

> what say you then to such a carriage for a cannon, wherby the peece with the helpe of two men only may be turned, mastered, and charged at pleasure in as good sort as ten men are able to doe at sea with their usual carriages? What thinke you of a portable boate, which one man may carie with ease, and yet will hold eight persons? And of a light, strong & sodaine bridge to be made by uniting these boates, and therby sodainly to convey each of a whole armie over a large river? What if an invention bee shewed how a serviceable vessel may chace with ten or twelve great peeces of ordinance as readily, and as aptly, as now any ship doth with two or with foure peeces only? Nay, what if such a Pinnesse were warranted to be made, as should upon her owne motion, without the helpe of any mariner to direct her, make a speedy way against all wind and weather upon the seas for one halfe mile at the least, and being laden with all kind of shot and fire-worke, upon the first touch of any other vessel, shal presently give fire to the traine, and so spend her selfe, and endanger such ships as are then next unto her? And what would you say to a peece of ordinance which one man may sufficiently manage, and yet twentie of them shal make five hundred Muskettiers to abandon the field? But to conclude these warlike inventions with a shot of the highest execution both for land and

[5] Bert S. Hall, *Weapons and Warfare in Renaissance Europe*, Baltimore, MD, 1997, pp. 201–234; Henry J. Webb, *Elizabethan Military Science*, Madison, WI, 1965.

sea. What if a bullet bee delivered that shal breake into a thousand parts, each part carrying both his fire, powder & shot with it, so as no garrison under the wals of any warlike towne or citie, no band of souldiers lying in the safest trenches they can devise to make, can possibly be free from the furie of this bullet: which because it may be shot compasse at anie reasonable distance, must needes force them to forsake their ground. Some of these new inventions the Author hath already shewed to divers of his honourable and private friends…

Plat left no detailed notes about many on the list, such as the self-propelled pinnace, others we will consider.[6] Although not included here, saltpetre, the major component of gunpowder, was a commodity that interested Plat greatly. A good supply was vital for warfare and, given Plat's enthusiasm for military inventions, it is no surprise to find him engaged in ways to improve its production. With the eclipse of bows by handguns, the use of cannon, mortars and mines in siege warfare, and the importance of seaborne guns, there was a constant need from the government for gunpowder. Large quantities were purchased throughout Elizabeth's reign. Despite growing demand and general price inflation, the wholesale price of gunpowder fell towards the end of the sixteenth century, leading A. R. Williams to suggest that saltpetre was being produced more efficiently. It is not possible to say if any contribution to this was thanks to Plat's ideas.[7]

Saltpetre is potassium nitrate (KNO_3) and occurs naturally through the action of microbes on decaying living tissue. 'Virtually any body of decaying organic matter will produce some nitrates, but not necessarily in useful quantities. Creating an environment in which conditions were most favourable was the task of the early saltpeterers.'[8] There were two stages to the making of saltpetre in Plat's time: the creation of potassium nitrate in earth and waste matter by microbial action, succeeded by its refinement from that earth. The first stage could either be left to nature in such places as dungheaps, old latrines, dovecots and tombs, or created

6   Plat, *A new, cheape and delicate Fire of Cole-balles*, 1603.
7   A.R. Williams, 'The production of Saltpetre in the Middle Ages', *Ambix*, vol. 22, pt. 2, 1975, pp. 125.
8   Hall, *Weapons and Warfare in Renaissance Europe*, p. 74.

INVENTIONS

> 101. *A Wagon to be drawn with Men, instead of Horses.*
>
> THe joints and other parts of this Wagon are so knit together with hooks and pins, as that it may easily be dis-jointed and taken in sunder, whereby many of them may be couched in a narrow room, & wil lie close together in a ship. It is to be drawn with six men, whereof two of them must labor at the fore-cariage thereof, and at either wheel other two, which must work by winding of the handles, (which are of purpose fastened both to the nave of the wheel and axletree) either forward or backward as occasion serveth. The use thereof is to convey their victuals and other necessaries from place to place, when the Mariners and souldiers have cause to land in some countries where the place affords no horse or other beasts that are fit for labour or carriage. I know not the author of this invention, but because it came so happily to my hands, and carries some good

Figure 7 (above). A wagon which can be taken apart and assembled quickly which Plat invented for military use, from The Jewell House.

Figure 8 (below). The portable pump Plat claimed to have invented and obtained a 'privilege' for.

> 146. *A Portable Pump.*

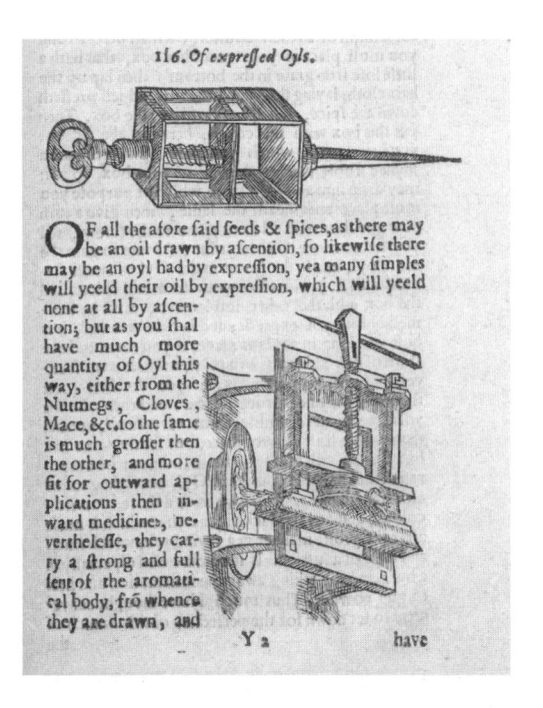

Figure 9 (above). Devices for expressing oils depicted in The Jewell House of Art and Nature.

Figure 10 (below). A copper press for moulding small objects as described in The Jewell House.

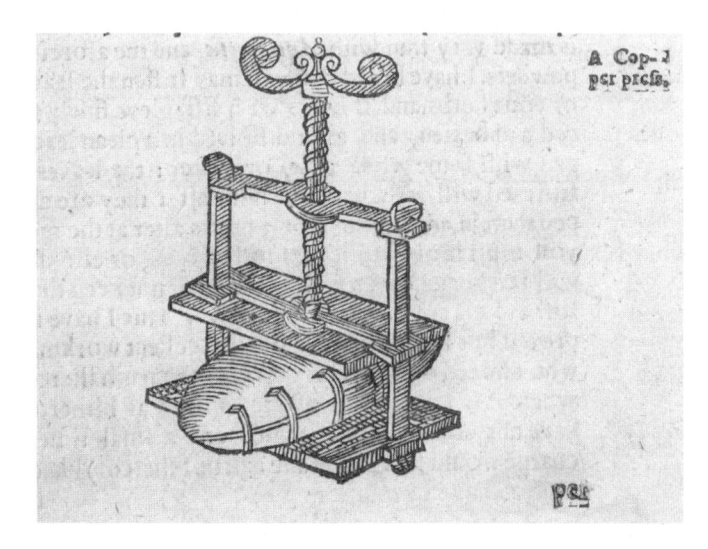

by 'nitre-beds' where dung, calcium-rich lyes, and urine were kept under cover to mature. In the second stage, when the earth was refined, it was placed in layers in tubs with lime and ashes, and water was poured through the mixture to dissolve the saltpetre. After settling, the liquid was drawn off the solid matter and evaporated to leave a useful residue.[9]

Plat issued three advertisements offering improvements to this process (this was the most heavily advertised of his military inventions). The first, in 1594, was for engendering the raw material, 'a multiplying earth, which would yeeld sufficient store of Peter, for the service of this Realme without committing such offenses, as are dayly offered, in the breaking up of stables, barnes, sellers, &c'. The offer in 1595 was to save 'the full halfe or moity of all such fewell, as is dayly, and grosselie spent in all the peterworkes of this Realme, …..and is now most vainly, & absurdly wasted & consumed'. In 1603 he enigmatically referred to 'my new & late discovery in Peter-works' which had led to a considerable saving in manufacturing costs and transport savings. The latter two inventions were probably both improvements in refining.[10]

Plat talked to at least one 'Peterman' (Wilkinson) and from him he took down a detailed description of refining saltpetre and a sketchy and confused account of a nitre-bed. Others, such as a dyer, a distiller and his Scottish acquaintance Nepper, provided piecemeal information on engendering saltpetre artificially. However, Plat never recorded a coherent description of the process to compare with the detailed recipe of the German expert Honrick, dated 1561, now lodged in the State Papers. Plat's inability to obtain a full recipe reflects the value of such a secret – Honrick's was to be sold for £300 to two English projectors planning to establish nitre-beds in England. It is not possible from Plat's notes to discern what his proposed improvement to nitre-bed technology was. In any case, he probably saw more opportunity in improving the refining of saltpetre because most of the English-produced raw material came not from nitre-beds (which had been in use for some time abroad) but from

[9] Stephen Bull, 'Pearls from the dungheap: English saltpetre production 1590–1640', *Ordnance Society Journal*, vol. 2, p. 5.
[10] Plat, *Jewell House*, 1593, pt. 3, p. 76; Plat, *A Discoverie of Certaine English wants*, 1595; Plat, *A new, cheape and delicate Fire of Cole-balles*, 1603.

naturally occurring, saltpetre-rich earth. Saltpetre production was of such strategic importance that it was a Crown monopoly. Commissions were issued to saltpetre-men authorizing collection of this earth in one or several counties. The patents gave the men, or their deputies, considerable power to dig in likely places – dunghills, dovecotes, cellars, stables, outhouses and the like, where decaying matter in cold, dry earth was likely to produce saltpetre. Their activities aroused fierce resentment when they arbitrarily dug up church floors, the interiors of houses and public buildings in pursuit of nitre-rich earth. The diligence with which the raw material was sought emphasizes the strategic importance of a good supply.[11]

Plat describes how he would improve the process of refining. Instead of boiling up all the liquor strained from the nitrous earth, he would,

> devide the first third part by it self wch will be both of great strength and deepe in collor, this liquor only wold I boile when I have sufficient therof, then I take the last two third parts wch will be clearer and weaker then the first, and infuse that uppon fresh earth, and of this liquor I will draw only so much as is equall in tast, strength and collor wth the first, the rest, either I will put of it self uppon more fresh earth, if there be sufficient for one or more water wth it, & then devide as before. but becawse I will loose none of the strength of my liquors remayning in the earth I powre fresh liquor uppon that earth wch hath had any stronge liquor in it before, and then I powre that liquor uppon fresh mold, and so gather all my liquors that I will boyle of one strength and coller as neere as I can ghess.

This method of reiterating, he thought, would mean that, 'in a ketel of 40 gallons I will make as much peter as they do comonly in a vessell of 60. gall wherby ⅔ of the fire is saved and ⅔ of the labor. & cariage of liquors only there is somme more laboring in often infusions in the beginning, wch is doble requighted in the ende'.[12]

This idea is likely to have been 'my new & late discovery in Peter-works' which in 1603 he claimed was 'the true foundation and groundwork

---

11 BL, SL 2189, ff. 89v, 109v; Williams, 'The production of Saltpetre in the Middle Ages', pp. 125–133; Bull, 'Pearls from the dungheap: English saltpetre production 1590–1640', pp. 7–9.
12 BL, SL 2245, f. 74v.

of the last letters patents granted for the same'. He was aggrieved that 'it bringeth in yearly & freely many 100 pounds to the Patentees' with 'my self not having received one half yeeres profit for the invention'. Presumably he had sold the idea, a genuine invention on his part, to the patentees but at a discount, because without a patent he could not exploit it himself. Cryptically, he concludes 'may be ere long I wil find sufficient reliefe in a strange maner'. He may here be alluding to his forthcoming knighthood.[13]

Plat has one suggestion for improving the process of making of gunpowder, suggesting a safe way of drying it (a dangerous process frequently resulting in explosions). His solution was to use hot water:

> you may make a vessell either of leade, latten or copper havinge 2 bottomes, betweene wch bottoms you may convey scalding water at a pipe, wch water may bee heated in an other Roome, for the more suretie of the fyre, and then you may lay your powder upon the uppermost bottom, and so yt will drie very well & wthout all daunger. & when the water beginneth to coole, you may let yt owte at a Cocke in the bottome of the vessell, & so lette in more scalding water into the vessell by an other cocke wch may bee fastened in the pipe that runneth into the vessell.[14]

He has a few thoughts for additives to improve the efficiency of gunpowder in guns, but Plat's interest in this explosive is principally as an ingredient in incendiary devices. The use of fire as a weapon on land and sea already had a long history. Greek fire, a thick combustible liquid which would burn on water, was used by the Greeks of Constantinople in the seventh century and by Plat's time many recipes were in circulation for this and similar combustible substances. Using mainly the secrets book of Isabella Cortese, *I Secreti* (Venice, 1574) and della Porta's *Natural Magick*, he collected together recipes for 'wild fire', described the wooden pipe the Greeks used to deliver the fire, and compiled a table of possible ingredients.[15]

13 Plat, *A new, cheape and delicate Fire of Cole-balles*, 1603. Plat was knighted in May 1605 and may have had hopes in 1603 that he was about to receive the honour.
14 BL, SL 2210, f. 59v.
15 BL, SL 2210, ff. 65, 108v; SL 2189, ff. 32, 130v; Isabella Cortese, *I Secreti...*, Venice, 1574. John Baptist della Porta, *Natural Magick*, 1568. Book 12 has many recipes for wild fire for land and sea use with which Plat was undoubtedly familiar. For instance, he copied the recipe for 'Fireballs flying in the air' into BL, SL 2197, f. 12.

Contemporary interest in incendiaries is indicated by recipes he collected from acquaintances, for instance this from Arthur Gregorie:

> Rx brimstone 3 partes, fine gonpowder 7 partes, salt fresh refined parts then take of turpentine, oleus benedictus, linseede oyle, oyle of Juniper, oleus petre, & aqua vite, and of all these put together ⅓ part, incorporate them all together. And iff you can not gett all theise oyles you may use only turpentine and linseede oyle making ⅓ of them two, but it is the better if you can gett them all, for that they are more pure & subtile to mix wth your .3. drie recite vz. powder, salt peter, & brimstone. for these liquid oyles in continuance do not harden nor alter and if you finde the receipt over moist you may wth drying it in a clos pipkin from the fier a prettie way or in an oven after the bread is drawne, amend it to your likinge./ yt is verie necessarie that all this compound be well mixed & incorporated togther, & that the resceit bee not more moist or fattie then the earth wch founders do use in their moldes, wch you shall finde by their clammines betweene your fingers and clodding together. this is Mr gregories resceit for fireworks, to cach and reteyne fier rounde abowte his great balls of execution to bee shott owte of greate peeces of ordinaunce, until it takes hold of the quick powder.[16]

Plat appreciated the value of these devices in naval warfare, offering recipes for fire-grenades to toss onto enemy ships and also several ideas for incendiary bullets for long-range use, both against shipping and to burn towns under siege. In addition, he included a self-propelled fire ship in his 1603 list of military inventions quoted above.[17]

Linked with incendiary devices were those whose principal object was to make smoke. These were battlefield weapons, aimed against the increasing numbers of infantry involved in warfare in Plat's time. A benign smoke to hide troop movements from the enemy was suggested:

> yf somme arte were founde how to raise a greate & soddain smoke, or rather diverse smokes betweene two armies, in everye smoke there might

[16] BL, SL 2189, f. 32. 'Rx' means 'mix'.
[17] BL, SL 2189, ff. 32, 131v; 2210, f. 109; 2197, ff. 12–13; Plat, *A new, cheape and delicate Fire of Cole-balles*, 1603, D2.

> bee placed a small nomber of shotte to discharge at the enemy, & on a soddayne to convey the greatest force of the army in an other smoke, wch shall not greatly bee galled wth shotte, becawse the enemy can not know where the greatest nomber are to ayme at.

However, most of Plat's schemes involving smoke were for poisoned smokes – hollow balls shot from cannon containing mixtures of gunpowder, arsenic, antimony, vitriol and the like.[18]

Warfare in Plat's time was marked by the increasing use of guns on the battlefield and at sea. But to judge from records of the campaigns in the Low Countries, Elizabethan armies made relatively little use of field guns. Artillery suffered from a number of problems in the field: guns which were inaccurate and with a potential to explode if mishandled, unwieldy carriages, lack of skill in aiming and loading, slow rates of fire, and vulnerability to attack by cavalry. Handguns, too, were having an effect on the battlefield – horsemen with pistols could inflict casualties on pikemen powerless to retaliate. Long-barrelled weapons such as muskets, calivers and harquebuses harried the enemy, sapping morale and inflicting casualties before the main bodies of infantry met in hand-to-hand combat. For both artillery and handguns, Plat notes recipes to improve the quality of gunpowder, and from military men such as Captain Bodley and Sr Romero he records tips on how to load handguns for maximum power, a French method of rifling a gun barrel to increase accuracy, an impractical suggestion for a curved gun barrel that will shoot round corners, and positioning a touchhole so as to minimize the kick of a recoil. This latter idea, 'If the tuchole be made so far backward that no powder may lye behind it, such a peece will not requoile' was advocated by Humphrey Barwick in a book on guns in 1594.[19]

Plat gives more than one recipe for hardening bullets to increase their penetration of armour or ships by adding other metals to the lead:

> Melt downe 2 oz of Copper put [lead] therto by little and little, least you coole the Copper vz. 14 oz [lead] to 2 oz [copper], caste the same into musket

18 BL, SL 2210, ff. 100, 166v; SL 2189, ff. 32, 130 –131v.
19 Webb, *Elizabethan Military Science*, pp. 78–107, 124–147; BL, SL 2210, ff. 136, 167v; SL 2189, ff. 32, 130.

or caliver bullets, and it will pierce much deeper in wood (qre of armors and targets of metal) then the ordinary bullets doth. This is serviceable against Shipps that are thought to bee musket free [i.e. musket-proof].[20]

To make field artillery more effective against large bodies of infantry or cavalry Plat had some fearsome ideas such as, 'A misceiffous bullet or cheyne shotte to destroy infinite nombers of men wth'. He provides full details of the manufacture of this fragmentation bullet:

> Make a Shotte of leade in this manner [circular diagram in text], first make a circle of leade of the iuste bignes of the Saker, Culveringe or Cannon &c. & of half an inche or more or lesse according to the greatnes of the shotte in thickness, devide the samme in partes as you see in this figure, & let everie peece therof bee severall & filed so trew as that when they are all put together they may make a iuste circle of the bignes of the Gunn wherin you meane to use them. make diverse of these circles of the like largenes and thicknes & devide them into the like partes. & so pile them in a large stronge or doble paper one uppon thother till you have the iuste weight of a Bullet for your peece wch they muste serve for, and then wrapp them in your paper and tie yt faste wth somme threede & so put it into your peece. you must first know the iuste weight of the bullett & then devide the samme into so many partes as all the circles put together may make a iuste weight wth the whole. when you pile the pieces, you must lay them as they use to lay bricks in a binding manner topped the Joyntes. also having once made one ball in this manner by hande, you may take one peece therof, and make a molde therof of copper, or copper and lattin mingled together …. and so in one molde you may caste 12 or 16. peeces at a tyme. a Canon shott devised in this manner will beare 200 peeces or more and every one of the doble weight of a muskett bullet. the midle circle of this bullet as you see in this figure wch conteyneth 1.2.3.4. may ether bee devided into partes as the uttermost circle is, or els you may convey any manner of wildfire in the midste of this bullett.[21]

This shot was of use on land and sea: 'yt serveth very well to keepe a

20 BL, SL 2189, f. 131.
21 BL, SL 2210, ff. 143v–144.

breach, to spoile men in galleys, to cutt downe shrowdes taking sayles, and to destroy men and all at one shotte. for the peeces therof being after a sort square, Do cutt and rende wonderfully where they light.' No less fearsome was his chain shot for use against horsemen, 'a leaden bullet devided in halfe and chained wth wier, so as the same may spread twice or longer as when it was first put in to the piece, is very serviceable against horsemen, because yt maymeth the horse whersoever it lighteth taking away a legg or a sholder from him'. These projectiles, together with several ideas for poisoned bullets, reflect Plat's interest in applying technology to warfare as well as a dispassionate attitude towards the subject not touched by gentlemanly chivalry.[22]

More effective guns of all calibres meant there was a need for any number of countermeasures. The guns themselves, in the shape of field artillery, needed their own protection for they were slow and unwieldy weapons, in danger of being overrun by horsemen. Gun emplacements were one use of sandbags suggested by Plat. He spent some time refining his ideas on the military use of sand – early manuscript notes on 'Winning of Breaches' involved filling holes in fortifications or river embankments with heavy elm chests filled with earth, sand or gravel. He later realized that bags were less cumbersome and fitted together better, suggesting what is essentially the modern sandbag, 'Heere wee shal neither have neede of nailes or timber, stones or mortar, but linen cloth & needles to make our strong defence against the furie of the cannon or the scourges of the sea, the whole art whereof consisteth in bags or sacks of linnen to be filled upon any present occasion either with sand or earth; and these suddenly layd or sunke upon any cause of service or irruption of waters.' For military use, 'A breach of 30 yards made by the Canon may be substantially filled upp in one hower'. The genesis of this secret came from Turkish military engineers, 'For so was the Golletto won by the Turke: a fort otherwise impregnable, whereby that honorable and glorious victorie of don Iohn de Austria obtained against the Turke by sea, was mightily eclipsed by a miserable overthrow at the same time given to the Christians by Land.' Plat states that had the commanders taken empty sacks on the expedition against Lisbon in 1598, the landing party 'might have found lining enough

22   BL, SL 2210, ff. 144–144v; SL 2189, f. 131.

to have raised a fort even upon the sands, and suddenly planted the cannon that should have commanded the towne itself'.[23]

Another ingenious military use of canvas and earth was to make an instant musket-proof wall with a large pocket of earth-filled fabric propped up by wooden stays. This was a suggestion of the Spaniard Sr Romero and may well also have come from Turkey:

> A trench or wall of muskett proofe yt is very portable and soddenly sett upp in the fielde to entrench a whole army. This is made by cordinge of 2 pieces of thinn wainskott together wth stayes in the topp and bottom, much like the bottom of a corded bedd. then applie sayle canvas to the cordes that hath 2 eylet holes at everie cros to bee tied wth tapes, this muste bee filled wth earth, and a stay or propp behinde in the midste of everie peece of trench being of 6 foote broade and 4 foote high. each peece of trench may bee hasped to the other wth hookes/ you may make the sides as broade as you liste.[24]

Plat's oft-repeated civilian application of the idea was to repair a breach in the riverbank at Erith on the Thames below London. He wrote,

> I doubt not that the great breach at Erith might bee won in a monthe ... And though al the dutch marsh-men have hitherto pusled themselves about the inning and winning of the forsaid breach, and have given it over as impossible to be won at the Thames mouth (because they find it in some parte to be nine or tenne foote in depth underneath the low water-marke, before they come to any firme ground) yet I doubt not but by the sinking of sacks of earth the workeman shal soone find or make a foundation sufficient to beare a strong marsh-wal, which may also consist of sacks of earth worke-manly placed, and after wel backed; which before the sacks be throughlie rotten wil closely couch and knit together, and likewise be so fronted and filled up with ooze, as that in a short time you shall have a firme and substantial marsh-wal against al wind and weather whatsoever.[25]

23 BL, SL 2210, f. 45; SL 2172, f. 5; Plat, *A new, cheape and delicate Fire of Cole-balles*, 1603, D2; The sea victory was the battle of Lepanto, 1571.
24 BL, SL 2210, f. 165v.
25 Plat, *A new, cheape and delicate Fire of Cole-balles*, 1603, D3; Henry VIII created a naval dockyard at Erith so there are military implications for this application too.

INVENTIONS

Soldiers needed better personal protection, armour and shields, against the fire-power brought to bear on them. The best armour in the sixteenth century was made from hardened steel and was 'proved' by subjecting a small part of it to assault by the weapon it sought to protect against, which by Plat's time was a bullet shot from a handgun. The effect of firearms was to make such armour of 'proof' (marked with the dent from the bullet) thicker and prohibitively heavy. The weight may have been the reason that infantry at the end of the sixteenth century wore less armour than at the beginning, having already discarded shields as too unwieldy.[26]

Plat's answer was armour made from lighter, non-metallic materials. Consulting Sr Romero, he was informed somewhat optimistically that 'a Dromme [drum], strongly braced, will hold owte a pistoll shotte' and this set him thinking about alternatives to metal for armour and 'targets' (shields). Following on from Romero, he wondered 'if somme stronge leather may not so be fastned uppon wooden targetts, that they may be blowne into the shape of targetts & whether the samme having quilles in the topp, may not be blowne or lefte lancke at ones pleasure.' Or 'qre of a targett made of thicke past boords, wth 2 or 3. tynned plates inserted amongst them/ qre of a targett pricked full of wyres like a hachell, of 1. 2 or 3. inches longe & sharpe at thends./ qre of wyres layd clos one by one uppon pastbords, v.2 or 3. layres of wyres first neled uppon 2 or 3 layers of pastboords.' He thought of leather armour: 'Take leather that is half tanned, owt of the Tanners fatte, streyne it hard uppon targett or Curat moldes of wood, let it drie and clap on an other wth some apte & holding substaunce between them. this will bee of good weapon proff: qre if horshides, bulhides or old boarhides bee best for this purpose.'[27]

He then thought of papier mâché and he spent some time elaborating on this idea. His recipe for this 'excellent cheape & light mixture for Armour & Targettes of weapons of Pistoll proffe' was as follows:

> Lay rowles of Isinglas in stronge Alliger [alegar, vinegar produced by the fermentation of ale], till they grow softe, devide the yellow & rotten parte,

26 Alan Williams and Anthony de Reuck, *The Royal Armoury at Greenwich, 1515–1649*, 1995, pp. 8, 12; Webb, *Elizabethan Military Science*, pp. 89–90.
27 BL, SL 2210, f. 125; SL 2244, f. 33v. Hatchel, a wire comb; curat (*cuirass*), a breast plate and back plate which buckled together to protect the upper body.

& breake the rest into small peeces, sett them on the fire wth a litle faire water mixed therwth, and when all is dissolved then streyne the size. and take that streyned liquor, thrust therin wth a small slice, sheete by sheete as many sheetes of paper as the size will beare, even till it grow to a thick & stiff substance yf you wold coller it yellow mingle somme saffron in ye water, worck this uppon your armors or woodden targetts that are thin & light, lay them in the Spring or Autumne to drie uppon canvas or packthread fraimes in a light roome then divide them from the Armors when the substance is drie. but that which is layed uppon the wood must remaine wth the Target. varnish them over wth oyle of turpentine, or cover them wth oyled Leather to hold owt rayne.

Whether layers of paper glued together with isinglass would withstand pistol shot is a moot point but there is no doubt that his ideas for laminated shields such as that of papier mâché he described would have had both lightness and a degree of strength surprising to those who knew what it was made of.[28]

Despite Plat's experiments, armour was still usually made of steel and subject to rust (something which also also attacked guns both large and small). Keeping armour clean therefore was a constant problem. Medieval armourers used sand and vinegar to burnish the plates, but by Plat's time the preferred cleaning materials were the abrasives pumice-stone and emery, with oil. Other options to preserve armour were painting or varnishing, as was adding a layer of tin, although these methods obscured the beauty of the shining steel. Plat thought that if he could invent a method of preserving armour in a clean, bright state, he had a product which he might sell at a profit. Selling the preservative, however, risked that purchasers might discover its ingredients. To avoid disclosure, he considered obtaining the position of officer in the Tower charged with keeping the armour there clean, or perhaps an exclusive contract from the Armourers' Company to maintain its armour.[29]

[28] BL, SL 2245, ff. 40–41.
[29] Charles ffoulkes, *The Armourer and his Craft*, 1912, pp. 78–82; Williams & de Reuck, *The Royal Armoury at Greenwich, 1515–1649*, pp. 47–8; BL, SL 2172, f. 21; SL 2197, f. 18. See more below on his attempts to make money from this invention.

INVENTIONS

Plat consulted the secrets books of Cardano and Lupton for ideas on this knotty problem of keep armour shining bright, and he talked as well to his friends Romero, who suggested powdered lead and salad oil, and Auditor Hill, who offered long-boiled salad oil or lard. He also noted varnish as a possibility and that a proven method was 'the cleere and white oyle of turpentine layd thinly on armour first well dried' which 'will keepe it 2 or 3 yeere from all shew of ruste'. By 1595 Plat had perfected his 'defensative' for armour and, citing a naval hero's approval, boldly proclaimed:

> This is a fit secret for all hir Maiesties armories, and all the artillery of her ships, wherein the beauty of the armor is still preserved notwithstanding this defensative upon it, it suffereth no dampe either of fresh, or brackish water to prevaile against it, this is an inestimable secret for the sea in the opinion of sir Fr Drake.[30]

His recipe was relatively simple, although notes added to it show he was still not totally decided on the right mix of oils:

> To a pinte of Sallet oyle, add one pinte of neates foote oyle, 1 ounce the bigness of a walnut of yellow wax and .3. handfulles of Leade small scraped or filed. prepared in this manner. First take the sallet oyle and boyle it until it leave cracking. Skomme ye same wth a feather as longe as any Skomme ariseth & when it seetheth smothelie then put thereunto one handful of yr leade therin, and doo in all points as before then set that likewise aside. Then take somme sheepes suet from the kidney .v. twice the quantity of a walnut, pull of the skin and clarifie it purely (Qre iff deeres suet bee not better) wth boyling in a potte a reasonable tyme, then sett that also aside. Then boyle the wax and clarifie it likewise. Then putt all thafforsaid substaunce together, and putt therto one handful more of Leade, and when you have boyled them well together, take them of and strayne the oyle from the Leade, and keepe it in a tynn or leaden vessel, gally pot or jar glasse until you have occasion to used it. Take a fine ragg being well dried and warme at the fier, & when your armour, weapon, pece &c is made very cleane and well dried also by the fier, rubb it over wth the saide cloth being

30 BL, SL 2244, f. 29; BL, SL 2210, ff. 136, 177; Plat, *A Discoverie of Certaine English wants*, 1595.

moistened before wth thafforsaide oyle and let it hange in the aer or where you please and it will not ruste in many yeeres. yea it is thought that it will not canker though it lay 14 dayes covered in the water. And iff the yron worke which you wold clense bee very rustie, then lay good store therof by the fier side upon it, and in 2 or 3 dayes it will eate of all the ruste, until it comme even to the mettall.[31]

To ease the problem of moving armies over rivers and dykes – a frequent occurrence in the Low Countries campaign, Plat tried to devise 'a portable boate, which one man may carie with ease, and yet will hold eight persons', and, for large numbers of troops, 'a light, strong & sodaine bridge .. [could] be made by uniting these boates, and therby sodainly to convey each of a whole armie over a large river'. Alternatively, he speculated (somewhat optimistically) that corks or bladders strung along two ropes might support a temporary bridge of canvas or fir boards.[32]

Finally, if the country was in extreme peril, Plat had a secret device exploiting the psychological impact of new weapons on the battlefield. The invention, included in the manuscript draft of *A discoverie of Certaine English wants* of 1595 but excluded from the published pamphlet, offered 'A philosophical and martiall invention not to bee disclosed much less to bee practised, but upon a Spanish Invasion', and only then 'when the Common pike and musket & other ordinary and commendable meanes doo fayle'. He claimed of this terrible weapon: 'The secret the Author will not willingly disclose but upon the command of his sovereigne'. Was this a 'poisoned smoke'? It is described at one point as 'chimicall'.[33]

It is perhaps stretching any definition of inventions to include within it the topic of waste products. I have done so because Plat was interested in the use of waste and he brought together his thoughts in a paper with 38 ideas for utilizing waste, either intending it for publication or as a petition for some royal reward. The treatise, written after October 1599, was developed with the experience he had so far gained from his investigation of trades in London, with many cross-references to other

---

31   BL, SL 2216, ff. 28–28v.
32   Plat, *A new, cheape and delicate Fire of Cole-balles*, 1603, D2; BL, SL 2189, f. 113.
33   BL, SL 2197, f. 4.

notes on trades.³⁴ It is axiomatic that people in Plat's time wasted few artefacts or materials. Many a house was made or repaired from ship's timbers or stone from old monastic buildings. Pedlars sold new crockery and clothes in part-exchange for old. Second-hand clothing was big business and clothes too far gone to wear might become rags for the paper industry, fertilizer when left to rot on the fields and even, if linen, bandages for wounded seamen. Durable materials, such as metals of all kinds, were recycled, either by melting them down or patching up the artefacts into which they were made. Human and animal waste was scraped from city streets and dug out of privies and laystalls for use as manure in gardens and suburban fields. Some industrial processes used waste as an important ingredient – urine for saltpetre, dog-dung for tanning, and rags for paper. Natural resources were similarly exploited. Edible wild vegetation and wild animals were food resources, while wood, furze, straw, peat and the like found use as fuel, building materials, animal bedding, and a wide range of other functions. In Donald Woodward's words, 'early modern society was characterized by the ability to find some purpose for virtually every natural material and agricultural by-product'.³⁵

In his treatise Plat was not particularly interested in using natural materials collected from the wild or in recycling artefacts – he wanted to

34   BL, SL 2189, ff. 163–165, has numbered paragraphs on the following (including duplications): Heat from domestic fires; waste liquor from starch-making; waste liquor from steeping barley in brewing; orange pulp discarded by confectioners; various fertilizers from waste; egg whites; spent Spanish black from dyers; sheep with rot; dead sheep; petrifying and alum springs; 'vile and contemptable thinges in phisicke'; blood and brains of slaughtered animals; wormwood as substitute for hops; molasses as basis for drink at sea; brine from salt pits and seawater; carrion; sawdust; woad roots; fallow woad ground; waste brine from salt makers; waste woad juice from dyers; pulp from crushed oilseed; urine; coal ashes; Lord Dudley's brine pit; Lord Shrewsbury's alum spring; household waste brine; soap ashes; waste iron filings; salt extracted during saltpetre making; shreds of parchment; ashes of beanstalks and kelp; sheep's dung; earth from graves; waste heat from industry; waste candle heat; use of dead bodies; tile stones as whetstones; garbled spices and indigo.

35   Donald Woodward, 'Swords into ploughshares', EcHR, XXXIII, 1985 pp. 175–191; Beverly Lemire, 'Consumerism in Preindustrial and Early Industrial England', Journal of British Studies, 27, 1988, pp. 7–9; Mark S.R. Jenner, Early Modern conceptions of "cleanliness" and "dirt" as reflected in the environmental regulation of London c. 1530–

SIR HUGH PLAT

take advantage of by-products of human activity. Six of his ideas are for the use of agricultural waste, and as we have discussed many of these in the chapter on agriculture, they will be passed over here. Apart from half a dozen ideas which are not really waste at all – wormwood as a substitute for hops in beer, molasses as a flavouring for a drink at sea, etc. – the ideas concentrate mainly on waste from trade and industry (20 paragraphs) and domestic waste (6 paragraphs). The industrial ideas reflect Plat's interest in the trades of London and its suburbs – soap boilers, glass houses, saltpetre works, brewers, starch-makers, dyers, confectioners, butchers, sawyers, oilseed crushers. But he does not ignore – no doubt thanks to discussions at Court – the extractive industries on noblemen's lands such as alum works, saltpits, and seacoal. He is aware of noisome waste, both industrial and domestic, such as slaughterhouse offal, dead dogs and cats, urine, and discarded domestic brine, which is produced by London and seeks to put this material to use, thereby justifying the expense of collection. Similarly he has seen that some industries produce large quantities of by-products which, although not particularly noisome, need to be disposed of usefully: ashes from soap-boilers; waste dye from dyers; sawdust; salt-maker's brine; oilcake; iron filings; and waste brewers' liquor.[36]

Plat's love of ingenuity is evident here, perhaps no more evident than in the ideas for the use of waste heat from industry and domestic fires:

> A man might make good use of his chamber fiers, especially all the winter if in either of the insides of the chimney, there were cubbards erected an ell or yard high as the chimney wold beare it and there either wth turned pillars latised or plates full of holes wthin which you may hange whites of egges in blathers and make gomme therof, or els you may place litle roundletts filled wth water infused upon a good quantity of rosemary, roseleafes, lavender, Isopp, sweetbrier thyme violets &c.

The cupboards might also be used to dry hearbs, mature hippocras or flavoured vinegars or, somewhat dangerously, dry gunpowder.[37]

---

c. 1700', D. Phil. thesis, Oxford, 1991, pp. 87–89, 95–99; Donald Woodward, 'Straw, Bracken and the Wicklow Whale', *Past & Present*, 159, 1998, pp. 43–76; Thick, *The Neat House Gardens*, pp. 101–2.

36   BL, SL 2189, ff. 163–165.   37   BL, SL 2189, ff. 147–147v.

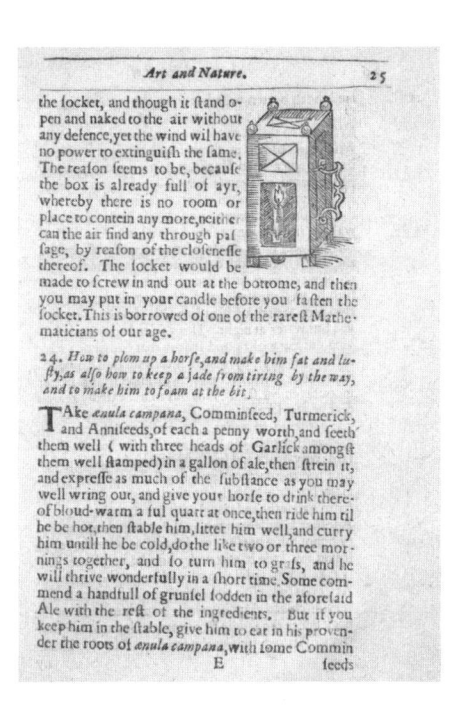

Figure 11 (above). A new and cheap lantern designed by Plat.
Figure 12 (below). Plat's coalballs, neatly piled up in a grate as he advised.

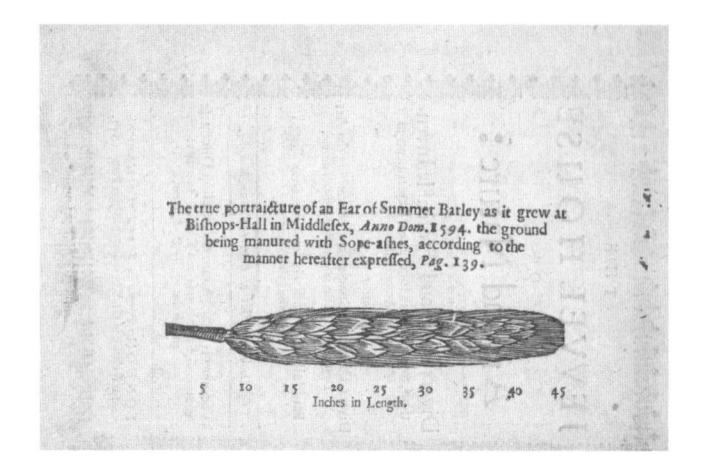

Figure 13 (above). An fine ear of barley that Plat grew in his garden using soap-ashes as fertilizer.

Figure 14 (below). Plat's bolting hutch, a mechanical device for sifting bran from flour which he exhibited in his London house to sceptical bakers.

Bigger kitchen fires produced enough heat for hot-air central heating. Plat suggested iron plates over a large fire with holes into which a funnel was fitted from which narrower pipes emerged to convey heat to other rooms in the house. To ease cold limbs, 'if your pipe run alonge under your chamber, you may pierce diverse holes thorough the floor iust over the fonnell where a woeman or man may stande till they bee thoroughly warme'. He speculated whether iron (thick or thin), lead, or clay pipes were better for this system. Plat recognized that open fires were wasteful, much heat escaped up the chimney, for 'you gathereth this greate heate, & yet your fier remayneth in his full strength either for seething or roasting'. More fanciful was his idea for a luxurious hot bath: 'yf you hange a pott of water upon the fier wth a clos cover and a place to let in a pipe of leade therin this pipe may bee conveyed through an other roome into the side of a bathing tubb wth a false bottom full of holes wherin you may sitt and bath yr self. you may putt sweete herbs in the potte to sweeten yr bath'.[38]

Industry also produced a good deal of waste heat; Plat identified 'glasshowses, sugerhowses, diers howses & sope howses', brewers and saltpetre works as particularly profligate. From such enterprises, the hot furnace gases, conveyed as above into another room, could be used to dry vegetables and meat or keep delicate plants and trees through the winter; or a large kettle might be hung in the flue to use the heat to raise the temperature of liquids. Waste sawdust was a combustible substance Plat thought might be used in these heating systems.[39]

Some of his ideas for using butchers' waste – the blood and brains of slaughtered animals – were quite simple. 'In these hard and extreeme yeeres of dearth and famine [they] wold wth a very little cookerie bee made a wholesomme and norishing victuall for the releyving of many thowsands of hungrie bellyes that of this present are ready to sterve for want of foode.' The idea for converting blood to chicken feed, however, which was set out in the *Jewell House* (originally the secret of a Dutch immigrant), exemplifies Plat's love of ingenuity. 'If you take the bloud of beasts, wherof ye Butchers make no great reckoning, filling stone pottes therwith, whose covers may be full of such holes, as that the flesh flies in

38 BL, SL 2189, f. 147v.
39 BL, SL 2189, f. 19v; SL 2210, f. 59a–60; SL 2189, f. 164v. Coalballs, see below.

sommer time, may easily get in and out at the same, you shall finde the bloud by the meanes of the flie-bloes and putrifaction together, wholie converted into white and glib worms (which the anglers call Gentils) which will fatten them exceedinglie & make them eate most tenderlie.' More prosaically, the blood might be boiled with bran and fed to fowl.[40]

As evidenced from his cookery writing, Plat had a good deal to do with confectioners. He noted that 'diverse Confit makers make no reconinge of the pulpe of civill or portingall orenges, but only of the pills or rindes wherof they make either orengadoes, or comfits … but if you wringe owte the uiyce of them', filter it, and store it well, 'you shall by this meanes have a very serviceable sawce all the yeere longe, for veale, capons, suwet &c.' A similar domestic saving involved egg whites. 'Rather then they shold bee caste away or merely lost wee may hange them in blathers wthin our kichin chimneys where in 9 or 10 dayes they will harden and become a gomm very serviceable for the pennsill but not for the pen, or els (yf you have the arte to keepe them) they will serve for a size either to lay on gold or silver upon paper, parchment, or any banqueting stuff, or to grinde or mix diverse colors wth for thart of limning or paintinge.'[41]

Many of the suggestions for using waste were not originally Plat's. Urine collection for industrial use, iron filings as an ingredient of Spanish Black dye, Ely tiles as whetstones, and Lord Shrewsbury's alum spring as a source of alum-substitute for dyers: these and other uses of waste were already in operation and he merely publicized them. In the case of waste soap-ashes, a by-product of the soap industry, Plat was particularly keen to point out the benefits already exploited and suggest new uses for them. He complained that in England soap-ashes were 'trampled under foot and contemned of al men', being mostly given away by soap-boilers to those who would carry them away, whereas until recently the Dutch had given three shillings a load for them, carried them to the Low Countries and used them as fertilizer. In England many hundreds of loads had been used to surface streets and bowling alleys (hence the 'trampling'). Plat recommended them to scour wooden vessels and floors as well as glass windows, and as a bleach for linen, but he was most excited by their

40 BL, SL 2189, f. 164; Plat, *Jewell House*, pt 1, pp. 12–13.
41 BL, SL 2189, ff. 163–163v.

potential as building material. Two soap-boilers, Mr Bronsede and Mr Meggs, had used soap-ashes successfully in building. Bronsede, 'dwelling without Aldgate, hath for the better encouragement of others, long since erected a faire and stately edifice of brick for his owne habitation, upon the good success whereof he hath also verie latelie built one other house of some charge and good receipt, the mortar wherof did consist of two loades of waste sope ashes, one load of loame, and one load Woolwich sand.' Meggs used 'onlie loame and sope ashes tempered and wrought together instead of mortar, whereby he hath laid both the foundations, chimneys, and their tunnels in his dwelling house in Southwarke and they have endured those stormes already which have overturned manie other both new and olde tunnels that hath beene built with the ordinarie morter.' Plat claimed in 1593 that at least 60 other houses had been build with soap-ash mortar, using many hundreds of loads. He acknowledges that soap-ashes were 'sharp' and could hurt bricklayers' fingers, but if soaked in water this problem was ameliorated, and they could always use gloves.[42]

Plat's invention of a portable pump was mentioned, without detail, in his 1593 pamphlet. Apart from a note of a method of hardening leather for pumps by soaking it in water with iron filings, there is nothing in his manuscripts describing how this invention was constructed or whence he got the idea. There is an illustration of the pump in action in *The Jewell House* and details were provided of its construction in 1600: 'yt is not above ¾ of a yard in length and half a yard in hieth'; it is light, made from wood, and portable by one man; it has holes for water at the bottom; the action involves two strokes; one man can operate it for 5 or 6 hours at a time, delivering 4, 5, or 6 tons of water per hour; it can be made for 30 shillings. The pump worked by drawing water into a tube on the upstroke of a piston through holes in a flat base, then, on the down stroke, a flap closes the holes in the base and water is forced up and out via a second tube set at an angle to that with the piston in it. Such a pump, operating

42 Woodward, 'Straw, Bracken and the Wicklow Whale', p. 70. Soap manufacture required considerable quantities of wood or fern ash or the like, from which, with the addition of lime, alkali was extracted, leaving waste to be disposed of. BL, SL 2210, f. 189; Plat, *Jewell House* pt. 1, pp. 77–8. (By 'tunnels', Plat means chimney stacks.) Plat, *Jewell House* pt. 2, pp. 50–51, 56–58.

as a garden spray, is illustrated in Thomas Hill's *The Gardeners Labyrinth* of 1577. So, Plat did not invent this pump and anyone who knew of the illustration in the gardening book could probably have constructed it.[43]

The 'Cole-ball' – a mixture of crushed coal and loam formed into balls – was an innovation given much prominence by Plat in his public offers. It was first mentioned in print in 1593 and detailed in his pamphlet of 1603. Posterity has endorsed his enthusiasm. Proposals were made to make this fuel commercially by the Thames in 1716, while William Hanbury reported to the Royal Society on their manufacture in Liège in 1719, recording that he had used the fuel for upwards of ten years: 'It is a most excellent Fire for Roasting, for heating of Irons, or warming a Room: I use it in my Kitchen, Laundry, Parlour, and Library'. Coal briquettes made from powdered or fine coal pressed into shapes with various substances to aid cohesion are manufactured in many parts of the world today: in the 1990s they were made in China using clay mixed with crushed coal which had been soaked in water for 24 hours to reduce the sulphur content.[44]

In 1803 Alexander Hunter found them made out of waste coal near Bristol, in other parts of the West Country, and in Wales. 'The way in which those balls are prepared, is to take a certain quantity of culm [i.e. small coal], to which they add an equal quantity of sleech, or mud.... after mixing them well with shovels, they blend them with their hands more perfectly, and mould them into balls of six inches in diameter; and, in making them up, they work as much culm into the sleech with their hands as they possibly can, without making them crumbly'.[45]

This description is very similar to Plat's method, although he suggested the best seacoal, crushed and mixed with loam watered to a thin pap:

> Spread these coles abroad some handful thicke, or thereabouts, equally upon the floore, then sprinckle some of your thinne pap all over the heape:

43 Plat, *The new and admirable Arte of setting of Corne*, 1600, D4–5.
44 Plat, *A Briefe Apologie of Certaine New inventions*, 1593; Plat, *A new, cheape and delicate fire of Cole-balles*, 1603; *Philosophical Transactions*, vol. 41, 1739, pp. 672–674; 'Greenhouse gases and other airborne pollution from household stoves in China', J. Zhang et al., *Atmospheric Environment*, 34, 2000, pp. 4537–4549.
45 A. Hunter, *Georgical Essays*, vol. 3, York, 1803, pp. 149–50.

then turne them with a shovel or a spade, and spread them againe as before, throwing more of your lomy liquor upon them. Continue this course till you have made the whole masse or lumpe of your coles soft enough to be wrought up into balles, betweene your hands, according to the maner and making of snowbals; then place them one by one, so as they touch not ech other til they be thorough drie, which will be in a few dayes.

In perfecting his method, Plat speculated on substituting all or some of the coal with sawdust or tanners' bark and using clay rather than loam. Both Plat and later commentators praised the lack of smell and smoke from coalballs and the heat produced by a fire of them.[46]

One might suppose that Plat was the originator of this invention, later copied by many others, but he was given a recipe in November 1590 by his Spanish informant Sr Romero, who may have come across the balls in the Spanish-controlled Low Countries. Following his own published recipe in 1603 Plat comments, 'And so you have seacoles wrought up into bals simply of themselves, according to the manner of Lukeland in germanie: which forme of firing hath been in use with them for many yeares past, and doth as yet continue to this day, as I am credibly informed.' Plat claimed to have taken the idea of coalballs and, by his own ingenuity, radically improved it, creating 'our new and English fiers, such as neither germany nor any other forrain king-dome or country, did ever to my knowledge as yet use or injoy'. He described among other things the advantages of various combinations of coals in the coalball mixture; making a sweeter-smelling ball with the addition of deal, fir, oak or elm sawdust; and the advantages and disadvantages of using cowdung, straw and tanners' bark. Some of these alternatives, however, were mere speculation: of mixing turf with coal, 'I have made no experience'; and 'what oozes will do either for the multiplying or binding of our colebals, I can not certeinly determine.'[47]

Plat described in detail how to prepare a fire of coalballs: 'I do first lay bricks edge wise on my hearth one by one, each bricke distant a full inch from the other, according to the breadth or compasse of the fire

46   Plat, *A new, cheape and delicate fire of Cole-balles*, 1603.
47   BL, SL 2210, f. 131; Plat, *A new, cheape and delicate fire of Cole-balles*, 1603. 'Lukeland' is Liège, where manufacture of these balls was well established in Plat's time.

which I intend to make... then do I place a rowe of faucon or saker shot for the neathermost rank ... and then an other row of these bals upon the neathermost, and so I frame my fire to what hight and compass I thinke best.' He admits his readers 'without direction might easily aime at' such an arrangement, but one feels he enjoyed the pleasing symmetry of the piled balls, as the women of South Wales still did in the 1840s when they arranged their oval-shaped balls: 'the good housewives not infrequently display their taste by the fanciful way in which they place these balls edgewise in the grate, each row being inclined at a different angle; and under the active influence of that passion for whitewashing...they are not unfrequently...whitewashed also.'[48]

Plat's shoe-blacking was advertised in 1595 as, 'A licour to keepe either boot, shoe or buskin made of drie leather, both blacke in wearing, and defensible against all raine, dew, or moisture, wherof there hath bin already a sufficient triall had by divers gentlemen, and others, this is to be had of the Author in severall kindes'. He canvassed his friends for recipes for this commodity. Auditor Hill suggested clarified butter to keep out rain; Thomas Gascoigne advised, 'The blacking of a lampe, tempered wth oyle of almonds, or somme other sweete oyle ... to keepe the glosse of spanish leather shoes or Buskins'. The recipe Plat settled on was quite complicated, and arrived at by several experiments. We know this because under the heading, 'A liquor that doth farr exceede blacking both to keepe boots and shoes in a good coollor & also to make them hold owte all weather', we find not a definitive recipe but a number of suggested ingredients: 'Qre of Neatesfoote oyle insteede of trayne oyle to bee used in this secrete. qre of sweetening the same wth oyle of spike or oyle of turpentine. qre if rosen fume bee not also sweeter then lamp blacke for this purpose. qre of Neatsfoote oyle and Rape oyle to bee mixed in a good proportion together/ qre if the fume or blackinge may not bee mixed sufficiently wth a heare pensill (putting but a little oyle to the blackinge) without any fier'. He concludes the note, 'for so I made somme in July 1595'. No doubt various combinations of dyes and oils were used in keeping leather in good condition in Plat's time (a literary reference in 1611 to shoes 'stinking of

48 Plat, *A new, cheape and delicate fire of Cole-balles*, 1603; Anon, *The history and description of fossil fuel*, 1841, p. 397.

blacking' reflects the problem of finding a liquor which blackened whilst being 'sweet') but he does seem to have invented his own special mixture and sold this to others, marking this out as one of his genuine inventions.[49]

'A vessel of Wood, to brew or boile in' was first introduced to the public by Plat in his *Briefe Apologie* of 1593. However, few details are provided of its construction in this or two subsequent advertisements. He chided those who speculated that outlandish materials were used: the vessel was not coated in salamander hair, 'whose bodies are rare and hard to come by', he wryly comments, nor was it coated with 'that chargeable and incombustible oyle of Talcum, fitter to bee thinly laid upon the face of a ladie, then grossly spent or dawbed' upon the sides of a wooden vessel. The chief benefit of this invention was its fuel economy – a saving of 50 per cent on fuel was promised: the vessel had been 'forced to boyle with a fewe flaming stikes' when demonstrated to dignitaries at Plat's London house. Other advantages were that it was cheaper than a copper vessel, would last at least 20 years (being not subject to burning or damage from boiling water), and if repairs were needed they could be carried out simply by the owners themselves. In 1603 he promised to reveal full details of this invention: 'within a terme or two, I purpose (God willing) to make a publike shew thereof to all commers, unto whom the secret it selfe shall also be revealed and made good: and therefore I do here labour to prepare their minds to a kind and probable conceit therof, least when it shall be offered to a publike view, it may happily bee taken for the second part of M. Venners Tragedie, lately acted at the Swanne on the bankside, with better profite to himselfe then pleasure to the beholders'.[50]

Plat never made good his revelation (not unlike Venner?), keeping the secret in his notebooks. He recognized that wood was cheaper than metal if it could be treated to withstand the heat required to warm large

49 Plat, *Discoverie of Certaine English wants*, 1595; BL, SL 2210, ff. 3, 135v; SL 2216, f. 38v; Middleton, *The Roaring Girl*, 1611. 'Heare pensill' is a hair brush.
50 Plat, *A Briefe Apologie of Certaine New inventions*, 1593; Plat, *Jewell House*, advertisements at the end; Plat, *A new, cheape and delicate fire of Cole-balles*, 1603. Venner, like Plat a member of Lincoln's Inn, advertised a special performance by gentry actors of a play at a Bankside theatre in 1602, took entrance money from playgoers, then decamped with the takings without providing a performance, T.S. Graves, 'A note on the Swan Theatre', *Modern Philology*, vol. 9, no. 3, 1912, p. 431.

quantities of liquid, whether in the workshop or the domestic kitchen. He tried wooden vessels with various coatings : 'Rubb over the owtside of your tubb, or boole, either wth oyle, butter, the fatt of powder beoff broth &c', but fire might still damage the wood, even if the vessel was suspended above the flames. Frequent re-coating would be required. He also speculated that such vessels might be suspended over a flue of hot gases rather than the fire itself but in the end he abandoned coatings. Instead he suggested a wooden vessel with a funnel of copper inserted into the centre, narrowing upwards, succinctly described thus: 'Cut owt a large hole thorough the midst of the bottom of any tubb. fasten close therin a fonnell of yron or copper, wth a grate underneath, & make your fier in the fonell.' Such a vessel would have no wooden part in direct contact with the fire, which would be wholly enclosed by the copper funnel. A tall funnel might carry two vessels, a second, smaller tub with a smaller funnel being inserted on top of the main funnel. A modern Storm Kettle, used by troops in the field in the last century and still made today, is based on the same principle – a jacket filled with water surrounds a flue with a small hearth. Today's advertising copy proclaims, 'A few twigs or pieces of dried grass are placed in the base, the hole is turned towards the wind and the thing is lit. The funnelling effect concentrates the heat against the inside walls of the kettle, and within just a few minutes the water is boiling. If in the meanwhile more fuel is needed, it is simply dropped in through the top.' As with so many of Plat's inventions, the idea was not original. 'It was described in print by Conrad Gesner forty or fifty years before and is figured in most works on chemistry in the sixteenth century.' [51]

Plat's invention of rainproof fabric was announced in 1593 and advertised a year later in the *Jewell House*. The range of fabrics he considered waterproofing show that he was well acquainted with the many types

---

51   BL, SL 2210, f. 51–51v; SL 2212, f. 16v; Plat, *A Briefe Apologie of Certaine New inventions*, 1593; www. hawkin. com – advertising copy for a Storm Kettle; *Mechanics magazine, museum register journal and gazette*, no. 1145, 1845, p. 46. Another of his inventions, the mechanical bolting hutch, was not original to Plat: Emperor Charles V granted a patent for a similar machine in 1547 and the grant is described in Cardano's *De Subtilitate*, 1550, a work Plat had read. See Jeremy Phillips, 'The English patent as a reward for invention', *Journal of Legal History*, vol. 3, 1982, p. 75.

of cloth available in London to a courtier with the money to indulge in fashionable clothing. He tried camlet, watered camlet, grogram, worsted, kersey, rash, taffeta, sarcenet, serge, calico and 'Scottish cloth'. Most of these cloths were expensive, many contained silken threads; some had angora wool or mohair in them.[52] To judge by his notes, he considered this matter over a number of years and tried several solutions before finding a method which inspired confidence. His earliest thoughts were to line fine clothes of silk or velvet with soft leather oiled with perfumed almond oil or a mixture of oil, wax and rosin, but he did not know in what proportion to mix these substances and these early thoughts may well not have been tested. He then speculated that weatherproofing cloth could be 'donn by thin layinge on of amber vernish' to protect 'clokes or other riding garments' as well as thin parchment covers for ladies' hats. He even wondered whether a cloak of 'a clos wrought chamlett', or one so lined, might be waterproof on its own. He finally settled on treating fabrics with oil, although it is not possible to know which of several methods he hoped to exploit commercially. One involved simply turpentine: 'Streyne the stuff [i.e. cloth] upon frames like tenters, and with a pensill lay on lightly so much oyle of Turpentine, till it appeere on the backside. in 3 or 4 days it will dry. I cold wish a grograin garment thus prepared to bee lined wth taffeta prepared in the same manner'. To get rid of the smell of turpentine he suggested scenting with wormwood or cloves during the oiling, or afterwards with damask powder.

Another method was more complicated. First powdered Benjamin was steeped in spirit of wine for 24 hours and to the liquid drained off was put an equal amount of oil of freshly crushed 'small nutts' (possibly hazelnut oil). Taffeta was dipped into this, then 'chafe the silcke well betweene your handes till it bee well entred abowte ½ of an hour, then hang it up to drie', and repeat the process the next day. This was apparently Auditor Hill's recipe and a scarf so treated had kept him dry 'in good store of rayne', although when Plat poured a spoonful of water onto the fabric it was only two minutes before the water penetrated it. Another of Plat's recipes fared better. Treating an ell and a quarter of 'taffeta sarcenett' with

52 Plat, *A Briefe Apologie of Certaine New inventions*, 1593; Plat, *Jewell House*, 1594; BL, SL 2210, ff. 143, 174, 180; SL 2244, f. 28v; SL 2245, f. 46; SL 2216, ff. 108, 128v.

8 spoonfuls of oil of turpentine and 4 spoonfuls of oil of spike he found 'it will hold in any place where there is no gall or pinhole a sponfull of cleane water for ¼ of an hower before it dropp through'.[53]

As well as moulding armour in papier mâché, Plat speculated on the use of this substance (and sundry other powders) mixed with glue to make objects in wooden moulds. Plat learnt to mould in gold and silver from London artisans but here he seems to be experimenting himself, trying possible combinations of substances in the moulding process and suggesting uses for the finished products. As well as glue, wax, rosin and isinglass are considered as possible binding agents and he also suggests mixing with the paper pap powdered jet, box, sawdust, sand, ebony, and ivory. Doing away with the paper altogether, he wonders if glue mixed with fine powdered hardwood, brick, tiles, horn ivory, alabaster, brazil wood, steel, pin dust, lead or tin might produce cast objects with interesting finishes. The casts could be painted with water- or oil-based paints, or coloured by mixing pitch or vermillion in the pap. Oil or varnish coatings made the casts able to withstand the weather. Weather-proofing allowed outside use: one of his father's tenants told him 'the Italians garnish the owtside sides of theyre howses wth fansies made of this [papier mâché] mixture', and Plat thought one of his recipes 'a good mixture to make signes for mens howses becawse yt abideth all weather'. Inside, the use of the mouldings was almost infinite:

> It is a pleasing and commendable practice, by this Art to mold of those excellent counterfeites, of carved or embossed faces, dogges, Lions, Borders, Armes, &c from toombes, or out of noblemens galleries: as also of pillers, balles, leaves, frutages, &c, wherewith to garnish beds, tables, court-cupboards, the Iawmes and mantletrees of chimnies, and other stately furnitures of chambers or galleries.[54]

Robert Boyle, a couple of generations after Plat, suggested the use of papier mâché for such 'curious moveables' as picture frames. From

53 BL, SL 2210, ff. 143–143v, 174; SL 2245, f. 46; SL 2216, ff. 108, 128v. Camlet – a costly fabric made from Angora wool, *OED*. Benjamin was the nut of the Benjamin tree, an import from the Far East and spike oil was derived from French lavender.

54 BL, SL 2210, ff. 148–148v; SL 2197, ff. 16–17; SL 2245, ff. 41–42, 106–7; SL 2216, ff. 111v, 125v; Plat, *Jewell House*, pt. 4, pp. 67–8.

the middle of the eighteenth century it was in vogue as an architectural material for genteel interiors. 'Papier mâché was often used as sculptural relief for ceilings and walls in imitation of plasterwork and was generally painted white. It was also used as decorative border or fillet in imitation of carved and gilded wood to border the "surbase" (chair rail), doors, windows, chimney pieces, cornices, or anywhere else it might look attractive.' In eighteenth-century England papier mâché was used for decorating rooms in much the way Plat envisaged and architectural historians have found some dating to Plat's time.[55]

Plat set much store by his inventions; notes and recipes for them occupy a good deal of his surviving manuscripts and, as we have noted, he frequently lists those of which he has high hopes of success or of which he is particularly proud. Inventions are also ubiquitous in his publications. Setting aside his youthful excursions into poetry and philosophy of 1572, every publication from 1593 until 1608 was either a pamphlet advertising his inventions or a work with an advertisement appended to it. Why? We have seen how he thought progress both possible and desirable and associated progress with technological advance, in other words inventions. On a personal level, he was curious and ingenious, and took pleasure in perfecting a recipe, mastering a technique, or making a machine work, as well as in coming across someone else's improvement on a craft or skill. One can sense his pleasure at the thought of a house with hot-air central heating, of being able to lead cold visitors to the chamber where he had installed warm air vents in the floor, or the brave show his symmetrically-piled coalballs made on a hearth. The satisfaction on testing his waterproofed fabric and finding it much better than his friend Hill's is evident. Making papier mâché casts to adorn the wainscot of a room was 'a pleasing and commendable practise'. He admired the industrious soapboilers who built fair houses with mortar made of waste ashes, and the ingenious Dutchman who invented the 'biological' way of turning waste blood into chicken feed.

55 Jonathan Thornton, 'The history, technology, and conservation of architectural papier mâché', *Journal of the American Institute for Conservation*, vol. 32, no. 2, pp. 165–166.

In his advertisements for various inventions, Plat is careful to emphasize the benefits his and others' ingenuity would bring to the realm as a whole, summarizing in one sentence that inventions, 'if they were brought into some generall, and common use, would procure great love and securitie to the rich, sufficient maintenaunce and reliefe to the poore, some credit to the Author, and no small benefite to the whole realme of England.' The new bolting hutch and his other inventions advertised in 1593 were 'to be put in use for the service of the common wealth'. Coalballs would bring wide economic benefits to England, helping 'Newcastel men' sell their small coal, bringing down the price of wood and charcoal, and preserving timber for shipbuilding. He put a general case for state support for inventors in the *Jewell House*, 'why should not little England ... seeke to raunge her selfe in the foremost rankes and troupes of all Minervaes crew.' With royal favour these men can join others 'that dailie perform their best endevours for the weale publike' for, without encouragement 'so yet there remain a secret number of choice wits, who being full fraught of more necessarie, yea more invaluable commodities, then either the east or west Indies are able to affoord, are nevertheles forced to consume their daies in melancholie, & ... to burie their talents in the bottomles pit of oblivion.'[56]

As well as the common weal Plat, a moderate Puritan gentleman, mindful perhaps of his good fortune in inheriting wealth and possessions for a comfortable life, had a genuine concern to help the poor. The first of a list of reasons for inventions in his early pamphlet of 1593 was to the help the poor 'in their present distress'. One benefit of coalballs was, 'It wil employ many thousands of maimed souldiers, and other poore and impotent persons, in the making thereof'. Plat also proposed that one tenth of the profits should be used for poor relief. Plat's piety and concern for the poor is manifested in the *Sundrie new and Artificiall remedies against Famine*, hastily written 'upon thoccasion of this present Dearth' in 1596.[57]

A more calculating reason for Plat's interest in and publicizing of inventions was the furtherance of his own reputation. In tracing his

56 Plat, *A Briefe Apologie of Certaine New inventions*, 1593; Plat, *Discoverie of Certaine English wants*, 1595; Plat, *A new, cheape and delicate fire of Cole-balles*, 1603; Plat, *Jewell House*, 1594, B2–3.
57 Plat, *A Briefe Apologie of Certaine New inventions*, 1593.

biography we have noted the often bitter comments in the pamphlets of the 1590s; how his attempts to interest courtiers and London worthies in his schemes for improving the common weal were met with 'sundrie and sharpe calumnations … by ignorant and yet malitious enemies of all ingenious devises'. His frequent mentions of the exhibition of his inventions: a 'publicke viewe taken in London by some of her Maiesties privie Counsell, and diverse other Gentlemen and Citizens of good worship and account', probably in 1593, shows how earnestly he hoped his reputation would be enhanced by these efforts. He complains bitterly that while talk of his coalballs was 'hot in every mans mouth for a while', his coals 'now lie raked up in the ashes of oblivion, suspected of some, condemned of many, and trampled under foote of all'. His inventions, 'to my great griefe, losse, and discredit (I know not by what forward fates or misfortune) hath bene staied, crossed, or coldly commended in Court, and most injuriously disgraced, and contumeliously handled in my native soyle at home'. He fervently believed in his inventions and sought to persuade his readers that it was malice rather than genuine doubts which motivated his detractors.[58]

His later advertisements are not so feisty and argumentative (the last barbed remarks are against Sir John Harington in 1603). His inventions seem to have met with more approval in King James's Court. In 1606, as well as the approval of the French ambassador for Plat's home-made wines, he gained similar plaudits from other courtiers, including Sir Francis Vere, Arabella, Countess of Cumberland, Lady Anne Clifford, '& most of the Maides of Honour'. It has been supposed that his knighthood in May 1605 was in recognition of his inventions and scientific achievements, although there is no documentary basis for this assumption.[59] A major reason for Plat's interest in inventions was economic – he thought he could make money from them. His moneymaking activities are the subject of the next chapter.

---

58 Plat, *A Briefe Apologie of Certaine New inventions*, 1593.
59 ODNB, entry on Sir Hugh Plat; Plat, *Floraes Paradise*, 1608, O 8.

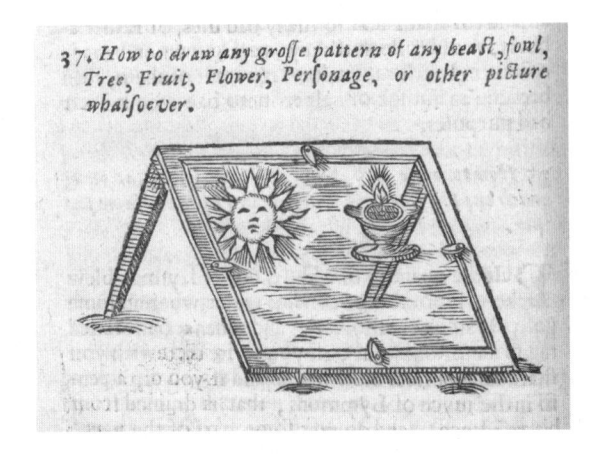

Figure 15 (above). A glass desk Plat describes which enables outlines to be traced using light underneath the desk.

Figure 16 (below). ABC dice, Plat's idea for a way of teaching children the alphabet.

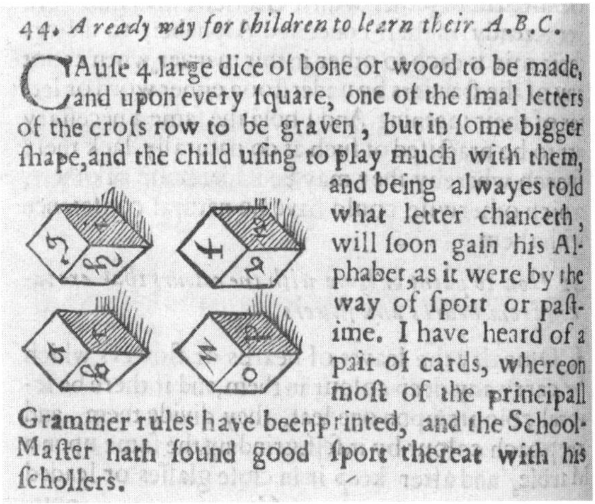

CHAPTER TEN

Moneymaking

In the words of Sir Thomas Smith, Sir Hugh Plat was of the class 'who can live idly and without manual labour, and will bear the port, charge, and countenance of a gentleman'. Exactly what he lived on is difficult to establish: as R.G. Lang remarked in a study of London aldermen at the beginning of the seventeenth century, 'We simply do not know what many of London's richest citizens did for a living and it is hard to find out'. As with most gentlemen, it is likely that the bulk of his income came from property. His will is not very informative: he declared that he had already settled unspecified property on his children, and his wife Judith is left the lease of a brewhouse together with copyhold lands and tenements (i.e. agricultural land). Judith's inheritance was said to have included land near St Albans in Hertfordshire, and also at Cowcroft and Kentish Town to the north of London. Plat mentioned in 1593 that he had tenants at St Albans. The brewhouse at Southwark yielded Plat a rent of £127 per annum at his death. In 1600 Plat's father Richard had left Hugh a good deal, including several holdings of freehold agricultural land near St Pancras in Middlesex and at East Greenwich, as well as copyhold land in the Manor of Tottenham.[1]

It is likely that the bulk of his rental income came from urban, not agricultural property, including the lucrative brewery. His father left him much London property: messuages and lands in Birchin Lane (off Lombard Street) and in Thames Street; more in the parish of St James, Garlickhythe;

---

1 Smith, quoted in Peter Earle, *The Making of the English Middle Class*, 1989, p. 6; R.G. Lang, 'London's Aldermen in Business: 1600–1625', *Guildhall Miscellany*, vol. 3, p. 242; TNA, E134/8 Jas 1/East 2; Guildhall Library MS 5485; TNA, PROB 11/112/114; BL, SL 2245, f. 114; Plat, *Diverse new sorts of Soyle*, 1594, pp. 33.

other freeholdings in the city of London and adjoining suburbs; lands on lease in the city of Westminster; and various tenements occupied by Edward Auborn in which Hugh was given a life interest. Plat had a house at Garlick Hill in 1600 and at least one property in Cornhill. London rents were high: the city was a magnet for adults from the south-east of England hoping to find work at good wages, business opportunities, or charity. Population rose steadily despite a high rate of mortality. A royal proclamation against new building in London of 1580 told of 'great multitudes of people brought to inhabit in small rooms'. The rents of these subdivided tenements and houses in back alleys allowed urban landlords like Plat and his father to enjoy incomes rising faster than inflation.[2]

Plat also had income from medical activities and although he was called to the bar, no evidence has been found of receipts from work as a lawyer. He made a list of suits he might profit from, which included 'the relief of Brewers against promoters' (presumably defending brewers, of whose Company he was a member, from the monopoly granted to William Carre in 1596 to brew and sell beer in London and Westminster for seven years) and the 'St Albons Case' – maybe a reference to the charter gained by St Albans under Elizabeth to erect three wine taverns, the profits of which were to maintain a free school. The last item on this same list is a note about 'printing of myne own books by priviledge'. These seem to be referred to in a later 'spreadsheet' of ideas for further consideration where he wonders if he can rely on friends in various parts of the legal establishment 'For the dispersinge off Coppies of my Law bookes into all the Inns of Court and Chancery'. Apart from a stray page of legal commentary, nothing survives in print or manuscript of Plat's legal writings, so we have no way of knowing what he had written or how saleable it was.[3]

Court papers from the dispute over his will mention several substantial debts and it is likely that, like other city gentlemen, Plat borrowed when

2   Ian W. Archer, 'Material Londoners?', in *Material London ca. 1600*, ed. Lena Cowen Orlin, Philadelphia, PA, 2000, p. 183; Guildhall Library MS 5485; TNA, PROB 11/112/114; BL, SL 2209, f. 40; SL 2172, f. 28.
3   BL, SL 2172, f. 13; SL 2223, f. 60v; Edward Porritt, 'Five centuries of liquor legislation in England', *Political Science Quarterly*, vol. 10, no. 4, 1885, p. 618.

necessary and lent spare money at interest in the city. Also in the case papers is a claim that he derived income of £200 per annum from a patent which had several years to run, evidence of an aspect of Plat's financial affairs which he probably did not share with many other city gentlemen, a sustained attempt to make money from inventions.[4]

Fifty years or so after Plat, in 1648–9, another aspiring gentleman, William Petty, made a list of ways to gain money:

> 1. By procuring privileges for public designs of universal use 2. By making models and directing private works and engines 3. By selling some works made by myself and friends 4. By giving, selling transcripts of secrets [and] artifices… 5. By printing the same afterwards… 8. By honorary pension from great persons 9. By the practice of physic, setting up and maintaining a *Noscomium Academicum*.[5]

Petty rose faster (and higher) than Plat, from younger son of a clothier to riches, a knighthood and offer of a peerage, occupying important government posts and becoming a founder member of the Royal Society, but his wide range of interests, including a medical practice and a restless interest in invention, identify him somewhat with Plat. It is therefore instructive to find him as a young don at Oxford listing ways of making money in a manner similar to Sir Hugh half a century before.[6]

We have encountered Plat's numerous lists of inventions already. Many were compiled as memoranda of ways of making money from his research. Headings such as 'The Secrets wch I accompte most proffitable in this Booke'; 'The Secretes wch I meane firste off all to putt in practise'; 'The choice & moste principall secretes off all my longe Travell & experience'; 'Proffitable conceipts', begin lists of potentially profitable secrets. The last of those above contained many items described as 'good cheape', that is, marketable. A list headed 'Matters of most royall and present Expectation' is explicitly involved with profitability, as is 'artificiall secrets to gaine by'.

[4] TNA, E 134/8 Jas 1/East 2.
[5] BL Add. Mss 72, 891, Petty papers, f. 8v, 1647–8. Quoted in Linda Levy Peck, *Consuming Splendor*, 2005.
[6] Toby Barnard, 'Petty, Sir William (1623–1687)', *ODNB*.

In the heading 'Certein odd observations for the gayning of money. Deus time', there is no ambiguity; this list has ideas for making money. 'When wines are deere, and reasons Curreines or figges cheape, then make wine of them' and, 'when there is store of beech mast & corne is deere, then make bread, aqua vite, drincks, viniger &c. of them', both involve some effort of manufacture. But he also has the idea of making money purely from buying cheap and selling dear, acting as a manufacturer's wholesaler: 'Buy black vernish for 6d the l., in Bucklersbery, & sell it againe for 8d to Cutlers and Armorers'. An alternative route to riches depends on cornering the market and sharp practice: 'when owtlandish wormseede is deere' buy up all the 'english wormseede which you can gett and collor, or finde owt amongst the herbwiffes somme seedes like unto them and coller them'. Such thoughts might have been written by a city merchant rather than an aspiring courtier and are a reminder of Plat's family background in commerce. He largely devoted one whole notebook to lists of possible gains from secrets, with one folio for each idea and notes of varying detail under each heading. Included in the book is also the draft of an advertising pamphlet.[7]

Probably written in the 1590s (a reference to the Queen's arms dates it before 1603), a hastily written 'spreadsheet' covering four pages brings together Plat's thoughts on the marketing of secrets and is further evidence of his methodical moneymaking plans. Headed 'All the necessary and helpinge means that I can imagine for the present gaininge of money either by artificiall wares, secretes or otherwise', the paper lists goods, notes on how to sell them, and the names of people he hoped might assist his efforts. At the end of the second page he finishes with a catch-all category, 'all my other Secrets and Wares not mentioned in any other particular here', but he obviously had more ideas later because he added two further pages of notes, albeit that the last consists only of headings.[8] We will

---

[7] BL, SL 2210, ff. 6–7, 161v; SL 2216, ff. 182, 188v–189; SL 2223, f. 61; SL 2189, f. 1v; SL 2197, ff. 1–44; SL 2172 ff. 28–29a. One of the lists of medicines in SL 2197 contains many that Plat proposed to sell to a Leicestershire gentleman in 1600. Bucklersbury was a street in London heavily populated by apothecaries.

[8] He makes detailed if brief notes on marketing the following: sweet oils and waters; casts to adorn borders, furniture, etc.; shoe-blacking; publishing law books; privileges and pensions for various schemes; cinnamon water; 'usqa bath'; macaroni for

discuss many of the details in this document later, but here is an example of the contents: his notes on 'Macaroni for the sea' in which he outlines his strategy to try and interest a government minister and commercial shipowners in this food.

> first by Sr John Fortescew to get the victuallyng of the Queenes ships. reasons. first, for lasting, secondly, the cheapnes, thirdly saving of fier .4. being brought home it serveth for an other voiage and may bee layd upp till cawse of employment. Then to use Bradbury, to harken after all the Captains and officers of long voiages. for the streights merchants etc.

His proposals for medicine sales follow a similar format, listing twelve medicines he hopes to sell, and retailers and practitioners he hopes to sell to:

> vente these by barbers , Surgeons and Phisicians D. Wyle.d.Hippocretes, Mr Clowes./ Mr Ren./Kemish./ Mr Westee./Bashford./D.Lester. D.Momford. Mr Messinger/

Amongst those listed we can recognize the surgeon William Clowes and the chemical apothecary Kemmish.[9]

This document shows the seriousness with which Plat pursued moneymaking. Despite the jibes of John Harington, his social standing appears to have been unaffected by any taint of trade. In any case, there was no rigid social code which forbade gentlemen, especially in the busy world of London, from engaging in commercial activity. Indeed, the hazy dividing line between a gentleman and a tradesman was blurred by the frequency with which younger sons of the gentry were apprenticed to London traders, their daughters were married to tradesmen, and elder sons of tradesmen (such as Plat) elevated themselves by wealth and education into the gentry. Ian Archer has noted the increasingly positive portrayal of merchants in print in the period when Plat was active. The secretary of the Merchant Adventurers' Company argued in 1601 that 'nobles could trade without derogating from their nobility'. John Wheeler, in his *Treatise of Commerce* of 1601, thought all classes pre-occupied with

---

seamen; turnsole; trinkets made from *Lignum Rhodium* (an imported aromatic wood); medicines; perfumes, plus another 14 ideas.

9   BL, SL 2172, ff. 13–14v.

commerce – everybody was engaged in it: 'The Prince with his subjects, the Maister with his servants, one friend and acquaintance with another… the Husband with his wife, women with and among themselves'.[10]

We do not know the value of Plat's rents, or revenue from medical practice or from the law. Life at Court was not cheap and he may have been forced to supplement his income by marketing his inventions. He would not have been the first at Elizabeth's Court to find that the expenditure necessary to support his status far exceeded his income.[11] He may, however, simply have been tempted to exploit his inventions by the rich opportunities for commerce in the London of his day. The city's size and wealth made it an expanding market which dwarfed all others in England.[12] A population of 40,000 to 50,000 in 1500 had risen to 200,000 in 1600, and continued to grow rapidly thereafter. In Europe, only Naples and Paris were larger in 1600 and London was sixteen times bigger than Norwich, the next most populous English town. Mortality in London was high, therefore London sucked in people, mostly young adults, from southern England to sustain its growth. Most of these economic migrants were poor people wanting work or charity (especially in the bad harvest years of the mid-1590s) but a good few, like Plat's father, came as apprentices hoping for a successful career. Such a large concentration of people created a market economy for everyday goods bigger and more complex than anywhere else in the kingdom. Food, drink, clothes, footwear, bedding, furniture, cooking and eating utensils, lighting, heating, water supply, waste removal, transport, and housing, all had to be brought into the city and sold at markets or provided at a cost to the individual. In the case of food and drink, the flow had to be constant to satisfy so many hungry mouths. Plat's schemes and

10   Peter Earle, *The Making of the English Middle Class*, 1969, pp. 1–16; Ian W. Archer, 'Material Londoners?', p. 181; John Wheeler, quoted in Deborah E. Harkness, '"Strange" Ideas and "English" Knowledge', in *Merchants and Marvels: Commerce, Science, and Art in Early Modern Europe*, ed. Pamela H. Smith and Paula Findlen, 2002, p. 142.

11   David Harris Sacks, 'London's Dominion', in *Material London*, p. 28.

12   For a detailed view of London's economic importance, see *London, 1500–1700*, ed. A.L. Bier & Roger Finlay, 1986; *Londonopolis*, ed. Paul Griffiths & Mark S.R. Jenner, 2000; *Material London*, ed. L.C. Orlin, Philadelphia, PA, 2000; F.J. Fisher, *London and the English Economy*, 1990.

inventions – a mechanical bolting hutch, ways to improve and cheapen the production of beer and ale, cheap alegar, food preservation, cheap candles, cheaper vessels to boil in, coalballs and new uses of waste heat, a new fertilizer and other inventions to improve agriculture, a cheap mortar for house building – are all ideas drawn from his experience of the great market for basic necessities and the many industries and trades that catered for this demand.[13]

London's role as the centre for litigation, seat of national government, chief port for international trade, financial heart of the country, and important provider of gentry education, drew in large numbers of the richer sort to stay for some part of their lives or as regular visitors, enriching the lawyers, teachers, merchants and traders who served them. Gentry enriched by rising agricultural rents came to London, often with their families, for pure pleasure. Their demand for London accommodation in turn benefited Londoners rich enough to let out property. As a courtier, Plat would have seen and participated in the conspicuous consumption that was part of that life. As an avid visitor of the traders who catered for such demand, he was acquainted with the bewildering array of specialists supplying luxury goods. Many artisans involved in luxury trades were on Plat's books as medical patients. His practice-notes record trades such as artist, painter, goldsmith, gold wire-drawer, gold-lacemaker, diamond-cutter, feather-dresser, gunmaker, upholsterer, and pewterer. He visited shops in the Royal Exchange where many luxury goods were on display and he also sampled the entertainments London offered both rich and poor, pleasure gardens and playhouses, wine shops and taverns. His coach, gold chain and the jewels his wife wore in her hair display a personal participation in the luxuries London could provide.[14]

His first father-in-law was a senior customs official and Plat would have seen the constant flow of expensive foreign goods through the port of London – including many imported from Italy and other parts of the Mediterranean – which influenced élite tastes and fashions. As a trained lawyer he probably participated in the legal business centred on London,

13   David Harris Sacks, 'London's Dominion', pp. 22–24; E.A. Wrigley, 'A simple model of London's Importance', in E.A. Wrigley, *People, Cities and Wealth*, 1987.
14   BL, SL 2209, ff. 21–25; SL 2172, f. 13; TNA, E134/8 Jas I/East 2.

and as an author he was familiar with the London book trade, another activity concentrated largely in the capital. Much death and disease made London a fruitful market place for medical practitioners such as Plat: as we have discussed, all sectors of society were eager consumers of medicines and medical advice.

How consumption was stimulated in this period is debatable but the range of goods on offer is not in dispute. Nor is the size of the workforce involved in supplying luxury goods and services (and cheap imitations of such products for those with some cash to spare and a desire to copy the rich). So it is no surprise to find amongst Plat's lists of profitable secrets many luxuries: new ways of making wine, interior decorations, silk dyes, perfumes, marbled paper. Also present were consumer goods to tempt the casual shopper: rainproof clothes, children's spelling-dice, trinkets made from exotic woods, strong drink, shoe-blacking and strong shoe-leather. The luxury goods made cheaply for those who wished to emulate the rich included silver and gold ornaments cast hollow from a small amount of metal, and ornamental architectural mouldings made from papier mâché and other cheap materials.[15]

Plat used various strategies to gain income from the commodities he produced or intended to produce and secrets of various kinds he discovered. He tried selling goods in retail and wholesale markets or to particular individuals and institutions. Then he attempted selling the secrets of his inventions by negotiation. Two more alternatives were to procure rewards, patents or office from the Crown for his secrets, or secure contracts for the supply of goods to the Crown.

In the long lists of secrets Plat hoped to make money from were many that he hoped to produce for sale to the public. In the case of pasta and other preserved foodstuffs, we have already examined his activities in some detail. Plat had high hopes of profits from other commodities, none more than from mouldings in papier mâché or some other finely powdered substances mixed with glue, for exterior and interior house decoration. He was always conscious of the commercial potential of this invention. An early description of the technique is headed, 'A Devise for cheape & fayre

[15] Sara Pennell, 'Consumption and Consumerism in Early Modern England', *Historical Journal*, 42, 2, 1999, pp. 549–564.

Mantletrees all kinds of embossed woorckes or pictures, pillers, Boles, Compartementes for Armes, Carved Antike Borders of carved Woorcke, Lions, dogges, all Beasts and fowle'. He later produced a full description of all anticipated uses for the 'Moldinge in paper stuff, sawdust, plaster, alabaster, sande', beginning with 'Borders or freises for chambers galleries &c. of all kindes of fruits and flowers knott together by their stalkes and trayles, or standinge severally upon their owne stalkes or slippes. theise beinge placed over hangings of tapestrie, dournick, sea &c will grace the roome exceedingly the same also beinge made in a dew proportion will garnish bedsteedes corte cubbors tables &c & serve insteede of a valence & especially they will adorne the rownde hoopes of a Canopie'.

These mouldings would also make, 'Valences for beddes and the toppes of Cannopies, made of sea velvet, satten taffeta or calico or other linnen cloth and collours and then garnised wth this stuff'. Or they could garnish mirrors: 'Borders of antique frutage or morisco worke being cutt according to the proportion of looking glasses, will serve to make such variety of frames for them as doe usually comme from Venice'. Furniture might be enhanced : 'if you have excellent moldes sharpely cutt of brasse you may mold owt of them ether bordes for virginals, or works to garnish cabinets, diskes, or jewell chests, whose ground being taken away wth googes and formers, then perforce an other work may bee placed upon a ground all guilt, and so will grace the same exceedingly'. And ceilings might be embellished, 'Large stately and rancke Compartiments wth blanke Schocheons' being filled with casts of various designs and similarly 'iff you choose apte work, you may bewtifie the Iames [jambs] and manteltrees of chimneys wth this worke, and make stately chimney peeces and other termes and pilasters according to the excellency of the moldes'.

Being light and insubstantial, 'This arte in paper serveth well to garnish pageants and play howses & banqueting howses'. Equally, 'Vizares and armors and targets for players might bee made of this paper stuff and signes to hange at mens dores'. The papier mâché casts could either be covered with marbled paper or silver leaf, or painted.

Plat's use of papier mâché was probably not his own invention but an idea he had borrowed from Italy. Indeed, the embellishment of houses, inside and out, with ornate plasterwork and papier mâché was introduced into England by Italian craftsmen and taken up by English artisans from the 1550s. At Hampton Court, in parts of the building built by Cardinal Wolsey and Henry VIII, ceilings moulded in the Italian manner survive. In Cardinal Wolsey's Closet, the ceiling is divided into panels, 'each panel is filled with Italian decoration modelled in papier mâché'. Italian style in art and architecture in general, as well as in other luxury goods such as mirrors and glassware, remained in vogue in Plat's lifetime.[16]

Plat may have been innovative in his use of materials but in his choice of designs he reflected prevailing tastes. We have already seen one idea, 'all kindes of fruits and flowers knott together by their stalkes and trayles, or standinge severally upon their owne stalkes or slippes', a style carved on a tomb in London in 1594 but more common in the mid-seventeenth century. Did Plat visit this tomb in St Helen's Bishopsgate and follow his own advice that, 'yt is an excellent and proffitable Course, to mold of, ffaces, doggs, lions, borders, & Armes &c. from Tombes or noble mens galleries'? I have already quoted a list of other possible designs, but a further idea was to cast 'Capitall Romaine letters being cutt large & ranke' to 'set upon borders,' for 'theise letters will serve fitly either to make any prose, verse, or contrived speech, or els to then serve for mens names and theyre wiffes to place in the corner of theyre mantletrees, or upon the pillers of theyre beddes or corners of theyre cubbords, tables, formes &c'. For a more uplifting display the letters might form, 'divine or Philosophicall sentiments, poeticall verses emblems, and adages &c.' in sumptuous colours 'the grownd azure & the letters gold'. (For an idea of such 'Philosophicall sentiments' in precisely these colours, see the frieze of the Selden End of the Bodleian Library, Oxford.) Plat helpfully adds that 'Yff you cutt loose Romane letters, cutt diverse of one sorte together, upon large tables, always havinge care to cutt most of those letters wherof you finde the moste woordes in a dictionary as the .C. the .S. the J. the R. the P. the V. &c'.

16   BL, SL 2210, f. 148; SL 2197, ff. 16–17; SL 2216, f. 111; Jourdain, *English Interior Design, 1500–1830*, 1950, pp. 21–24; *Material London*, pp. 64–5, 268–289; John Schofield, *Medieval London Houses*, 1995, p. 119; J. Alfred Gotch, *Early Renaissance Architecture in England*, 1914, pp. 197–9; Linda Levy Peck, *Consuming Splendor*, 2005, pp. 14–15. 'Termes': terminus, a post with a carved head at the edge of a piece of ground.

Yet another bright idea he had was that within the compartments of coffered ceilings casts of gentlemen's arms might be placed, or for more general consumption, 'A square frame wherin all the 12 Companies armes of London, or if yt please, the Q. armes and the Cities armes also may bee graven, will bee likely to afford tables that will sell well.' Demonstrating a flair for selling, Plat continues, 'these armes though they bee all molded together, may bee devided and sold to every Citizen severally to garnish the corners of his mantlettrees, or placed in Court cubbards, bedsteads &c.'[17]

Courtiers were the innovators in this new taste for interiors ornamented by fruits, flowers, animals, birds, figures, heraldic designs and improving texts. Claire Gapper has traced the spread of the fashion for another form of plaster decoration, strapwork, from courtier's houses to other well-off householders in the early seventeenth century and then to quite modest dwellings. Influential courtiers built (or renovated) houses in London or residences just outside the city. Their sumptuous interiors were meant to impress visitors so it is not surprising that decorative details were copied in other London houses. Wooden moulds for the ornaments, once made (they were purchased from woodcarvers or joiners), became part of the stock of plasterers and were reusable – an easy way to spread fashion.[18]

Many of Plat's ideas for moulded ornaments would have come from the great houses and royal palaces he visited in London. Book illustrations were further inspiration. Plat advises, 'Buy printed bookes in Collors and without collors, and make choise of the most speciall work for this purpose.' He recommended a work by Conrad Gesner, for 'beastes, fishes, &c whose colors & counterfeites may bee founde' therein. He also points to woodcuts from 'Munster', that is Sebastian Munster, whose *Cosmographia* went through many editions after 1544, and 'the choysest of whitneys emblems', illustrations from the popular emblem book of Geoffrey Whitney, *Choice*

17   *Material London*, p. 310; BL, SL 2210, f. 104v; SL 2197, ff. 16–17; p. 000.
18   Claire Gapper, 'The London Plasterers' Company and decorative plasterwork in the 16th and early 17th centuries', *The Journal of the Building Limes Forum*, vol. 9, 2002, pp. 8, 13, 19; *Cambridge Cultural History of Britain*, ed. Boris Ford, Cambridge, 1992, p. 282.

of *Emblemes*, first printed in 1586. The use of printed sources as patterns for moulds was common: at Blickling Hall, Norfolk, the long gallery ceiling of *c.* 1625 has eleven central symbols copied from plates in Henry Peacham's *Minerva Britannia* of 1612 and the gallery at Little Moreton Hall has devices copied from the *Castle of Knowledge* by Reynold Wolfe, printed in 1556. Engravings of likely subjects were available as single sheets at reasonable cost: many moulds were made of some popular engravings. Plat was thinking of such sheets as sources for his mouldings of likenesses of the Queen and her close relatives, former kings of England and other notables such as Drake and Essex.[19]

The ready availability of patterns and the use of cheap wooden moulds, coupled with a spread of the fashion for moulded ceilings, make Plat's enthusiasm for selling these castings made out of cheap ingredients understandable. Despite being architectural features which one would normally think had some permanence, these were fashion items, cheap substitutes for plaster mouldings, ready-made and presumably capable of installation by unskilled labour (undercutting members of the Plasterers' Company). Plat's care in listing many designs for his mouldings and imaginative ways in which they could be used reflects an understanding of market conditions in London for cheap 'luxury' goods. John Sykes emphasizes the sophistication of the London market: 'The density of information networks and personal interactions in which Londoners were enmeshed, educating them as consumers by exposing them to fashion and novelty in a particularly intense way, was unmatched in even the largest provincial towns'. Early modern England was receptive to new consumer goods but sellers had to create the right conditions, tailoring products to new markets with a balance of novelty and familiarity, making unfamiliar products, such as these mouldings, less daunting by carefully choosing popular designs, and recognizing the market for cheaper versions of new products for the 'middling sort' of customer.[20]

19   BL, SL 2197, ff. 16–17; *Cambridge Cultural History of Britain*, , 1992, p. 283; Jourdain, *English Interior Design, 1500–1830*, p. 23; Gapper, 'The London Plasterers' Company and decorative plasterwork in the 16th and early 17th centuries', p. 20.
20   John Sykes, 'Product innovation in Early Modern London', *Past & Present*, no. 168, 2000, pp. 124–169.

For no other invention did Plat make such detailed notes on how to present the product to the public. He has a few ideas for artificial wines, such as, 'Qre of making rennish wyne all the Sommer and selling the same to the vintners as the stilliard men use to doo'. And he jotted down more brief thoughts on marketing preserved fruits, and gilded 'Armes, Banners, Creasts &c. good cheape' which he believes will sell well – 'when the easines of the price shall bee generally knowne, there will want no Customers'. He probably intended more memoranda in this vein, to judge by the many headed but blank pages of his notebook devoted to selling techniques.[21]

Plat's most intriguing project for selling wares created from his secrets was to display them either in a shop, or what he called a 'Jewell House'. There are several references in his manuscripts, such as to 'My shop of Arte' (i.e. stocked with manufactured goods); a plan to sell sweet oils and perfumes in 'myne own shopp'; and a long list headed 'Wares for my new shop'. Repeated use of the possessive establishes he envisaged being the proprietor rather than just supplying a trader. The list of 'Wares' contained some specific inventions – ink powder, multiple roasting-spits, oil for armour, couscous and macaroni – as well as general categories: 'Bewties for woemen', 'All the oyles of spices', 'conceipted sugar knacks', or 'Vinegars of severall flowers & severall colors'. It is a reminder of Plat's many inventions and interests: enough to stock a shop.

Plat's cousin Henry Davenport wrote to him in 1595, telling him he and his wife were planning to move from Coventry to Hoxton, a village just east of London and near his home. 'This may encourage you to proceede cheerfully in furnishing one shop or Jewell House without any fear of not having it applied diligently', he observed. Maybe Plat had asked Davenport to manage the establishment. He certainly thought he might help him sell goods at the nationally important Stourbridge Fair in Cambridgeshire and at London's Bartholomew Fair, and that he might also advise on selling a new type of shoe leather. In all probability Davenport was a merchant, experienced in buying and selling a variety of goods. If Plat was planning a shop, the site which springs to mind is the Royal Exchange, which was mentioned by him as a possible outlet for various perfumed goods.[22]

21   BL, SL 2197, ff. 20v–25.
22   BL, SL 2216, ff. 188v–189; SL 2172, ff. 13–14, 21; SL 2197, ff. 14–15.

SIR HUGH PLAT

The Royal Exchange, completed in 1569, was part 'bourse' or meeting place for domestic and overseas merchants and moneymen, and part shopping mall. After a slow start, it gained fashionable attention when the Queen visited in 1571. By the time Plat was writing, there was a waiting list for shops. Its success can be gauged by the number of foreign visitors who described it, the many literary references to the Exchange as a place of assignation or intrigue, the countless prints depicting it, and the general acceptance that for high quality goods or just to mingle with fashionable company, one went to the Exchange and perused its 120 shops.[23]

Haberdashers, mercers, painter-stainers, merchant tailors, grocers, leathersellers, and clothworkers were early tenants, with other luxury trades such as goldsmiths, girdlers, milliners, stationers and upholsterers well represented. These were vendors of the sorts of goods in Plat's list of items to sell. The shops themselves were on the upper floors. A French visitor described the layout in 1571: 'The third walk is a gallery, which is above the others and is excellent, beautiful and rich. You climb to it by 25 or 30 steps which are arranged in flights of seven. That gallery has all around it 150 stalls of rich merchandise, most notably of all sorts of mercery'. 'Stalls' was only partly accurate. There were probably boards at the entrance to each shop displaying wares, for inside every inch of space was needed as they were little more than booths, measuring only 2.28 metres by 1.52 metres. Some of Plat's list of goods: 'A waggon for the sea men to draw theyre vessels on land', 'my Engin to take any dogg or beast in', would never have fitted the space. Others, such as his trinkets and perfumes, rainware and children's spelling-dice, were small fashionable goods which might tempt rich browsers.[24]

As well as listing goods to sell in his shop, his plans for 'all my other Secrets and Wares not mentioned in any other perticular here' were 'to sell the wares in my Jewell house'. His Jewell House was, therefore, a physical place, not a reference to the book of 1593, and in his manuscripts his Jewell House is differentiated from his shop (although Davenport implied they were the same thing). A jewel house, in common parlance, was a strongly

23 The many essays in part 1 of *The Royal Exchange*, ed. Anne Saunders, 1997, provide much detail on this building and popular reaction to it.
24 *Royal Exchange*, ed. Saunders, pp. 48–9, 59, 89; BL, SL 2197, ff. 14–14v.

made building to secure jewels. In more metaphysical terms, it was a place of innermost secrets, a cabinet of curiosities. It may be significant that Plat anticipated visitors to his Jewell House by invitation; it was not open to the general public. So, to see his 'Secrets and Wares' he proposed asking ' Sr Charles Cavendish, Dr. Doile. Mr Brooke' to 'procure all the Innes of Court and Chancery to repayre unto mee'. Other select groups of individuals may also have been invited to the Jewel House, for he placed a good deal of faith in his ability to inform gentlemen directly about his products, often relying on friends for suitable introductions. To sell blank papier mâché escutcheons he hoped 'To deale with Noblemen and gentlemen … to put in their Coates' using his friend the herald. 'Mr Segar will bee a fitt man to recomend these blanke Schochond to all gentlemen', and his 'Cosin Robertson may bring gentlemen'. Another idea was that 'Some particular persons of each of the Companies might bee dealt withall, to take a certein number of the Armes of theyre owne Company and by that meanes many meyst bee vented'.[25]

The idea of a shop may have come to nothing but the Jewell House did exist; it was the exhibition of artefacts for sale at his London house. He makes a note to have papier mâché adornments put on furniture and 'To have a table, Coort cubbard and deske in myne owne howse ready finished for all gentlemen to beholde'. He placed great emphasis on the exhibition of his inventions. In his *Briefe Apologie* pamphlet of 1593 we are told repeatedly of his display of artefacts at home to various classes. He had 'called some of the choicest of you [bakers] with diverse other Citizens of good worship and account' to his house, and also a 'Court of Aldermen' had attended, as well as 'one of the Lords of her Maiesties privie Counsell, with three other Gentlemen all well born'.[26]

For wares that did not lend themselves to display in shop or exhibition, Plat intended to sell through agents. To sell macaroni for sea voyages he

25   Plat, *Certaine philosophical preparations of foode and beverage for sea-men*, c. 1607; BL, SL 2172, f. 13; SL 2197, f. 17v.
26   BL, SL 2172, f. 13; Plat, *A Briefe Apologie of Certaine New inventions*, 1593. Deborah Harkness finds much public interest in viewing mechanical marvels in London at this time. She highlights the exhibition in 1588 at the Guildhall of an 'artificial motion' devised by a Dutchman, Deborah E. Harkness, '"Strange" Ideas and "English" Knowledge', p. 147.

hoped to use the agent 'Bradbury, to harken after all the Captains and officers of long voiages'. Gentlemen might be approached by intermediaries to buy his home-made wine (we have already seen that he organized a wine-tasting for a group of courtiers and the French ambassador), and he hoped to gain the attention of 'all the fine dames of London' to advertise his cinnamon water to them 'by somme speciall meanes'.[27]

Another way of selling was to approach institutions which might buy many items at one time. To sell his papier mâché blank escutcheons he proposed: 'To deale with my L. of Canterbury for the furnishinge of all the Churches in England wth the Queenes Armes. To deale first with the Fishmongers of London to furnish soundry armes for them, & then all other Companies of London. To deal with both the Universities for all theyre College armes.'[28]

Selling at the major annual fairs in England might also shift a good deal of stock. Plat suggested selling his castwork at 'Bartholmew fair and Stur bridge fayre'. Bartholomew Fair, a summer fair held in London at Smithfield, was a big trading and social event, the setting of one of Ben Jonson's most famous plays, while Stourbridge Fair in Cambridgeshire was the largest fair in England at the time. 'According to the charter of 1589, Stourbridge Fair "far surpassed the greatest and most celebrated fairs of all England; whence great benefits had resulted to the merchants of the whole kingdom, who... sold their wares and merchandises to purchasers coming from all parts of the realm". With its elaborate layout of streets and squares, its booths for jewellers, perfumers, silkmen, haberdashers, drapers, upholsterers, cabinet-makers, potters, pewterers, braziers, gunsmiths, and ironmongers, Stourbridge Fair provided for sixteenth-century provincials much of what a department store in the West End of London does today.' One might add that in 1611 the reverse comparison was made, London was described as a fair which lasted all year and, indeed, the many Royal Exchange booths resembled a permanent fair.[29]

27   BL, SL 2172, f. 13; SL 2197, f. 23v.
28   BL, SL 2172, f. 13.
29   *Agrarian History of England and Wales*, vol. iv, ed. Joan Thirsk, pp. 535–6; John Davies, quoted in David Harris Sacks, 'London's Dominion', in *Material London*, ed. Orlin, p. 20.

Wholesaling was another means of disposing of his goods. Plat wondered about supplying artisans with materials: joiners and upholsterers might buy his ornamental castworks, so might 'all new builders for chimney peces, manteltrees, freises etc'. Builders might show interest in his artificial stone and mortar of soap ashes, and painters might buy his verdigris. Shopkeepers might be induced to retail his products. 'Sweete oyles & waters to perfume with' could be placed 'in somme Apothecaries or Milliners shopp' and apothecaries might also stock cinnamon water and other distillations of strong waters. Other items could be sold to grocers including shell gold and silver, artificial opium, and refined myrrh. 'Cheynes and Braceletts' made of aromatic wood could find their way to the customers through 'goldsmiths, phisitians and Exchange shopps'.[30]

Plat's sales methods were sophisticated, and showed a keen appreciation of new markets. The warfare in Ireland at the end of the sixteenth century brought increased numbers of Irish to London, some of whom were street traders, and Plat saw a market for his home-made 'Usqua Bath or Irish Aqua vite' which 'might bee made knowne amongst the irish Costermongers and they wold soone make it knowne to theyre fellows, that it is to bee hadd at 8d the pinte'. Even more ingenious were his proposals to sell shoe leather strengthened according to his secret recipe and his shoe blacking-oil:

> yff somme duch man didd sett upp bills uppon poastes, to serve a payre of soles for 12d and if they will have them sett on, then for 14d, which shall last as longe againe at the leaste as ordinary soles doo so peradventure hee might soone gayne stoare of Customers. The soles might bee delivered to such strangers first prepared wherby there cold no disceyt bee shewed toward the Author of thinvention. this offer might bee published at play howses. The secrete of my blacke shew liquor might also passe herwth.

The use of Dutch cobblers may have been a way of avoiding conflict with London guild regulations. Many Dutch immigrants traded from Blackfriars, a liberty outside the control of the city by virtue of its former

30 Patrick Wallis, 'Apothecaries and Medicines in Early Modern London', in *From Physick to Pharmacology*, ed. Louise Curth, 2006, p. 6; BL, SL 2172, f. 13–15.

status as a monastery. Being careful to protect the secret of this strengthened leather may have led Plat to prefer Dutch craftsmen. In the absence of guild protection they were very protective of their crafts and would not have divulged any secrets they obtained to other London artisans.[31]

The proposals for selling shoe leather involved advertising, an art Plat used extensively. The simplest method was the bill or poster, so named because they were commonly fixed to posts. Tiffany Stern has conclusively demonstrated the importance of the playbill in advertising theatrical productions. Such bills, already the subject of a jest concerning a fraudulent advertisement in 1567 and made the monopoly of one printer from 1587 onwards, were printed in some numbers, from 100 to 1000 per announcement. They were set up on posts of all types – tethering posts, pissing posts, door posts, the outsides (and insides) of public buildings, and must have been a sight akin to the poster-strewn spaces near many modern students' unions. Not only plays were advertised by these ephemeral papers; bills survive for archery contests, fencing matches, bear baiting, and a rope-dancing entertainment. Booksellers (whose livelihood was print) festooned their stalls with title pages of new works as advertisements. In another reference to selling shoe leather, Plat thinks 'it will not be amisse to printe the place of repayre & to set upp the bills in every place'. In the crowded, bustling streets of London this was shrewd. Bills were fast and flexible. When the plague was rife, Thomas Lodge complained of flyposters for plague remedies 'that have bestowed a new Printed livery on every olde post' and, because some posters near his house were anonymous 'because at the first he underwrit not his billes', people thought they were his and pestered him for the promised cure. A French medical practitioner, Charles Cornet, was punished by the College of Physicians for trading illegally, advertising by putting up bills 'on all corners of the City'.[32]

31 BL, SL 2197, ff. 18v, 23v; N. Goose and L. Luu, *Immigrants in Tudor and Early Stuart London*, 2005, pp. 69–70; D.J. Bryden, 'Evidence from advertising for mathematical instrument making in London, 1556–1714', *Annals of Science*, vol. 49, 1992, p. 303.

32 Tiffany Stern, '"On each Wall and Corner Post"', in *English Literary Renaissance*, vol. 36, I, 2006, pp. 57–89; Paul J. Voss, 'Books for sale: Advertising and patronage in late Elizabethan England', *Sixteenth Century Journal*, vol. 29, no. 3, 1998, pp. 737; BL, SL 2172, f. 14; Thomas Lodge, *A treatise of the plague*, 1603, A3–4; Harkness, '"Strange" Ideas and "English" Knowledge', p. 141.

Plat's broadside of *c.* 1607 advertising medicines for the sea was most probably designed as a bill to be set up near wharves. Its confident title *Certaine philosophical preparations of foode and beverage for sea-men, in their long voyages: with some necessary, approved, and Hermeticall medicines and Antidotes, fit to be had in readinesse at sea, for prevention or cure of divers diseases* was set in eye-catching type beneath a decorated headpiece within a double-ruled border. The title leads the reader into a detailed description of the food, drink and medicine which 'I am bold to offer and publish for the benefit of seafaring men'. Plat undertakes to supply all the goods on offer and boasts of distinguished clients, 'With this the Author furnished Sir Francis Drake and Sir John Hawkins, in their last voyage.' Since both died on their joint last expedition, how wise a boast was this? In Plat's papers is another document, written in his very best hand, headed 'Waters, Oyles, Electuaries, & other Compositions necessary for the Sea'. Without preamble, it lists many medicines, sold by the ounce, pound or pint, together with 'Certaine other necessaries for the sea'. Again the name of Drake is used, here to sell macaroni and oil to prevent armour from rusting. All the items offered are priced, for example:

> An excellent plaister for greene woundes, & old ulcers,
>   the pound 0–16–0
> A pill for a violent coughe & procuring sleepe to be taken
>   severall times oz 0–10–0
> Rose water the pinte 0–1–4

Either this paper is a manuscript bill to be handed to likely customers or a draft for another printed poster.[33]

The main media for Plat's advertisements were the pamphlet and the book. Beginning with the pamphlet *A Briefe Apologie* in 1593, a year later he printed advertisements in *The Jewell House*, followed by another pamphlet in 1595. After this flurry there was a gap until 1600 when a book and a pamphlet with advertisements appeared, then another in

---

33   BL, SL 2172, f. 30; see Appendix IV for a full transcription of this advertisement.

1603, finishing with *Floraes Paradise* in 1608 shortly before he died. Over the years Plat developed his persuasive techniques. Like other authors, Plat was confronted with a sharp decline in the generosity of literary patrons in the 1590s. It is significant that only one of the works from his whole output from 1590 to 1607 has a personal dedication: *The Jewell House* of 1594 was dedicated to the Earl of Essex. Paul J. Voss has demonstrated that the publishing industry combated this decline in income by advertising. 'Instead of appealing to a wealthy patron or aristocrat, the printing press allowed investors direct access to the buying public.... Moreover, these advertisements served a distinct number of functions, including promoting reputations, establishing expertise, advancing knowledge and encouraging investment' – to which we may add, in Plat's case, selling goods to the public.

The 1590s were a period of increasingly creative advertising and this is seen immediately at the front of books and pamphlets by the use of 'front matter' (title page, preface, frontispiece, dedicatory poems and the like) as an advertisement for the work's contents. Printed last, this section could be used as an up-to-date advertisement for the new book. As many books were sold unbound, title pages and the like caught the eye of the browser in a stationer's shop or even fluttered in front of him as an advertisement stuck on the doorpost of the shop.[34]

Plat and his publishers became practised in this sort of advertising. While the title pages of his books and pamphlets up to 1600 were unexceptional, in that year the title page of *The new and admirable Arte of setting of Corne* displayed more verve. A very full title takes up half the page and invites the reader to look at the contents page: 'the particular titles whereof, are set downe in the Page following.' The bottom half of

34 Paul J. Voss, 'Books for sale: Advertising and patronage in Late Elizabethan England', pp. 733–756. Patronage had benefits over and above any largess from a patron. Stephen Pumfrey and Frances Dawbarn comment in general, and with particular reference to *The Jewell House*, 'Given the inferior status and uncertain reputation of the typical author of a work of natural knowledge, it was the dedicatee who first guaranteed the authority of a work – who was, in the sense of authorization, the principal author; a serious work, especially a novel one, without an authoritative dedicatee risked lacking credibility.' Stephen Pumfrey and Frances Dawbarn, 'Science and patronage in England, 1570–1625', *History of Science*, vol. 42, 2004, p. 152.

the page is illustrated by a dramatic sheaf of barley growing from a single root and a spade with 'Adams tool revived' on a scroll entwined around it. In 1603 the title page of *A new, cheape and delicate Fire of Cole-balles, wherein Seacole is by the mixture of other combustible bodies, both sweetened and multiplied* is set in bold type, above a woodcut of a pile of coalballs blazing on a hearth. The page goes on to describe the advertisements at the end of the pamphlet: 'Also a speedie way for the winning of any Breach: with some other new and serviceable inventions answerable to the time.'

This schema is repeated in *Floraes Paradise* of 1607: a title in relatively restrained typeface surmounts the advertisement 'With an offer of an English Antidote, (being a present, easie, and pleasing remedy in violent Feavers, and intermitting Agues) as also of some other rare inventions, fitting the times', continuing, '*Huis fruere, et expecta meliora*' (enjoy this and expect better things). On one occasion, in the front of *Delightes for Ladies* (1602) Plat uses the increasingly popular device of an introductory poem to entice browsers further into the book, summarizing in verse his main achievements to date as well as the contents of this book.

The relatively new conceit of inserting advertisements into the actual text of a book is employed in *The Jewell House*. For instance, Plat promises more on rose-growing in his next impression and that more will also be found in 'my new conceyted booke of gardening, wherein I have set downe sundrie observations, which neither M. Tusser though hee have written sharpely, nor Master Hill though hee have written painfully, nor Master Barnabe Googe though hee have written soundlye ...[have covered]... All which are readie for the Presse, and doe onely attende to see if Noahs Pigeon will returne with an Olive branch, seeing his Raven hath as yet brought nothing with her'. (No second impression of *The Jewell House* appeared, as far as is known, and the gardening book he was extolling took until 1608 to be printed.) Later, he tells readers he has further navy victualling secrets to reveal, pending 'some honest pencion etc.' He advises mariners requiring portable stills for fresh water at sea that, in these instruments 'I have found maister Sergeant Gowthrowse, the moste exquisite and painfull practizer and performer of our times'. In *Sundrie new and Artificiall remedies against Famine* (1596) he recommends *The*

Jewell House to readers, adding 'The Book is to be had at the Greyhound in Paules churchyard.'[35]

But Plat's most sustained publicity occurs at the ends of his pamphlets and books, where secrets are set out in varying degrees of detail. The form varies from a bald list at the end of *A Discoverie of certaine English wants*, to a section at the end of *The Jewell House* advertised as a part of the book in the table of contents and containing eye-catching woodcuts of the main secrets advertised. The advertisements at the end of *Floraes Paradise* also occupy a specific section with a new title. The initial resort to print to advertise his secrets was prompted, apparently, by the poor reception of the inventions he exhibited at his London house. In 1593 he berates the 'ancient and rich bakers of London, which thinke your selves both too old and wise to learne' because they did not applaud his new bolting hutch, and the 'senslesse blockes and penifathers of our time' who scorned his wooden vessel to boil in. In the pamphlet on coalballs of 1603, although he reveals the full details of that secret and promises to reveal others, he further suggests he has many military secrets of national interest to disclose: 'Some of these new inventions the Author hath already shewed to divers of his honourable and private friends, and the rest upon reasonable reward shall bee made good for any publike service'. In 1607, in an appendix to *Floraes Paradise*, he advertises a totally new set of secrets although, significantly, these are all luxury goods aimed at the gentry. Maybe his presence at the Court of James I was more conducive to successful selling than his attempts to sell to London artisans and the Crown under Elizabeth.[36]

Certainly there is a marked difference in the type of product advertised and the language used between the appendix to *The Jewell House* in 1594 and that of *Floraes Paradise* in 1607. In the first the secrets offered are utilitarian: coalballs, pasta, bolting hutch, saltpetre-making, a portable pump and the like, the majority of use to artisans and ordinary Londoners,

---

35 Plat, *Jewell House*, pt. 1, pp. 5–6, 9, pt. 3, p. 17; Plat, *Sundrie new and Artificiall remedies against Famine*, 1596, f. C1. Plat's reference to Noah alludes again to the cool reception at Court of his early pamphlets and exhibitions of secrets (the raven).

36 Plat, *A Briefe Apologie of Certaine New inventions*, 1593; Plat, *A Discoverie of Certaine English wants*, 1595; Plat, *A new, cheape and delicate Fire of Cole-balles*, 1603.

whereas in 1607 various ways of making gold and silver objects more cheaply, new types of wine and new medicines are on offer. The copy differs too. In 1594, long lists of reasons to buy the product or secret are set out, with emphasis placed on cheapness, utility, durability and profitability to the purchaser, as well as the benefit to the poor and commonwealth in general, themes which run through Plat's other advertisements in the 1590s and early 1600s. The tone is sometimes aggressive: malicious slanderers and doubters of his wooden vessel to boil in are challenged to 'wage a competent sum of monie, as may countervail the discoverie of the secret' and Plat will then demonstrate the vessel 'before anie indifferent Iudges' and pay out double the sum wagered if he loses. In contrast, the tone of the 1607 advertisements is smooth and gently persuasive, beginning, 'And now it is time to solace our selves, with some pleasing and extraordinary secrets… that I may both give you a taste of some variety of skill and also furnish such as desire the same, with the delicate fruites of some new inventions, which I know the nature of man doth earnestly affect.' He introduces his secret of thinner and lighter silverware to gentlemen who wish for 'all sorts of dishes, basons, sawcers, and trencher-plates of Silver; which shall be every way as serviceable, as faire, large, & beautiful, and in fineness truly answerable to the Standert and touch of Goldsmiths hall'. These vessels are 20 per cent cheaper than the norm but Plat is appealing to those who want a high degree of luxury first and a cheaper price second. The copy emphasizes pleasure, no mention is made of utility, profit or the common good. He moves on to pieces made of hollow cast gold, saving 80 per cent of the gold used to make solid pieces, 'fit for such as have money enough, as wel to satisfie their pleasures, as their necessaries'. Emphasizing the nobility of the secret, he reveals that similar pieces had been made for 'that worthy Prince, the Lantzgrave of Hessen'. The nobility is invoked when he advertises his artificial wines: the French ambassador came to his house expressly to try the wine and proclaimed 'he never drank any better new Wine in France', and a string of other leading courtiers are claimed to have tasted the wine with approval.[37]

In the preface to *Floraes Paradise* is a similarly sophisticated advertisement for a medicine, a 'spagiricall Antidote' of wide application,

37   Plat, *Jewell House*, 1594, part 4; Plat, *Floraes Paradise*, 1607, appendix.

marketed again at 'the Nobility and Gentry of this Land'. Plat reels off lots of medical conditions helped by the medicine along with detailed instructions for taking it, to give the reader confidence in his skill as a physician. Then he tells of many successful cures already effected by the medicine, 'neither have I seene it faile, but in one particular person to this day'. This last is a nice touch, admitting it is not perfect while emphasizing it is almost so. In a final piece of salesmanship, he bemoans the fact that the raw materials to make the medicine are in short supply and the preparation is long and tedious, encouraging customers to buy quickly. But never fear, if they are too late and 'my store shoulde happen to faile me, I have also another extract for intermitting Agues' which is not unpleasant to take. Plat summed up the rationale for advertising in print in his 1600 pamphlet on setting corn, 'because I know no better means to give a publicke notice thereof unto all my country men then by this pamphlet, which taking the wings of fame unto it, is like to disperse both in itselfe and his companion abroad in a most speedy and sodaine manner through all this little Iland'.[38]

We have so far concentrated on Plat's attempts to sell goods based on his inventions to the public. As important, if not more so, were his efforts to sell the secrets themselves, and to gain reward or protection from the state for his inventions (he also tried to sell goods to the state). Only occasionally does he advertise in print that he is willing to sell secrets to individuals. In 1594 he was willing impart the secret of improved saltpetre production, 'if the Author might receive a condigne reward for his profitable travell'. Two years later he announced, 'all those whome the author shall finde willing and worthie of the same, may uppon reasonable composition, become owners of the skill' of making his new compost. In 1607 he offered the secrets of his artificial wines to an individual, 'Or if there should fall out any competent number of gentlemen, that were desirous of the secret, onely to make whites or clarets, for their own private houses, I might happely be drawne upon reasonable termes, and under reasonable conditions, to impart my skill'.

38 Plat, *Floraes Paradise*, 1607, Preface to the Reader; Plat, *The new and admirable Arte of setting of Corne*, 1600.

MONEYMAKING

The problem with selling secrets to many people for their own use, Plat recognized, was the danger that the secret would soon leak out to the public at large and become worthless. The danger might be less in the case of wine-making secrets sold to gentlemen because they were honourable persons. A secret might also be safe with companies or other groups of mutually interested parties. Thus he remarked of the secret compost recipe, 'But now I see not how the Author shold reape any proffitt answerable to his discovery, unlesse somme greate person or somme whole shire wold compound for the same'. London companies might buy secrets for the use of their members: 'The Company of the bakers might compound for this secrete [more efficient use of wheat in bread] for they shall gayne by my computation, at the least 6s 8d uppon every quarter more then they doo when wheate is sold for 40s the quarter'. The vintners might buy his wine recipes and painters, or the Grocers' Company, might pay for his verdigris, and compound either for 'all that shold bee made or for the secrete'. Sometimes, on first use a secret would become obvious to all. Such would have been the sandbags used to fill a major breach in a dyke. Plat's solution was to share in the profits of the first user. 'This may bee discovered for the gayning of somme one greate breach reserving a reasonable parte for him that shall discover the secrete'. But the most efficient way of marketing secrets was to sell them outright for a lump sum, 'to sell the secrets to such as will compound for them' and this Plat tried on several occasions.[39]

We know he succeeded in disposing of at least two. In 1603 Plat lamented that he had been forced to sell at a discount 'my new & late discovery in Peter-works' to the holders of the saltpetre patent. Four years earlier, Plat sold to a London grocer the secret of extracting and refining the juice of logwood, *Haematoxylon campechianum*, a tree native to Central America and the West Indies. Used as a dye for black, grey, violet and dark red, logwood was also a medicine. Harvested by the Spanish from marshy ground on the Gulf of Campeche in southern Mexico, it was highly prized. Plat's sale is recorded on a small piece of paper in his hand and may be the sole survivor of a number of such agreements:

39   Plat, *Jewell House*, pt. 4, p. 76; Plat, *Sundrie new and Artificiall remedies against Famine*; Plat, *Floraes Paradise*, appendix; BL, SL 2189, ff. 18–25; SL 2172, f 15.

Memorandum that I Arthur Blackamore Citizen and grocer of London doe testifie by these presents to have receyved at the hands of Hugh Platt of London, aforesaid gentleman, before thensealing and delivery of certain Articles Indentured made and agreed uppon betweene the sayd parties bearing date the xviiith of February & in the two and fortith yeare of her Majesties reigne, one Secret, skill or Arte for thextracting or drawing owte of the force, strength or vertew of Logwood, alias Blockwood wherby the same may bee reduced into a new and lesser forme or boddie then naturally it haveth. In witness whereof I the sayd Arthur have sett to my hand the xxviii of February aforesaid Ao Do 1599.

Beneath the agreement, also in Plat's hand, is as short description of 'A trew and plaine discoverie of the above named Secret or Skill', a modified version of the recipe in one of his notebooks where he opines that logwood will serve as a substitute for turnsole 'and make an excellent rich secret for the diers'.[40]

From his surviving papers, we find Plat trying to sell medical secrets on two occasions, in the summer of 1600 and in the summer before he died in 1608. In 1600 he offered seventeen of his most successful medicines to Thomas Elkinton, a gentleman of Wellsborough in Leicestershire. He included in the deal advice on how to practise as a medical man, with Elkinton shadowing Plat for a time to gain experience. Letters between the two men display elaborate courtesy but Plat also demonstrates a keen sense of salesmanship. Unfortunately we only have one exchange of letters and the outcome of negotiations is not known.

The medicines he offers are all, writes Plat, best quality. Of a gout remedy he remarks, 'I never knew this to fayle in the ffoote'. 'An excellent extract for the running of the Reynes', is followed by an 'excellent & pleasing diet' for 'Rhewme' and 'a most rare water to cure any skin, filme, knot' etc. (in the eye). Finally, he lavishes superlatives on a 'philosophical quintessence' which will make Elkinton an Ajax amongst English doctors and has found favour amongst German princes. Unfortunately, Plat has

40 Plat, *A new, cheape and delicate Fire of Cole-balles*, 1603; BL, SL 2172, f. 12; SL 2189, f. 106v; Arthur M. Wilson, 'The Logwood Trade in the Seventeenth and Eighteenth Centuries', in *Essays in the History of Modern Europe*, ed. Donald C. McKay, New York, NY, 1936.

none of it left and does not know the secret of this medicine, valued at £10 the ounce (but he hoped to obtain more by next Michaelmas). This enticing list of medicines to whet Elkinton's appetite is followed by words of warning to remind him of the value of the secrets: 'I pray you of all loves and as you esteame my love and ffrendshipp, to bee very secrete even from all men saving your self in all these particulars ... and I pray you have care to seale upp all letters that you send mee in such sort as no men may open them before they come to my hands'. Plat puts pressure on Elkinton to close the deal, 'And as you were desirous of a speedy answere of your letters for I pray you let mee crave the same at your hands, ffor I have at this present somme great Schollers that are very earnest wth mee abowte these secretes in Phiscke which I meane to suspend till I heere your answere, so as it bee speedy and cary also a speedy expedition with it ... I pray you lay all ends together and then make choice of the best course both for your self and mee'. Elkinton was enthusiastic in his reply but we do not know if a deal was concluded.

The price Plat proposed to Elkinton is not recorded. The prospective purchaser offered 'Composition'. Plat offered to sell 'secretes of great woorth' but whether he proposed to sell the recipes or just supply the medicines is not clear. He would supply the 'philosophical quintessence' at a reasonable rate when he had some. In the case of a gout remedy he suggested, 'I can furnish you wth somme reasonable quantity therof at 10s the oz which is the lowest rate I can aford the same for', but he then writes that he would be loath the take £5 an ounce 'unles I have a reasonable share of your cleere gaines (the charge deducted)'. Such a profit-sharing arrangement suited Elkinton and the evidence suggests that, given worries about Elkinton's lack of secrecy, Plat probably favoured supplying medicines on this basis.[41]

A philosophical quintessence of vegetables, of great worth and efficacy, was offered for sale by Plat once more in August 1608. A few months before his death, this time he wished to sell secrets outright, probably to add to the value of his estate. To that end, he planned to approach Moritz, Landgrave of Hesse-Kassel to sell him medical and alchemical

41   BL, SL 2172, ff. 28–29v.

secrets. This may seem astonishing for a London gentleman but he had good reason in hoping to interest Moritz as the Landgrave was a leading patron of alchemical scientists. His 'Collegium Mauritanum' included a good deal of natural science in the curriculum and he gathered some distinguished alchemists at Hesse-Kassel. The college was international both in the student body and professorial staff and Plat probably hoped it would welcome his new ideas. Moreover, Plat was already acquainted with Francis Segar, a fellow-alchemist who was in the service of Moritz from 1592 until 1615. He led diplomatic delegations on behalf of Moritz to England in 1599 and 1604 and at the time of Plat's letter was in England as the Landgrave's agent.[42]

We have considered, in the chapter on alchemy, the letter Plat penned in his best hand in August 1608 to Francis Segar but meant for Moritz. Now we return to examine the letter as a carefully crafted attempt to sell, for considerable sums, Plat's most desirable chemical secrets. He begins by asking if Moritz still favours alchemy and true professors thereof 'which by your report at our last metinge [he did]'. He then goes straight into his sales pitch: 'I make no Question but that I shall fit his excellencye in his best desires, and give him that light into nature both in Chimicalls, Vegetables, and minerálls, as nether I dare valewe at anye price, nor yet can he make sufficient requitall of (in my poore conceipt notwithstanding his greatnes)'. He continues with an incentive for Segar to pass on his proposals, 'And as you have bene exceedinglye beholdinge to him for his honorable and kind Respect and recompance of your faithfull service towards him soe nowe a iust occasion I have shalbe offred him to reioyce et serio triumphare that ever a gentleman of your worth came under his cullors'.

After discussing who will be the Hessian representative to whom Plat will teach his secrets, he broaches the tricky subject of money: 'what honorable requitall he [Moritz] will make', bearing in mind he is buying 'suche particuler Secretts of Art as I have with great chardge in long time, not without great danger of my life and health attained unto', Plat is aware that alchemical and medicinal secrets are notoriously easy to fabricate

42 H.T. Graf, 'The Collegium Mauritanum in Hesse-Kassel and the Making of Calvinist Diplomacy', *Sixteenth Century Journal*, vol. XXVIII, no. 4, 1997, pp. 1167–1180.

and he is anxious to not be thought fraudulent by claiming to sell too great a discovery for too little a price. 'I must crave that I be not mistaken, for my meaninge is not to bringe anye invaluable skill suche as is the trewe openinge and exaltinge of all the minerall bodyes, to anye prince in valewation for then you myght well suspect me to be a great Impostor that would sell you manye millions for a fewe thousands if my knowledge should fall out to be sound and perfect.' He does, however, want a fair price. 'But my purpose is onely to valewe things valewable wherin I doubt not but that his excellencye shall have both good ware and good measure for his monye'.

The main secrets are set out, including an 'oyle of [mercury] irreduceable yet not trewlye philosophicall, curinge the pocks in an easie and delicate manner as alsoe that sweete oyle of Amber whiche Sr Walter Raleighe will not willinglye discover for 500li which cureth the dead palsie and all crampes and convulsions'. Other major secrets are offered: cinnamon water quickly made, quintessence of herbs, spirit of wine made without fire. These are practical, profitable inventions and may have been put forward early in the letter to persuade the Landgrave that he was getting something tangible for his money.

He next turns to intangibles. 'Now if his excellencies shall hold himselfe royally satisfied for such recompense as he shall yield in liewe thereof with these and such other skills as by these, he maybe Imagine that I am like to bestow [i.e. if the practical items are thought good value] then out of my love and kind affection, I will frelye discover whatsoever grounde and meanes that I have for the trewe procedinge in all philosophicall courses, concerninge minerall animal and vegitable bodyes which I doe here ingeniouslye acknowledge that I have now done or finished, (for then his whole estate though it were 10 times greater, could never counterpose) for I assure my selfe that either ther is no truth in philosophie or els that I can leade my ffriend by the hand into the secretest corner of her howse yea into the bedchamber of Nature her selfe'.

'And to give you a touch of my conceipt in philosophicall courses' he discloses the best way to make a quintessence, and thoughts on making the red stone: guarded words which imply he is a serious alchemist. His concluding remarks emphasize the potency of the secrets he holds. He asks Segar if the proposals are rejected, 'that you would consecrate those fewe and Idle lines to vulcan without reservinge anye note or memorye of them for I knowe not the man except your lord, for whome I would by writinge have adventured this farre to wade in the deepe streames of philosophie wherein manye witts better than mine doe dayle plunge and sinke themselves'. Further emphasis is given by asking for secrecy if negotiations continue, 'I must allsoe yf I proceede with your lord crave the like secresie at his hands as I was forced to yeelde unto before I could drinke of this Ambrosia'. A final inducement is the disclosure that another gentleman is after Plat's secrets. 'I must crave to have a spedie answere with a full and perfect conclusion for I assure you upon my creditte, I staye my conference with a great englishe Nobleman herein in respect of your lord whome I cannot long delaye'.[43]

The coincidence that here, as in the letter to Elkinton, another buyer is introduced might provoke the thought that Plat has invented him to heighten Moritz's interest. There is indeed a document confirming Plat was negotiating with someone else. This is a draft schedule with brief, one-sentence notes of fourteen secrets concluding, 'Yor Honor in all Dewtie Devoted./ H.P. Miles [knight]'. Several of the medical secrets are the same as those offered to Hesse, a speedy cinnamon water, spirits of vegetables extracted, herb- and spice-flavoured butters and sugars. Others are new to us. One is 'to make a Lion, beare, dogg, or any other beast of cleane gold wth ⅛ part of gould that any ordinarie Jeweller shall doe.' This was a secret Plat knew had been given to Hesse by a foreign artisan some years before 1608. The document was written after he obtained his knighthood in 1605, and the wording of the secret of making gold figures is very similar to that in the preface to *Floraes Paradise* in 1608. So this was not a schedule to be attached to the Hesse letter. Why would he repeat here, with no further elaboration, offers made in the body of that letter and why would he offer a secret he knew Moritz possessed? This was rather the offer made to the 'great englishe Nobleman' Plat mentioned.[44]

Plat did, however, intend to append a schedule to the Hesse offer, making clearer what he would offer in the case of recipes for three chemical

43 BL, SL 2172, f. 18.
44 BL, SL 2203, f. 112.

medicines, and his secret of extracting the brine from sea water.[45] This note, a draft in Plat's 'rough' hand, as well as seeking to assure the potential buyer of the novelty and utility of the recipes, broaches the subject of price. We find that Plat is asking a fortune for his secrets. A value of £500 is placed on 'a sure and cheape way to make store of rosewater, violet water, honysuckle water and gelliflower water &c.' with his 'Antidote of Antidotes' being valued as highly. But these sums are dwarfed by 'the Quintessence of Eyebright' which he considers 'alone to be worth 10000 L', a sum justified by the comment that the remedy is 'especially for noble men whose noses are not fit to carry spectacles, and therfore you may see that I doo greatly undervalue this secret'. Despite the valuation, Plat is willing to sell it for 1000 marks (£333) with that sum returnable if the secret prove false (the great reduction on what it is 'worth' and the sale-or-return terms are clearly marketing strategies). The secret of processing seawater is valued at £400.[46]

Almost certainly the plan to sell secrets to Moritz did not succeed. The letter to Segar now in Plat's papers is a fine copy, in his best hand, endorsed with an address to Segar, in other words this is the letter which was to be sent and was never dispatched. The schedule of secrets to go with it is still in draft. Was Plat told by Segar that the offer would only be entertained if he travelled to Hesse to disclose his secrets in person, rather than tell them to agents – to Dr Mosanus or Segar – as he proposed? Plat hints at this reason in July 1608, when he wrote 'the Lantzgrave of Hessen, that now liveth (From whom I could easily draw some honourable and bountifull pension, if I were disposed to travell)'. Plat was then ailing and not fit for the journey.[47]

Plat's attempts to make money from and with the aid of the English Crown met with mixed success. They were intertwined with social ambitions. As well as selling goods and services to the Crown and seeking

45 As evidence that the note was meant to complement this letter, is the phrase 'you may read at large in both the pamphletts', surely a reference to the two pamphlets mentioned in the letter. Additionally, all the recipes detailed in the note are also mentioned in the letter.
46 BL, SL 2172, ff. 20–21.
47 BL, SL 2172, ff. 18–21; Plat, *Floraes Paradise*, 1608, Preface.

reward for inventions seen as useful to the country as a whole, he might hope for financial reward as a courtier from the wide range of patronage available to him. Like many a well-dressed and well-spoken gentleman, he could try and achieve fame (and fortune) by being noticed and nurtured by those already in positions of power and influence, the ultimate aim being royal recognition. Such ambitions were risky; in trying to catch the eye of powerful courtiers by wit, intellect, sense of fashion, beauty, or just plain chance, one might be spectacularly successful, a quiet failure, or publicly humiliated and ridiculed.[48]

We have traced, however faintly, Plat's fortunes (or rather misfortunes) at Elizabeth's Court and the evidence is that many of the attacks on him, both at Court and from some Londoners engaged in manufacturing and trade, were aimed at belittling his ideas and inventions. On the evidence of John Harington's remarks, as well as Plat's own complaints of mistreatment, it seems that the inventions were derided not for their lack of merit so much as Plat's laying himself open to ridicule for his persistent advocacy and his demands for reward before he would put them into the public domain. Thus Harington, a dilettante whose famous invention was put forward as a satirical tract with the tiresome technical details as a mere appendix (written by a member of his household), poked fun at Plat's desire to sell the secrets of his inventions or seek a monopoly for them and alluded to the unpopularity of his father-in-law's starch monopoly. Harington mocks Plat's earnest endeavours: Harington assumes his own invention (the flushing water-closet) and Plat's fertilizer and coalballs all involve ordure, and he calls himself and Plat 'two such rare engineers', both of Lincoln's Inn. Harington makes it clear in his playful prose that he is an aristocrat with a lively mind, whereas Plat is a gentleman with the mind of an artisan, a new gentleman on the make.[49]

---

48   This tale, still remembered when John Aubrey wrote *Brief Lives*, sums up the silliness of Court life which could make or break reputations: 'This Earle of Oxford, making his low obeisance to Queen Elizabeth, happened to let a Fart, at which he was so abashed and ashamed that he went to Travell, seven years. On his returne the Queen welcomed him home, and sayd my Lord I had forgott the Fart.' John Aubrey, *Brief Lives*, 1999, p. 305.

49   Harington, *The Metamorphosis of Ajax*, ed. Donno, pp. 164–9.

Plat had to contend with more than personal antagonisms in his quest for patronage at Court. As Pumfrey and Dawbarn have shown, England had not the diverse sources of patronage for scientific study available to Continental natural philosophers. England was an island state, with no chance of cross-border intellectual traffic. Unlike the Catholic church in Europe, the English church offered no alternative source of patronage to inventors. Oxford and Cambridge taught a conservative curriculum not conducive to such ideas as Paracelsianism or practical mathematics, although some study of the latter took place outside the official instruction. There was no 'Court' university, or university at all in London, and the Inns of Court were no substitute. Court patronage, especially in Elizabeth's reign, was predominantly utilitarian when it did exist. It was more interested in practical solutions to problems of navigation, warfare, or new industries to reduce imports, than 'ostentatious' enquiry, astronomy, alchemy and the like, which might produce discoveries which boosted the status of the patron. This belies the fair comment made by Pumfrey and Dawbarn that, 'court culture everywhere was agonistic and competitive, governed by an aristocratic desire to gain and display power, including power over nature, in accordance with humanistic codes of etiquette.'

The one source of patronage, or at least intellectual stimulus, which Plat could rely on was the significant body of 'natural knowledge makers' working in London. These were primarily mathematicians, instrument-makers and medical men but, as we have seen, included a great many other enquirers, the majority of them artisans rather than gentlemen. As Deborah Harkness has demonstrated, this community embodied a formidable amount of knowledge. Plat's one advantage in his quest to interest the Crown, or at least the nobles and administrators who were Court brokers of patronage, was that his ideas were predominantly utilitarian. As we have seen, he offered solutions to current problems, 'aimed at the practical control of nature'.

Lack of evidence precludes any assessment of the harm to Plat's quest for patronage caused by the fall of Essex in 1601 but it may have severely dented any further hope of patronage until the accession of King James in 1603. If we are to judge by the lack of complaints of Court antagonisms after that time in his published works, and by his knighthood in 1605, he was more socially at ease in James's Court, if no further forward in his attempts at extracting significant financial gain from the state.[50]

The simplest way of gaining from the Crown was to sell it goods or services. Surviving evidence of Plat's efforts all relate to military goods, predominantly the preserved foods we have already discussed. Of these efforts, only the note to Lord Buckhurst dated Michaelmas 1598 is an unambiguous offer of goods in a commercial transaction.[51] Other notes such as the three lists containing ideas for munitions and military rations and medicines headed with the names of Thomas Arundell, Sir Walter Raleigh and the Lord Treasurer, and the memorandum to badger Sir John Fortescue about victualling the royal navy with pasta, were probably *aides-mémoire* for lobbying at Court rather than direct commercial propositions. Plat's shrewd choice of lobbyists included two military men: Arundell, a mercenary soldier who distinguished himself fighting the Turks in Hungary, and Raleigh, a naval hero; and two senior Crown administrators, both heavily involved in royal finance. Lord Treasurer Buckhurst may have had links to Plat via his first father-in-law Richard Young: like Young he was involved in suppressing Catholics in the London area and he acquired in 1598 the starch monopoly held by Young until his death in 1595.[52]

The note to Buckhurst, headed 'An offer of certen light, fresh, lastinge & portable victualls for the service of Ireland wch the author will, undertake eyther to make for the said service or uppon reasonable Composition to discover unto suche as shalbe thought meet for the same', is a brisk document sent to a busy administrator, giving detailed but brief summaries of the benefits of Plat's preserved foods. They are said to be ideal for military campaigns, albeit that the cooked meat 'is onelie serviceable (As the author supposeth) where it maie eyther be Carried by

---

[50] Stephen Pumfrey and Frances Dawbarn, 'Science and patronage in England, 1570–1625: a preliminary study', pp. 137–188. They point out (p. 153) how Thomas Harriot's 'appalling bad luck' in moving as client from the disgraced Raleigh to Northumberland, who later suffered a similar fate, badly affected his status and reputation in England; Deborah Harkness, *The Jewel House*, 2007.

[51] Although, as we have seen, there is strong evidence of some success in providing at least some victuals to naval expeditions.

[52] BL, SL 2189, f. iv; SL 2172, ff. 13, 21. See also relevant *ODNB* entries.

shipp on the sea, or by Carte or other Carriage on the land, and not by the souldier himselfe'. Buckhurst is assured that Plat can deliver any of these comestibles as required. As we have seen, Buckhurst was sufficiently impressed to mention these foods in a memorandum on preparations for the Irish campaign.[53]

Plat also considered how he could exploit his secret oil for preserving armour. He must have discussed his plans with his cousin Henry Davenport who wrote to Plat in 1595, 'Me thinks it were well for you to go forward with ye office in ye Tower of keeping the Armer bright, bycause it may be supplied by a deputy if you think it not good to exercise it your selfe'. Such a plan would yield an income from the Crown in the form of a stipend or pension and allow him to control access to the secret ingredients in the invention. Linking all these direct approaches to government and more circuitous ways of making money from patronage, via pensions, grants of patents of monopoly or other legislative favours, is the importance of lobbying, getting the ear of those who have the power to grant one's wishes.[54]

Since the mid-sixteenth century, patents granting monopoly rights over an aspect of economic activity for a number of years had been established as the way an inventor (or innovator of a foreign invention brought into England) might exploit his novel process or product with Crown protection. There has been disagreement over exactly how and when the system first began. Joan Thirsk concludes that the long-standing minister William Cecil copied a Continental practice, encouraged by foreigners who offered to bring new skills to England in return for protection from competition.[55] The initial impetus was thus economic, encouragement of new enterprises of all sorts so that they might substitute for imports and

---

53 BL, SL 2172, f. 21.
54 BL, SL 2172, f. 27.
55 This was just a new use of documents known as patents. Patents had for centuries been used by the Crown for many things. William Blackstone defined them in 1768: 'The king's grants are also a matter of public record... These grants, whether of lands, honours, liberties, franchises or ought besides are contained in charters, or letters patent that is, open letters, *litterae patentee*, so called because thay are not sealed up... and are usually directed by the king to all his subjects at large'. Quoted in Christine MacLeod, *Inventing the Industrial Revolution*, Cambridge, 1988, p. 10.

give employment to the poor. Early patents were granted for activities such as making glass, searching for and working metals, manufacturing hard white soap, and making saltpetre. They were framed to protect the entrepreneurs from competition when using the new techniques they had developed or imported, not to control the whole of an existing industry. These early patents could be revoked if the innovations were not swiftly brought into use. Often there was a stipulation that goods produced should be cheaper than foreign imports and of good quality. The government was careful to protect existing industries from the effects of the innovations. Such patents were granted with increasing frequency in the third quarter of the century.[56]

By the 1580s and 1590s, some patentees of new and renewed patents were courtiers, merchants or speculators who took on the financial risks of the enterprise, hiring skilled craftsmen to carry out the actual work. Because many were for new manufactures which produced goods previously imported, the Crown was concerned that customs revenues would decline. Prospective patentees, therefore, offered the Crown a share of profits as recompense.[57] This additional burden encouraged monopolists to be more vigorous in enforcing their patents. It also turned patents from purely instruments of economic policy into another way of raising revenue. Grants of monopoly patents were also used by the Crown as a way of paying off old debts and meeting obligations such as pensions to Crown servants. Those thus recompensed were often men of means already, with

---

[56] Joan Thirsk, *Economic Policy and Projects*, 1978, pp. 52–7; Macleod, *Inventing the Industrial Revolution*, p. 11. The early history of patents of monopoly is covered by Joan Thirsk and Macleod, ops. cit. and also E.W. Hulme, 'The history of the patent system under the prerogative and at common law', *Law Quarterly Review*, 12 (1896), 141–54; 16 (1900), 44–56; William Hyde Price, *English Patents of Monopoly*, 1906; Harold G. Fox, *Monopolies and patents*, Toronto, 1947; Jeremy Phillips, 'The English patent as a reward for invention', *Journal of Legal History*, vol. 3, 1982, pp. 71–79. The role of William Cecil in developing the system is explored in Felicity Heal and Clive Holmes, 'The economic patronage of William Cecil', in *Patronage, Culture and Power, The early Cecils*, ed. Pauline Croft, 2002.

[57] A patent monopolizing starch making was granted to Plat's first father-in-law Richard Young on payment of £40 per annum to the Crown. Hyde Price, *English Patents of Monopoly*, p. 142.

the energy and resources to enforce their rights against small producers who had previously used the new techniques unmolested. Other actual or perceived abuses of the system occurred. Patents were renewed after an innovation had been exploited for many years without hindrance, and the patentees merely fined those involved in the industry rather than promoting new techniques. Similarly, patents were granted based on no new industry or invention at all, merely enriching the patentees (and the Crown) at the expense of those producers involved. Patents for the production of commodities considered essentials, such as salt, caused particular resentment. The sweeping powers granted to the saltpetre patentees to search for that commodity were a source of much discontent and tales of extreme vexation. The floors of churches, council chambers and private houses were allegedly dug up without notice.[58]

Disquiet over the way patents were exploited surfaced in Parliament as early as 1571 and some in the government were also worried at that time. In 1597 an address was presented to the Queen on the subject. She rebuked the Commons for presuming to interfere with the royal prerogative but promised to examine existing patents to ensure they were lawful. Nothing was done and much more discontent was voiced in the Parliament of 1601 when several attempts were made to bring in an act to assert that the common law freedom to trade overrode the rights of monopolists (this may have been part of a concerted plan by common lawyers to defeat monopolies). The debates in the Commons saw an outpouring of discontent from MPs about monopolies. Their complaints combined frustration at the abuses their constituents suffered at the hands of patentees with the judgment of common law experts that monopolies were illegal infringements of the rights of individuals to trade. The tenor of the debates is encapsulated by the MP who sarcastically remarked that if nothing was done to control monopolies bread would be subject to a patent before long. The Queen was forced to defend herself in the 'Golden Speech' to the Commons in November 1601. While upholding the royal prerogative, she went some way to deal with complaints, agreeing to abolish a number of patents and subject others to legal scrutiny. Out of

58   Stephen Bull, 'Pearls from the dungheap', *Ordnance Society Journal*, 2, pp. 5–10; Thirsk, *Economic Policy and Projects*, pp. 57–65.

the latter assurance came the case of Darcy v. Allen in 1603 when the grant to the courtier Darcy of the right to transport and make playing cards for 21 years was challenged.[59]

It was in this uncomfortable climate that Plat sought legal protection for his inventions. We have mentioned that John Harington ridiculed Plat in 1596 for trying to make money from his inventions rather than publish them fully for the good of the country. Harington assumed that Plat was in pursuit of a patent of monopoly similar to that of his father-in-law's for starch – which aroused much discontent, being in effect nothing more than the right to tax and fine members of an established industry, not a reward for enterprise. There is some evidence that Plat sought to benefit from patents under existing laws. In his manuscript notes he vaguely hopes he may seek a privilege for 'Spanish silke' and for printing his law books. He claims to have obtained a privilege for his portable pump: 'I wil here either borrow leave, or commend without leave, a new, light and portable Pumpe, being of late graced with her maiesties most favorable priviledge, which I am bold to publish..... because I know no better means to give a publike notice thereof unto all my countrymen then by this pamphlet'. No trace has been found of a royal patent issued to Plat for the invention.[60]

Plat did obtain indirect benefit from the patent for saltpetre production. In *The Jewell House* he advertised , 'And if the Author might receive a condigne reward for his profitable travell, he would peradventure find out a multiplyeng earth, which would yeeld sufficient store of Peter, for the service of this Realme without committing such offenses, as are dayly offered, in the breaking up of stables, barnes, sellers, &c.' He apparently sold his idea to the holders of the patent, claiming in 1603 that 'my new & late discovery in Peter-works [was] the true foundation and groundwork of the last letters patents granted for the same'. The invention 'bringeth in yearly & freely many 100 pounds to the Patentees', but he was forced to sell an idea he could not himself exploit because of their patent, received

59 Thirsk, *Economic Policy and Projects*, pp. 75, 99; David Harris Sacks, 'The countervailing of benefits', in *Tudor Political Culture*, ed. Dale Hoak, Cambridge, 1995, p. 274; Fox, *Monopolies and patents*, pp. 71, 75.
60 Harington, ed. Donno, pp. 175–72; BL, SL 2223, f. 60a; Plat, *The new and admirable Arte of setting of Corne*, 1600.

under 'one half yeeres profit for the invention'. Plat may have exaggerated the meagreness of his return from saltpetre. In 1600 the monopoly was granted to a consortium headed by one Richard Harding and one of the interrogatories in the law case over Plat's will is, 'Item howe longe are ye letters patents to endure wch are mencioned in the Copie of an Inventory touching CC l. per Annum to be paid by Hardinge'. Two hundred pounds per annum was no small sum and, because Plat sold his idea to the patentees, he avoided having to seek one for himself. Such a course was slow and expensive, involving ten stages of bureaucracy from petition to enrolment with fees and gratuities to be paid at each stage. Perhaps for this reason, in 1594 he offered to sell his inventions generally to any who were willing themselves to try 'to procure some privilege for them'.[61]

Plat realized that all economic legislation to help inventors risked being condemned by many, 'crossed and swallowed up, I know not how in the bare and naked word of *Monopolies*' and he sought alternatives to the now much-hated patents which would reward inventors without public outcry. Optimistically, in *The Jewell House* he hoped his preservatives for food and drink at sea for the benefit of the navy 'may bee rewarded with some honest pencion' but we have discussed the difficulty of obtaining such patronage in England.[62]

Plat placed much emphasis on schemes which would be of demostrable benefit to the inventor, the people, and the Crown (which, indeed, was the aim of the early Elizabethan patents for invention). In the appendix to *The Jewell House* outlining several inventions we find the germ of an idea for a new scheme. Ten reasons for taking up his offer of coalballs include the cheapness of the new fuel, ease of production, setting many poor to work making it, a decrease in air pollution, less pressure on timber and charcoal supplies, and an assurance that the existing Newcastle coal trade will not be damaged. In short, these were the facts a cautious government minister would consider before granting a patent. The final

61 Plat, *A new, cheape and delicate Fire of Cole-balles*, 1603; Plat, *Jewell House*, 1594, pt. 4, p. 69; Fox, *Monopolies and patents*, p. 315; TNA, E134/8 Jas 1/East 2; MacLeod, *Inventing the Industrial revolution*, pp. 40–41.
62 Plat, *A new, cheape and delicate Fire of Cole-balles*, 1603; Plat, *Jewell House*, pt. 1, p. 11.

reason is novel: 'And if this secret might, (by anie carefull and provident meanes) be brought into the publike use, with the good contentment of the Authour: Then the tenth part of the profit that should arise therby, might be yearlie distributed amongst sicke persons, and maimed soldiers, or otherwise conferred upon such other good uses, as should bee thought most convenient.'

In 1595 he develops this idea of taxing the benefits of innovation as an alternative to monopoly patents. At the outset, he assures his readers he does not wish to extract money from them but merely to harvest some of the revenue consequent on his fuel-saving inventions of coalballs, saltpetre processing and a wooden boiling vessel. These benefits, he says, will go to help the poor. He estimates brewers and saltpetre-men will save half their fuel costs and all people living in areas using seacoal for fuel will save one third. Even in areas not using coal there would be savings from the more plentiful availability of wood as fuel, because nationwide demand for charcoal would fall. The accumulated economies would be vast. He feels that people will willingly give up one half of these savings in gratitude for the increased security of fuel supply and the reduction of vagabondage, as well as in charity to the poor.

He is vague about the administration of the scheme, suggesting that existing tax assessors collect the savings and assess it in the same manner as other taxes rather than use any strict mathematical formula to assess fuel savings. The poor 'shold scape skotfree'. He is also not specific about the benefits which might accrue to him as inventor, beyond saying that he would be entitled to 'an enita pars', the first choice in any division of gains. In 1603 he further suggested an Act of Parliament might be passed giving a monopoly 'for a time' to inventors and a proportion of gains to the poor, reflecting the continued feeling that patents were being abused under the royal prerogative.[63]

His ideas were not taken up. They were put forward tentatively and close scrutiny by contemporaries would have revealed many flaws. Taxation of any sort was never popular and in this case the uncertainty over the exact benefits, who should pay, and how much, would have made

63   Plat, *Jewell House*, pt. 4; Plat, *Discoverie of Certaine English wants*, 1595; Plat, *A new, cheape and delicate Fire of Cole-balles*, 1603.

the scheme unworkable. Plat grasped the idea that reform of the patent system must end existing abuses and substitute a scheme which rewarded innovators and inventors, helped or at least did not harm government finances, and had clear benefits in the form of new products or cheaper goods to the public at large, but he failed to produce a workable solution. The eventual system, which emerged over time, was a modification of patents of monopoly. Although legislation and court cases reformed patents to a degree in the seventeenth century, it was not until the first decade of the eighteenth century that they became recognized instruments for protecting the interests of inventors and entrepreneurs.[64]

Evaluating Plat's overall success in making money from his natural science and technological research is not easy. Apart from the odd document, no records of his finances have been found. Some pattern however, emerges from the evidence available. He had most success when selling directly to customers and producing goods of all sorts on a small scale, either himself or under his supervision. He was least successful when trying to manufacture on a large scale or sell ideas to others for large-scale exploitation.

Probably his most lucrative 'small-scale successes' which we can prove or infer from evidence relate to medicine. In his medical case-notes we found a few recorded instances of gratuities received – a ruff, sugar cakes and 2 pullets – but these were just tokens of gratitude. He could earn much more in fees. He tells Elkinton, a prospective buyer of medical secrets, admittedly as a sales pitch, that he has received gold 'Crownes enough' for one medicine from patients, another has yielded gold chains in profits and he is unwilling to sell another outright for under £5 an ounce. The prices he puts on selling formulas to the ruler of Hesse, even if we accept some exaggeration as an initial bargaining position, are surely some indication of the retail price Plat had previously received for these cures. Other possible successes were the cinnamon water he claimed to make in a new way which he may have sold to or sold via John Clark; and the broadsides advertising medicines to seafarers may have yielded some success.[65]

64 MacLeod, *Inventing the Industrial Revolution* p. 40.
65 See Chapter 7. BL, SL 2172, ff. 28–29v.

A rare glimpse of small retail sales of an invented product is afforded by the following notes in Plat's rough hand at the end of a notebook which probably record sales of Plat's shoe-blacking oil:

November/95
  Mr Jefferys of Mawlin in Suffolk ½ pint of black ½ pint of white.
    at xii d. & 3d.
  Sr Tho North L. Northes brother i pint white oyle xvi d.
    1 pinte black/ vi d.[66]

We also have evidence that he sold dried pasta for use at sea. He claimed to have supplied this to a ship which went to China and he received £18 for four barrels provided for Drake's last expedition in 1596. We have seen that he may well have supplied pasta to another expedition against Spain in 1589 when his putative kinsman Anthony Plat was slow in providing victuals to be shipped from Plymouth because 'friends and acquaintances' were tardy in supplying him. Being involved in supplying this expedition may have involved a large amount of pasta and Sir Hugh may have found difficulty in producing the required consignment with just one macaroni machine. If so, this was an example of the difficulty Plat encountered in exploiting an invention which required extensive equipment and manufacturing capacity.

A much easier course was to sell the idea to someone with the resources to exploit it and willingness to take the business risk. We have seen that he adopted this course with his new technique for processing saltpetre and that he likewise disposed of a method of extracting the 'strength or vertew of Logwood' in 1599 to the grocer Arthur Blackamore. The price of the secret is not recorded and this is the only agreement to sell a secret surviving in Plat's papers.[67] There may have been others but when faced with the task of trying to exploit secrets on larger scale, success usually eluded him. True, he may have obtained a patent for his pump but he does not seem to have sold on the rights of manufacture.

66 BL, SL 2197, f. 43v.
67 BL, SL 2172, f. 12; Deborah Harkness interprets this document as a sale *to* Plat *from* the grocer, I disagree; Harkness, *The Jewel House*, p. 217.

MONEYMAKING

He apparently had no success in selling the mechanical bolting hutch idea, there are no records of success with his new fertilizer, papier mâché castings, the coalballs or the wooden vessels to boil in. In fact, admitting defeat, he published full details of coalball manufacture and promised to do the same for the boiling vessels. His plan for a shop selling a range of his wares in the Royal Exchange seems to been a pipe dream. Nevertheless, he had the consolation of the social cachet of a knighthood, perhaps granted in reward for his inventive skills although no citation exists. He remained ever hopeful of future rewards. In the last pages of his last book, he advertised a medicinal 'crystalline Cawsticke' which he hoped to finish making 'next yeere'. He then concludes, optimistic to the end:

> And thus I have culled out a fewe choice flowers, out of a large ground: which if they shall worke either favour, credit, or profit unto the Publisher, I shall be readie to second them with a new supply of fresh Inventions; *ut semper novus veniam*, not against Antony, that famous Conspirator of Rome, but against Ignorance, the professed Enemie to all true Religion and Learning.
>
> FINIS[68]

---

68 Plat, *A new, cheape and delicate Fire of Cole-balles*, 1603; Plat, *Floraes Paradise*, 1607, advertisements at end of book.

CHAPTER ELEVEN

Conclusion

Deborah Harkness's book *The Jewel House* was published as I was completing my manuscript. Subtitled 'Elizabethan London and the scientific revolution', it contains a chapter on the vibrant state of research into natural science in Plat's London and the many people who engaged in it. Other chapters develop aspects of this theme, including one devoted largely to Sir Hugh Plat and his part in early scientific research. One chapter rather than a whole book on Sir Hugh inevitably leads to a concentration of ideas and reading her stimulating thoughts has enabled me to test my conclusions against her review of Plat's manuscripts and some of his published works. I am heartened by the many points of agreement between us, stimulated to review those areas where we disagree, and prompted to re-read Plat's written remains to see how I can sum up his work. I fully concur with Professor Harkness's view that 'What Londoners like Plat and his informants contributed to Elizabethan science was a practical experience that could serve, like the counterweight on a clock, to move the study of nature from its largely theoretical and medieval foundations toward something more empirical'. Plat himself was conscious of this, as his barbed comments about 'schoolmen' and their approach to learning show.

Harkness describes his wide circle of informants, mostly Londoners, supplying both information about their own trades and professions as well as general ideas and secrets. She is impressed by Plat's open approach to informants' knowledge: 'What is striking about Plat's encounters with this varied cross-section of London's science practitioners is that he appears to have treated the muskmelon man and James Garret [an apothecary] with the same respect as he accorded Sir Francis Drake'. A substantially true observation but an unfortunate choice of informants – the muskmelon man is always referred to by Plat as a gentleman, he was probably the 'Gentleman Keeper of the Queen's house at St. James'. And Plat did have extra respect for facts obtained by word of mouth from gentlemen, sometimes noting information from them 'per verissimo'. She rightly highlights Plat's delight in 'spectacular feats of useful ingenuity', he liked novelty and was at pains to decry those who thought progress impossible.[1]

Examining Plat's notebooks, Harkness is impressed by the amount of information they contain, notes made of innumerable conversations with fellow-Londoners, and the number of practical experiments Plat carried out. Looking closely at his workings she discerns 'a bare-bones anatomy of the scientific method ... including use of hypotheses, testing, the inclusion of contradictory information or results, the generation of further hypotheses and experimental procedures to test them, and evaluations of their success and failure'. She notes the ubiquity of 'qre' – that small abbreviation at the start of a new thought generated by what he had just written, usually leading to suggestions for a new experiment to be tried or variations of a tested recipe. 'Page after page of the notebooks include this mixture of observation, query and hypotheses as Plat considered his experimental options'. Harkness notes the willingness of Plat to publish material 'with a certain amount of hypothesis and contingency still present' (referring to the presence of queries in the printed text). This is true, although as I discussed in my chapter on gardening, in *Floraes Paradise* (the book most liberally sprinkled with queries) they are present because Plat had no time to polish the text before publication.[2]

My disagreements with Harkness are not radical, but on points of emphasis. She stresses the importance of Plat's earliest substantial work of 1594 : 'Plat's public reputation as one of Elizabethan London's most assiduous experimenters rested on his most wide-ranging and important work, *The jewell house of art and nature*'. I know of no great approval of

1   Harkness, *The Jewel House*, pp. 217, 221, 226; BL, SL 2210, ff. 161, 168; Gerard, *Herball*, 1633 p. 918.
2   Harkness, *The Jewel House*, pp. 227, 229.

this work in late sixteenth-century London and it was not reprinted until 1653 (although one part, *Diverse new sorts of Soyle*, was separately published in 1594 – perhaps then the best-regarded part of the work?). In contrast, *Delightes for Ladies*, probably published in 1600, was reprinted in 1602, 1603, 1605, 1608, 1609 and a further eighteen times up to 1656. It would seem that a book on cookery, cosmetics, perfumes, distilled waters and preserved foods, which was only partly Plat's own work, was much more to the public taste than secrets of nature. (Incidentally, who bought this book? Did men buy it for their wives or were women browsing the bookstalls in St Pauls' churchyard?) True, the distillation and food preservation within this little book are in many cases the fruits of Plat's experimentation but they are presented not as scientific work-in-progress but as recipes to be followed. Plat himself voiced strong disappointment at the reception of *The Jewell House* in his pamphlet of 1595, *A discoverie of certaine English wants*. In general, one has the feeling that his published works on natural science were not greatly lauded at the time. Here we may draw a parallel with his lack of success in promoting most of his inventions and his general failure to make much impact on Elizabeth's Court. I believe we should regard *The Jewell House* as the publication of only the first stage of Plat's researches into natural science and technology. As I have outlined in the appendix on his manuscripts, this book is based on BL, SL 2210, the earliest substantial body of information Plat collected. Subsequent manuscripts, and publications based on them, show him developing his ideas, his range of interests and his approach to research. One has the feeling that, had he lived for a further 20 years, his notebooks would have contained not only a great deal more information and experiments but much-improved methods of arranging his material and conducting work in the laboratory.[3]

I can agree that Plat's publications (like Gerard's monumental *Herball*) have given him prominence amongst historians and obscured the work

3   Harkness, *The Jewel House*, p. 235; Bent Juel-Jensen, 'Some uncollected authors XIX', *The Book Collector*, 1959, pp. 60–68. Apart from *Delightes for Ladies*, the only works reprinted in his lifetime were one edition of his little treatise on setting corn – another didactic piece – and *Diverse new sorts of Soyle not yet brought into any publique use*, which was issued as a separate pamphlet in 1594.

which was going on in London amongst all the other tradespeople and gentlemen of enquiring minds. And how many sets of manuscript notes of experiments, thoughts and queries from these researchers were consigned to the fire, kitchen, or privy, rather than ending up in the British Library?

I see Plat as much more than an early scientist and, even when considering his scientific work, view this as much a means to enrich and advance himself as an end in itself. His most disinterested research seems to have been into alchemy. This may be understandable in terms of his swearing to uphold secrets divulged to him by his teachers and undertaking to keep secret anything he discovered of this mystery, sharing information only with fellow-initiates. Though it might be pointed out that even his alchemical secrets were offered for sale for substantial sums in 1608. Alchemy probably introduced him to research, hence the emphasis in his early notes on explaining the world of nature in chemical terms.

He moved on from simple alchemical enquiry, seeking knowledge from friends and acquaintances, most significantly from the artisans of London. This was for a purpose: he wanted to improve existing techniques or find new ways of doing things in order to make money from the secrets. As was clear when charting how he advertised and marketed his secrets, disinterested enquiry was not his main motivation – he wished to sell much of what he discovered. This is nowhere better seen than in his medical practice. Harkness has highlighted his apparent success in curing patients of the plague in 1593, drawing the attention, Plat claimed, of the Privy Council. Referrals of friends by satisfied customers in subsequent years betokens further successes. We must be cautious, however, of his practice-notes: they only list successes and I suspect they were written in part as proof to prospective buyers of his practice that his medical secrets were worth the money he was asking, and would yield a substantial income. How many of his patients died or continued to suffer despite his ministrations?[4]

In contrasting the impact, during Elizabeth's reign, of Plat and Francis Bacon, Harkness sees Plat as a clear winner, establishing himself as 'a man capable of understanding nature and producing authoritative, experimentally based natural knowledge for the benefit of the English

4   Harkness, *The Jewel House*, p. 231.

commonwealth', publishing bestsellers and catching the attention of the Privy Council. This overstates his impact. As I have argued, 'bestsellers' is a conjecture: we have no indication his works (apart from one) were popular. The one mention of attention from the Council comes from Plat, I can find no corroboration in the State Papers. True, his ideas on victualling Irish troops were acknowledged by Buckhurst and he may have supplied some pasta to the navy, but this is hardly taking the establishment by storm. I know of no contemporary natural scientist who commented on Plat's work, only that some courtiers and citizens were not impressed with his early attempts to establish himself on the London scene. True, he fared better under James, to judge by the knighthood and his ceasing to complain himself about unfair criticism. What then was his legacy?[5]

I know much less about Bacon than I do of Plat so I will not comment on their relative merits. Harkness, and others, have seen in Plat's work the sort of method advocated by Bacon, and lamented that Bacon's ideas were posthumously adopted from the second half of the seventeenth century onwards as the starting point for scientific research, burying the work of Plat and the other London-based researchers of Elizabeth's reign. I do not doubt the influence of Bacon but would not consign Plat so readily to oblivion. I see a renewal of interest his work a couple of generations after his death.

His library and papers were thought valuable very soon after his death: one of the more intriguing interrogatories put by the Attorney General to witnesses in April 1609 concerning his estate was 'Whether did you knowe or have you seene of a library of bookes or anie other written bookes paprs or notes whereof ye said Sr Hugh did posses'. The question goes on to try and establish the actual or potential sale price of the collection, and although one deponent puts a value of only £10 on them, clearly the state considered them of some importance.[6]

By the 1650s Plat's kinsman Charles Bellingham considered Plat's work worthy of renewed attention. In 1653 he republished *Floraes Paradise*

5   Harkness, *The Jewel House*, pp. 241–3.
6   TNA, E134/8, Jas I/East 2; BL, SL 2243. Examples of low valuations of book collections at this time can be found in F.T. Levy, 'How information spread among the gentry, 1550–1640', *Journal of British Studies*, xxi, 1982, pp. 11–34.

CONCLUSION

as *The Garden of Eden*. The book was needed, 'there being not extent (in our language) any work of this Subject so necessary and so briefe' (i.e to the point). He continued, 'no English man that hath a Garden or Orchard can handsomely be without it, but at least by having it will find a large benefit'. Five subsequent editions, to 1675, confirm Bellingham's confidence. In that year *The second part of the garden of Eden* appeared, made up of previously unpublished notes by Plat on gardening, possibly also published at the instigation of Bellingham.[7]

If Bellingham controlled the manuscripts from which *The second part of the garden of Eden* was produced, he tells us in 1653 that many more of Plat's papers were not in his possession. By piecing together the peregrinations of these manuscripts from Plat's death until acquisition by Sir Hans Sloane, we can assess the importance they were accorded at the time. The many additions made in a bold, clear, if ugly hand to some of the manuscripts now in the British Library are, on the authority of Sir Hans Sloane, the work of Thomas Hodges. Hodges, who acquired the papers some time after Plat's death, was probably either the cleric of that name (*c*. 1600–1672) who became Dean of Hereford, or his son Thomas (died 1656). Thomas the Dean was said to have been interested in alchemy. Another of his sons, Nathaniel, was a noted physician, taught by Theodore de Mayerne, a fashionable physician who favoured chemical medicine. Thomas junior and Nathaniel were probably the translators of a Rosicrucian work, Michael Maier's *Themis Aurea*. Both father and sons would have been interested in Plat's papers. The latest date in Hodges' catalogue of his and Plat's books is 1655, suggesting the son Thomas as owner of Plat's papers (he died 1656). There is, however a recipe in Hodges' hand attributed to the Bishop of Chester, very likely John Wilkins, a noted natural philosopher from the 1630s onwards. Wilkins was first appointed bishop in 1668 and died in 1672, dates which imply the notes in Plat's manuscripts are by Thomas Hodges senior. Whichever Thomas Hodges owned the papers, they were closely read and extensively annotated by their new proprietor. As Plat had done with the notes by 'T.T.' that he acquired, Hodges used the manuscripts as the basis of his own research,

7 Plat, *The Garden of Eden*, 1653, p. 8–10; Plat, *The second part of the garden of Eden*, 1675; Juel-Jensen, 'Some uncollected authors XIX', pp. 60–68.

especially into chemistry and medicine. He carefully indexed Plat's notes and, if we assume a good many of the printed books he listed that date before 1609 belonged to Plat, he also acquired a useful library.[8]

The likelihood is that the Plat material owned by Thomas Hodges remained in his family until they passed to Sir Hans Sloane. Sloane directly attributes the manuscript list of books mentioned above to Hodges, and a letter to the historian Thomas Faulkner from Thomas Martyn published in the *Gentleman's Magazine* of April 1826, provides a link between the Hodges family and Sloane. Martyn (1735–1825) was Professor of Botany at Cambridge and writes that he is descended via his great-grandmother from the Dean of Hereford. 'Rhoda, the only daughter of Dr Thomas Hodges, born 1635, married my great grandfather, John Martyn, merchant of London, 1660'. Martyn continues, 'I beg leave to consider Sir Hans Sloane as one of my patrons. The condescension of the venerable and amiable old gentleman to me when a schoolboy, will never be forgotten by me… I usually carried a present from my father of some book that he had published, and the old gentleman in return always presented me with a broad piece of gold, treated me with chocolate, and sent me with his librarian to see some of his curiosities'. It is not hard to imagine in this close relationship the acquisition by Sloane of Plat's papers from the descendants of Thomas Hodges.[9]

Plat's kinsman Bellingham bitterly denounced those who possessed some of Plat's manuscripts in the 1653 preface to *The Garden of Eden*. He was republishing 'to check their forwardness who were ready to violate so usefull a Work. There are some men (of great name in the world) who made use of this Author, and it had been civill to have mentioned his name who held forth a candle to light them to their desires; but this is an unthankfull Age'. Manuscripts held by others included 'other pieces of Naturall Philosophy, whereunto he [Plat] subjoined an excellent Abstract of Cornelius Agrippa de Occulta Philosophia; but they fell into ill hands,

8   BL, SL 2210; SL 2242, front endpaper; SL 2244, f. 21; SL 2242; John Wilkins, *ODNB* entry. Hodges compiled an index to *his* library and, as many of the works on alchemy and medicine predate Plat's death, the core of this library may be that of Plat.
9   *Gentleman's Magazine*, April, 1826, pp. 291–2.

and worse times' (these documents are in BL, SL 2223). The 'Abstract' Bellingham mentioned was in Plat's best hand, a fair copy ready for the printer and approved by the Stationers' Company; was Bellingham worried the material would be published without acknowledgement that it was by Plat? Note too that he thought Plat's manuscripts had revealed secrets to their new owners. Plat's 'candle' provided them with illumination.[10]

Was Bellingham's bitterness directed against Thomas Hodges and his sons? Probably not. Samuel Hartlib, co-ordinator of much of the enthusiastic enquiry into natural science in the middle of the seventeenth century, records in his diary in 1650, 'Mr Hinshaw hath Sir Hugh Plats MS'.[11] In the previous December Hartlib had probably just been introduced to 'Hinshaw', noting a brief biography in his diary:

> One Hinshaw about Kensington a Gentl[man] of 2. or 300. a y[ear] a universal Scholar and pretty communicative. Hee pretends to have the Alchahest or a true dissolvent. Is skilled in the Coptical Language. Exercises hims[elf] in Chymistry. Brought over an Excellent Historie of China in Ital[ian] wherin are the Annual Letters of the Jesuits. wch is worthy to bee translated. Hee is to bee ranked in the number of Experimental Philosophers. Hee hath a good Optical Glasse and kn[ows] one that hath one wherin you use both your eyes. Hee hath a number of MS.

'Hinshaw' was collecting manuscripts such as Plat's and his activities in the 1650s and after shed light both on the value placed on Plat's manuscripts and Bellingham's unease about their possible exploitation.[12]

Hartlib's 'Hinshaw' was Thomas Henshaw (1618–1700). Educated at Oxford and the Middle Temple, in the Civil War he was an active Royalist, captured in 1642 and released on condition he did not fight again. He went to the Continent, spent some time at the University of Padua, conversed

10 BL, SL 2223; Plat, *The Garden of Eden*, 1653, pp. 7–9.
11 *Hartlib Papers* 28/1/60B–71AE, Ephemerides 1650 part 3. There are several attributions in Hodges' notes of recipes acquired from 'Henshaw' implying continued contact between the two owners of the manuscripts – and confirmation that the writer of the notes was Hodges, not Henshaw.
12 Hartlib, Ephemerides(1649), *Hartlib Papers*, 28/1/37a, as quoted in Donald R. Dickson, 'Thomas Henshaw and Robert Paston's pursuit of the red elixir', *Notes & Records of the Royal Society of London*, vol. 51, I, 1977, pp. 58, 73.

with Continental natural philosophers and collected manuscripts. He returned to England in 1649 and as a defeated Royalist had little chance of an active political or professional life. So instead he settled in a house in Kensington, devoting himself to the study of chemistry, in Hartlib's words, 'a universal Scholar and pretty communicative'. He knew the alchemists Thomas Vaughan and Robert Child, Elias Ashmole drew on his library for his alchemical publications *Theatrum chemicum* (1652) and *The Way to Bliss* (1658), and he formed a lasting friendship with John Evelyn. Later on he became a founding member of the Royal Society.[13]

I quote Donald R. Dickson on Henshaw's activities at the time of Bellingham's complaints: 'The engaging Henshaw, not surprisingly, made quite a mark during the Commonwealth period. In 1650, together with Thomas Vaughan (1621–1666), who was widely known in his time as an apologist for the Rosicrucian Brotherhood and an alchemist, Henshaw formed a research collegium of chemists known as the Christian Learned Society or the Chymical Club. A handful of people lived or worked in his manor house in Kensington, called Pondhouse or Moathouse… Prominent among them were Obadiah Walker and Abraham Woodhead, two Oxford dons, both recently ejected from their Fellowships, who had tutored and befriended Henshaw at University College. What they had in common was an interest in experimental science and mathematics that had been cultivated by the leading mathematician of the age, William Oughtred (1575–1660)'. When Henshaw met Hartlib he was about to set up his research establishment and at that time he was most interested in the secret of 'the Alchahest or a true dissolvent'. This secret Henshaw discovered in Sir Hugh Plat's manuscripts: it was divulged to Plat by the Low Countries physician and alchemist J.B. van Helmont when he visited England. Henshaw obviously valued Plat's notes, telling Hartlib, and probably others, of their worth, to the chagrin of Bellingham.[14]

13 Dickson, 'Thomas Henshaw and Robert Paston's pursuit of the red elixir', pp. 57–58; Henshaw, *ODNB* entry.
14 Dickson, 'Thomas Henshaw and Robert Paston's pursuit of the red elixir', pp. 58–60; BL, SL 2212 is a manuscript of Plat's largely concerned with alchemical and medical recipes; it includes many secrets acquired from van Helmont; *Hartlib Papers* 28/1/60B–71AE.

CONCLUSION

A collaborator of Henshaw's, Robert Paston, copied a recipe of Plat's into one of his notebooks in 1662, so Henshaw probably still had Plat's manuscripts then. Thomas Hodges senior was first lecturer (in 1641) and then vicar of St Mary Abbots in Kensington. He may well have been a member of Henshaw's research group, as could another of the men from whom Hodges notes (in an addition to Plat's manuscript) he obtained secrets: Baptist Noel, Viscount Campden, who also lived in Kensington. Another contributor to Hodges' notes, Stephen Skinner, physician and philologist, was an acquaintance of Henshaw; Henshaw edited his philological treatises for publication. Hodges also records secrets gained from Henshaw himself.[15]

The picture which emerges from these scattered links between Plat's manuscripts and the Henshaw circle is that these documents were valued as sources of alchemical knowledge and consulted with much interest. Later, in the hands of Hodges, they continued to be appreciated for their scientific information. They were certainly not forgotten in the second half of the seventeenth century, nor were his published works.[16] Many of those interested in advancing the nation by better agriculture, industry and technology, who corresponded with Samuel Hartlib during the Commonwealth, mentioned Plat's publications with approval. An early correspondent writing in 1645, Sir Cheney Culpepper, had heard of Plat's agricultural writing but did not have his books; he wrote requesting information from Hartlib: 'I have a farme this yeere turned into my hands & woulde willingly make some advantage of that inconvenience by makinge severall tryalls, yf either in Mr Plattes book or elsewhere you can informe yourselfe clothinge corne with a riche Composte I would willingly be a learner'. Hartlib may have sent him a copy of *The Jewell House* for he later writes, 'Mr Plattes hath five excellent hints' about using London waste.[17]

In the 1650s, Bellingham republished Plat's gardening book. As well as the 1653 edition, *The Garden of Eden* reappeared in 1654, 1655, 1659 and

15   Dickson, 'Thomas Henshaw and Robert Paston's pursuit of the red elixir', p. 74; BL, SL 2247, f. 46; *ODNB* entries for Baptist Noel; Stephen Skinner.
16   In addition to my remarks here, see above, pp. 41, 45.
17   *Hartlib Papers*, 13/66B–67A; 13/122A. The 'excellent hints' were probably extracted from *Diverse new sorts of Soyle*, part 2 of the *Jewell House*.

1660, and the 'Second Part' was published in 1660 and 1675. *The Jewell House* was republished in 1653 by Arnold Boate, a Dutch-born natural scientist in the Hartlib circle. And *Delightes for Ladies* continued in popularity, going through seven editions between 1640 and 1656. These new editions stimulated Hartlib's correspondents.[18] Hartlib noted in 1653 that the Earl of Salisbury's gardener had tried several of Plat's experiments, presumably from the freshly available *Garden of Eden*. From *The Jewell House* of 1653 Robert Child noticed that 'Sir Hugh Platts in his writings setteth downe divers ingenious ways of fattening Poultry, &c'. John Beale was enthusiastic about the reissued Plat books and set about trying experiments based upon them, writing to Hartlib in 1659,

> Sir H Plats garden of Eden is written without pompe. I would I could see another like collection of later discoveryes. Sometimes from five lines in him, I could deduce hundreds of pretty experiments, as (for ought I know) I shall shortly shewe you. I can reduce the tallest, & most glorious sunflower speedily to descend lesser & lower than the smallest single marigold, & make English yarrowe, knapweede, & field plants to challenge the tallest flowers of the Easte &c. This belongs to my 2d part of Transmutation of plants. I owe more of thiese pretty slights to Sir Hugh, than to any newe man.[19]

After the Restoration some writers continued to regard Plat's agricultural and gardening works favourably. The Oxford don Robert Sharrock mentions Plat several times in *The History of the Propagation and Improvement of vegetables* of 1660. Using the 1653 edition of the *Garden of Eden*, he tried an experiment: 'I have Cherryes that grow upon

18   Juel-Jensen, 'Some uncollected authors XIX', pp. 60–68. It is clear that some of the correspondents were reading the new editions of Plat: Henry Jenny mentions 'Sir Hugh Plat to his garden of Eden hath annexed a short Philosophicall garden'. This was at the *beginning* of the first edition, moved to the end of the 1653 edition, *Hartlib Papers*, 53/35/3; Robert Child writes that 'Lately divers small Treatises have been made by divers, as Sir Hugh Platts, Gab. Platts, Markham, Blith, and Butler'. Putting Plat's works with others produced in the 1630s, '40s, and '50s, he probably read the reprinted versions of Plat, *Hartlib's Legacie*, p. 89.

19   *Hartlib's Legacie*, p. 89; *Hartlib Papers*, Ephemerides, 1653; 62/25/1A.

CONCLUSION

Plum-stocks which is Sir Hugh Plat's experiment from Mr. Hill'. John Evelyn referred approvingly to Plat's agricultural knowledge in *Pomona* (1670). In *A philosophical discourse of earth* (1676) Evelyn described him as 'that industrious man' and lauded Plat's experiment described as 'A Philosophicall Garden' in *Floraes Paradise*. In 1675, another member of the Royal Society, John Worlidge, referred both to *Diverse new sorts of Soyle* and to 'His New Art of Setting Corn' a work not republished since it appeared in 1600. The anonymous reviewer writing in the *Philosophical Transactions* of 1675, quoted on an earlier page, while crediting Plat with stirring up interest in agricultural improvement, says it is within 'fresh memory' that improvements began, implying that the reprinted works of Plat had stimulated activity.[20]

Those who used Plat's work in the period 1645 to 1675 were mining his output for information. Henshaw and Hodges used his notes on alchemy and medicine whereas the correspondents of Hartlib and later writers took experiments and secrets on agriculture and gardening. His work, and the way he worked, commanded respect. Robert Child calls him 'the most curious man of his time', and Hartlib refers to him as 'witty Sir Hugh Plat'. I have quoted John Beale's enthusiastic comments on *The Garden of Eden* and it is noticeable that Beale is inspired by Plat's experimental method to conduct trials in his own garden.

Although they respected his achievements, these men did not think him a great theorist. Nor did they think his books in any way comprehensive. In 1655, Child lamented to Hartlib the lack of a complete work on husbandry, 'That we have not a Systema, or compleat Book of all the parts of Agriculture. Till the latter end of Queen Elizabeth's days, I suppose that there was scarce a book wrote of this subject... Lately divers small Treatises have been made by divers, as Sir Hugh Platts, Gab. Platts, Markham, Blith, and Butler, who do well in divers things; but their books cannot be called compleat books, as you may perceive by divers

---

20 Robert Sharrock, *The History of the Propagation and Improvement of vegetables*, Oxford 1660, pp. 51–2, 67, 72; John Evelyn, *Pomona* (Part of *Sylva*, 1670), pp. 12, 16; John Evelyn, *A philosophical discourse of earth*, 1676, p. 129; John Worlidge, *Systema Agriculturae*, 1675 pp. 35, 63; *Philosophical Transactions*, vol. x, 1675, no. 114; see also p. 113 above; John Evelyn, *Terra*, 1787, p. 28.

particular things; not so much as mentioned by them'. Child acknowledged the work of Plat among others on soil fertility but again could find no definitive work in print, adding 'The Lord Bacon hath gathered stubble (as he ingeniously and truly affirms) for the bricks of this foundation; but as yet I have not seen so much as a solid foundation'. Bacon, although he had not produced a comprehensive work himself, had set out a framework for doing the research. This, to my mind, explains the eclipse of Plat and his contemporaries by Bacon. Plat muses on progress in various scattered passages, and the need to experiment rather than rely on the precedent of past writers, but did not concern himself much with theories of research. He left no blueprints for others to follow, only a body of useful information on many subjects which have been mined by later researchers and continue to this day to be appreciated.[21]

21   *Hartlib's Legacie*, pp. 6, 38, 89; *Hartlib Papers*, 8/9/5A.

APPENDIX I

Plat's manuscripts

Unlike most writers of books and pamphlets of this period, Plat has left us many manuscript notes and working papers and in some cases recognizable drafts of his printed output.[1] It is possible both to trace the finished works from these manuscript beginnings and to see what he borrowed from earlier writers. The documents, however, contain more than mere preliminaries. Some complete works remain in handwritten form – intended either as scribal publications for a select group or, as in the case of two which were approved by the Stationers' Company censor, destined for but not delivered to the printer.[2] He wrote more notes and collected more material than was ever published. *The second part of the garden of Eden*, produced by his nephew Charles Bellingham in 1660 from his surviving gardening MSS, was an early recognition of the wealth still to be quarried.[3]

In the manuscripts are acknowledgements of ideas taken from books he had read. His library was said to be extensive and he consulted friends' books and manuscripts including probably the excellent library of John Dee. In some cases, such as his summary of Bernard Palissy's work in

---

[1] Unfortunately, virtually no personal papers remain and, apart from medical practice-notes, nothing to record his day-to-day activities. But the manuscripts are full of indirect references to his life: names of acquaintances, food and drink he liked, references to his family, house, and garden. His hopes and ambitions are also displayed in the papers through his attempts to gain money and status from his activities.

[2] BL, SL 2223, ff. 1–47. See Harold Love, *The Culture and Commerce of Texts*, Amherst, MA, 1993.

[3] Plat, *The second part of the garden of Eden*, 1660, Preface.

APPENDIX I

Diverse new sorts of Soyle, the debt to the original author is clear. Closer inspection of other publications, for instance *Delightes for Ladies*, reveals much that has been more subtly borrowed.[4]

Early writers on Plat based their opinions on inspection of the printed works alone. A more complete account of his ideas and working methods is possible from close examination of both the MS and printed sources. For example, previous critics have praised his willingness to cite his sources, an opinion based largely on *Floraes Paradise*. But Plat was not here demonstrating a generous nature. Illness drove him to print this work prematurely, before he had time to tidy his notes for publication, hence all attributions in the manuscripts were printed.[5]

Plat's surviving manuscripts are in the British Library, part of the Sloane Collection. Although now sturdily bound and safely housed, they suffered losses and damage between Plat's death, acquisition by Sir Hans Sloane and subsequent bequest to the nation in the mid-eighteenth century. Some manuscripts are now missing. Plat himself mentions on several occasions a book of medical recipes which has gone. It is possible he sold it when he tried to dispose of his practice. The manuscripts, as now bound, are sometimes jumbled: SL 2189, for example, is an amalgamation of two notebooks, with sections of each spliced together and some pages bound upside down.[6]

Later additions (mostly notes by Thomas Hodges, who acquired the manuscripts in the seventeenth century at some time after Plat's demise) have both augmented and disfigured the originals. But Hodges did posterity a service by creating a single index to many of the documents. Charles Bellingham had control of Plat's manuscript on gardening (SL 2210) in 1660 when he produced from it *The second part of the garden of Eden* and, as he claimed to be a relative, at least some manuscripts at that time were still under family control. But in his introduction to *The Garden of Eden* in 1653, Bellingham implies that an attempt to publish Plat's work without acknowledgement was underway, and he was publishing to forestall this. What else we know about the ownership and use made

[4] TNA, E134/8, JasI/East2.
[5] Eleanour Sinclair Rohde, *The Old English Gardening Books*, 1972, pp. 32–3.
[6] BL, SL 2210, f. 145; Sir Hans Sloane, ODNB.

of the manuscripts before their acquisition by Sloane I have described in my Conclusion.[7]

I have also noted in my Conclusion the interest of the Attorney General in the value of Plat's papers.[8] Note-taking on the scale undertaken by Plat was by no means unusual among gentlemen in his day, neither was his method of using commonplace books. Rosemary O'Day sums up the situation succinctly: 'The impact which common placing had upon the Tudor and Stuart mind is obvious indeed to anyone who has read widely among private papers of the period. These are strewn with commonplace books or notebooks. As early as March 1557 John Foxe was to publish (in Basel) a notebook of commonplaces for students at the university. The pages were blank for students to fill in; only headings and sub-headings were supplied. Common placing learned at school and at university had a profound influence upon the study habits of all professionals – clergy, physicians and lawyers alike.'[9]

The increasing number of the gentry who went to grammar school first learnt this note-taking method there and carried on the habit in later life: 'They learned the trick of compiling a commonplace book, under whose artfully devised headings they entered the "flowers" of their reading. Then, when occasion demanded it, in conversation or letter, in the law courts or parliament, they could search out the appropriate *topos*, in their memories or in their notes, and bring to bear the weight of classical (and even modern) wisdom.'[10]

Those, like Plat, who went to university had this habit strongly reinforced. Richard Holdsworth (1590–1649) a Puritan divine and long-time tutor at Cambridge from 1614, left in manuscript 'Directions for a Student in the Universitie' which contain a section on keeping a commonplace book. The interesting detail, and the similarity between

---

7 BL, SL 2243; *The second part of the garden of Eden*, 1660, pp. 7–9. See the discussion in Chapter 11 for more about the later ownership of the MSS. *Hartlib papers*; Ephemerides 1650 part 3, 28/1/60B–71AE.
8 See p. 370.
9 Rosemary O'Day, *Education and Society*, 1982, p. 159.
10 F.T. Levy, 'How information spread among the gentry, 1550–1640', pp. 11–12; Love, *The Culture and Commerce of Texts*, pp. 197–203.

APPENDIX I

Holdsworth's recommendations and Plat's practice, justify quoting it in full:

Of a Common place book.
I have observ'd in many Students a commendable endeavour to make Common place books, in wch they might recorde the best of theyr studies to certain heads of future use and memorie. But few of them either continue constant In it, or bring it to any perfection. Neither doe I much wonder at it, when I consider the toyle & the interruption it must needs create to theyr studies, to rise evry foot to a great Folio book, & toss it and turn it for evry little pasage yt is to be writt downe.

    I was told of one, who to prevent this toyle, caused a box to be made with as many partitions as he could have had heads in his booke, so that writing his Collection in any bit of paper, he might without more trouble throwe it in to its Topick, & look over each division on occasion. But perhaps you like this no better than I however I shall councell you rather then to trouble your self with any Voluminous Common place booke especially in these your rawer studies, to follow this order in your Collection. Cut some handsome paper bookes of a portable size in Octavo, & rulle them so with Inke or black lead that there may be space left on the side for a margin & at the top for a title: Into them collect all the remarkable things wch you meet with in your Hystorians, Oratours, & Poets. Ever as you find them promiscuously, especialy if out of the same book, in the title space set downe the name of the Authour with the book, or Cap: & after every Collection, the number of the page, or Section whence it is taken, These Collections you shall render so ready and familiar to you by frequent reading them over on evenings, or times set a part for that purpose, that they will offer themselves to your memory on any occasion. Which if you could doe would be far above the use of any Common place book, but if you finde by experience that your memorie is not either faithfull, or quick to doe this, you may at any time with a little pains reduce such bookes of Collection to a Commonplace book wch. shall be only a kind of large Index to them setting downe with every reference, a word or two of each Collection to put you in minde of the rest; wch: reference that you may the better marke you must page all your fore-sayd bookes of Collections (if you will continue the number from one to another or else at the least distinguish them one from another either by different names or order of the Alphabet).

A Common place booke ought to be fitted to that profession you follow, whether of Law, Divinity, Phisick or the like, wch. probably you cannot resolve upon when you begin first your studies, & therfore now Collect many things uselesse, heterogenus raw, Common, and Childish, wch. in a riper Judgment you would be ashamed vexed to have your Common places filled with, wheras In such an one as is here described when you are come to maturitie of Judgemnent, and have pitched upon your Profession, you may avoyd this inconvenience referring to your Commonplace booke only what you like in your former Collections, & Omitting all the rest.[11]

Holdsworth would have recognized Plat's manuscripts as being the product of a university-trained man and would probably have approved of his note-taking methods. Almost all Plat's pages are ruled, usually in red ink, with lines defining left and right margins as well as the head and the foot. Samuel Pepys had his paper ruled by 'an old woman in pannier ally' and perhaps Plat too had his paper ruled for him. Writing paper was expensive, largely imported from France and subject to excise duty and so it paid a gentleman to be neat and organized.[12]

I believe Plat used the two-stage method of commonplacing described by Holdsworth. Manuscripts such as SL 2210, SL 2216, SL 2189, (and probably SL 2203 after f. 88) form the primary stage, when he first made notes on any topics which interested him. Manuscript notebook SL 2210 was an important repository of original notes, although Plat originally planned it to be a neat, tightly organized volume. The book is ruled throughout with head-notes and is today incorrectly bound. The notebook which forms the bulk of the manuscript started at what is now folio 23 with a title page headed, 'GENERALL & SPECIALL Rules for graftinge,

[11] Richard Holdsworth, 'Directions for a Student in the Universitie', Emmanuel College, Cambridge, MS I. 2. 27(1), ff. 51–52. Reproduced as Appendix II of H.F. Fletcher, *The Intellectual Development of John Milton*, Ithaca, NY, 1961.

[12] *The Diary Of Samuel Pepys*, ed. R. Latham & W. Matthews, 1970–83, vol. vii, p. 98, quoted in Love, *The Culture and Commerce of Texts*, Chapter 3, which provides much detail on the paper, ink, and pens of the time.

plantinge, settinge sowinge &c. off trees, cions, buddes, seedes, rootes, slippes. &c.' followed by two pages headed 'EXPERIMENTES in trees, fruits, Flowers, herbes, seedes etc./' with 'Experimentes' crossed through and 'Necessarie Rules 'substituted. After two unheaded pages which contain numbered paragraphs on trees, fruits, etc., and a blank leaf, comes a page headed 'Questiones, chimica, & earum solutiones' and this heading is carried across a few subsequent pages. Then another section 'Diverse Conceipts for the strange plantinge or forwardinge of all vegetables' begins at f. 29. The pattern continues with headings for other topics: 'Conceiptes in buildinge or bewtifiyinge of Howses, Galleries &c.' (f. 31), 'The Arte of Limminge uppon Glasse' (f. 32a), each with several numbered paragraphs of observations, secrets, etc., relevant to the heading.

So far, this book conforms to the highly organized pattern of a commonplace book, but after a page and a half devoted to hop-growing, a change takes place. The next heading: 'Approved & coniecturall experiments in trees, fruits, flowers, hearbes, seedes, Rootes, &c. Elements, Animalls, stones, & mineralls' (f. 35), covers a multitude of topics. There are no further headings to the end of the original book (f. 3a of the present binding) and almost 400 pages of notes, in 720 numbered paragraphs, follow on many subjects. Sometimes Plat finds more to add on a subject he has written about some pages previously. Having no space to augment the original text, he continues with the topic thus: 'Residuum secreti ante no. --' (f. 32a). As we go through the manuscript these residuums get more frequent as he continues to find information on matters he has started to investigate.

On many pages the head and side margins are crammed with additional notes, usually small variations or queries relating to the original paragraph on that page. They are particularly noticeable on topics in which Plat had a deep and continuing interest. Paragraph 349, 'An excellent kinde of victuall both for rich and poore & good cheape & good for the Navye' (his original note on the nature and production of macaroni in Italy), has margin notes galore: 'make your past wth creame or new milcke & some suger et vide quod fiet. qre of sacke muskadine or spirite of wyne...'; 'Ryce rosted egges, & otemeale sodden & buttered are very good meates for the Navye/ sonne drie stockfish upon a Kill, & then powder yt, & temper the powder therof wth oyle or butter./'; 'qre iff starch bee not good to make bread off see the popes booke of Cookery for the making of and dressing of a delicate macaron./'; 'make thinn large cakes of somme good paste, and cut the same owte either in losenges, or longe narow peeces of this foorme [figure of a long oblong] & so you may ridd quantity wthout any presse, then drie them./'; 'they make cuscusow in barbary most like to this/ it is made like small comfitts/'; 'qre of dried Coucumber the powder of dried Carretts & Parsnippes./ made upp into Macarro./'. As well as telling us a good deal about the possibilities Plat saw in this food as a naval victual, the notes show that he re-read this notebook on many occasions, adding ideas as they occurred to him or he came across them.

Several indexes break up the text of the book. That at the end, arranged alphabetically by keyword, aims to be comprehensive. The others are largely lists of subjects worthy of more study and/or potentially worth money. To sum up, while started by Plat as an organized (secondary) commonplace book, SL 2210 quickly becomes a preliminary notebook, only loosely organized, into which Plat poured thoughts, ideas and recipes culled from many sources on many topics. He must frequently have re-read the manuscript, adding notes and listing what he thought was of importance.

SL 2189 is another 'primary' notebook, completely unruled, again incorrectly bound at present. There is one comprehensive index and a few lists of the sort found in SL 2210. Plat used fewer and more widely spaced headings in this book but again there are frequent 'residuums' and some marginal additions. Here he employed side-notes to draw attention to important parts of the text, a technique used in the parts of this manuscript which are not by Plat but, as explained in the chapter on *Delightes for Ladies*, probably written by a priest in the reign of Mary. A third 'primary' notebook is SL 2216, a small, fully ruled book sharing the characteristics of those discussed above. Finally, SL 2203, after f. 203, is another such book. The original pagination starts at folio 281, so what remains is a fragment. With running titles reading 'EXPERIMENTA RERUMQUE METAMORPHOSEIS', the fragment is mainly concerned with alchemy and distilling.

It is difficult to date these notebooks with any accuracy. The scattering of dated recipes in them makes it likely that SL 2203 (after f. 203) was one of Plat's earliest primary notebooks, with dates over the narrow period

APPENDIX I

1580 to 1584 recorded. Although the recipes are more specialized than in the other primary books, there is no sign that they have been transferred from elsewhere and some have been crossed through (see below) indicating they have been copied to a secondary book. As I discuss elsewhere, Plat may have been attracted to alchemy before broadening his interests and this may be his earliest notebook. After SL 2216 Plat appears to have commenced notebook SL 2210, which has one date as early as 1581 and a scattering of others up to 1592. With its multitude of topics, this book may have been used by Plat as his intellectual interests broadened from the study of alchemy. The earliest date noted by Plat in SL 2189 is 1588, which may be around the time he acquired the manuscript from the 1550s which forms the heart of this book. While he filled up SL 2210 with his own schemes and ideas, so he also annotated SL 2189. Other dates that occur in SL 2189 are 1595, 1597 and 1598. The notebook SL 2216 contains many more dates from 1595 to 1597. It would therefore appear that he was using three commonplace books more or less in parallel. This would not be an unnatural course of action if he was intending in due course to transfer the recipes and secrets into more highly organized manuscripts.

On the front paste-down of SL 2216 Plat affixed a quarter-leaf on which is a note explaining the relationship of this primary notebook to the secondary commonplace books I turn to next. He writes, 'Note that where I [have made] a X over all the secret there the whole secrete is poasted over to my booke of Titles, and where you find but half a crosse as this / there the title of the secrete is only posted over and not the substance./' Plat has mined this book for secrets to arrange neatly into one of his secondary commonplace books, aptly called a 'booke of Titles' of which SL 2247 is a typical example. Small and neat, it is ruled without a footer. After a couple of pages of lists, headings appear, all concerned with different dyeing operations. Ff. 39–42 are headed 'Table'. After that the heading of 'Wines Artificial' covers several pages, with the remaining leaves left blank. No additions are squeezed into margins of the pages, no 'residuums' carry additional thoughts and there are many pages headed but otherwise blank, including those intended for the Table. In contrast to the primary notebooks, this is a well-spaced, planned commonplace book which remains unfinished.

SL 2244 and SL 2245 are also small books, and have the same rulings and basic characteristics. SL 2244's first heading is 'Naturall Wines' and headings follow every few pages on diverse topics: 'Bread', 'Houses', etc. Many headings have no text beneath, e.g. 'Borrace', 'Verdegrece', 'Prospective Glass'. The blank page headed 'Cheap Watch-candle' may have been intended for recipe number 207 (f. 93a) in SL 2210, 'cheape & sweete kinde off Candles for all sorts of men & easie to bee made'. 'Pich Multiplied' was copied from SL 2216, f. 19: 'How to multiplie either pich or Tarr', the original being struck-though with a diagonal line, as explained on the front paste-down noted above. SL 2249, another little book, has no headings, just the titles of individual recipes spaced one to each page. Most have no text, where there is a recipe or secret it is very briefly described and this book seems to have been an *aide-mémoire* of some kind. SL 2197 (after f. 16) is also an unusual commonplace book, being largely composed of lists and notes on how to exploit secrets for money.

SL 2195 is another sophisticated variation on the secondary commonplace book. It is fronted by a list of 46 alchemical authors and/or their works, including well-known treatises such as Norton's *Ordinal*, Ripley's *Twelve Gates*, the works of Roger Bacon, Charnock's *Breviary*, and a reference to Plat's own compositions. The rest of the book is headed with alchemical topics, the headings changing every few pages, in alphabetical order, beginning with 'Alchimie Anatomia' and ending with 'Vsus [Usus] Medicinae'. This set of alphabetical headings was insufficient, and there are more on subsequent pages in random order. The whole book is completed by a neat alphabetical index giving page references.

Plat was sophisticated in his use of these secondary commonplace books. They could be tidy and highly organized receptacles for notes taken from his primary notebooks – easy to read, well arranged and indexed – a reminder of his most important recipes, to be flicked through quickly, maybe, before he went off to lobby someone at Court for recognition of his worth as an inventor. Or they could form the basis of a concerted strategy to exploit his secrets to make money.

APPENDIX II

Plat's alchemical poem
(BL, SL 2195, ff. 119–123v)

C To the trew chimicall Reader.

    No ey so quicke but dazeleth at our Arte
    No witt so sharp whose edge is not rebated
    No science heere gives light enough to learne
    And pallas with her Muses all are mated.
    For colde or durst the pen of mortall wight
    The secretes of this secrete arte unfolde,
    Saturnus age longe since had wheelde abowt,
    And made our Tropheis all of massive gold.
    Yet Nature hath her trew adopted Sons
    Though seldom seene, now caring to confer
    Wth vulger witts, which to vaine-glorious endes
    Their lives and Labours whollie do refer.
    Then heere a Noves whoe presents himself
    In native Robes amidst the learned Crew
    A Knight, a Scholler, not mured in Cell,
    Nor treyned up within a monkish mew.
    He gives the matter wheruppon to worke
    In playner termes then ever he cold find,
    The grosse and subtile preparation
    Within the lightes of arte he hath confinde.
    Proportions he hath prized to thie hands
    A knot of Nature never yet untied,

The shape and nomber of thie vessells apt
Supprest by all, yet hath he not denied.
More inward fiers he hath disclosed to thee
Then Authors all in all their workes reveale,
And owtward treates of trew Experience
From Hermes Children he wold not conceale.
Something in Collors, (if thow markst him well)
He glanceth at, and giveth farther light,
How thow wth Phoebus heaven must rubifie
The Stone itself when thow hast made it white.
Yet must he still wth auncientes all conclude
No pen so plaine that ever penned all,
But that in fine when wearied are thie witts,
Thow must be forste for Maisters help to call.

C of the matter of our Stone.

Cap.1.

C When God began both heaven & earth to shape,
And Creatures fit for each of them to frame,
He found a chaos a rude unformed masse
From whence he drew and did derive the same.
Even so, an other chaos thow must seeke
In which a Microcosmus closely lies
That may bring foorth the .4. contending seedes
And turne the wheele of our celestiall skies
But if a chaos be no where to be had
Then ransack mines till thou an Adam mettalline
Hast found, whose Eve lies lurking in his side,
This doble sex, our matter will define
Or els review the spangled orbes above
Cawse Son and Moone to shine wthin thie glass.
Let fleetinge mercurie raunge up and downe
Till Lion greene from foorme to forme do pas
But if thy wit can none of theise attaine

APPENDIX II

> There is a roote from whence three braunches spring,
> Which if thow canst well rott and putrifie
> Into a Tree of gold thow shalt it bringe
> Let Man & Weif in trew commution meete,
> Unite the fixed wth his flyinge Mate,
> Prepare a bed for that incestuous broode,
> Whose lawles luste doth strive to mend their state
> Yf thow by nomber woldst obtaine the same
> One, two, three, fowre, which makes iust 10 in all,
> Is that which all the Sophies in their bookes
> Do count the nomber philosophicall.
> Yet one alone which is our Mercurie
> Our animated Azot if thow hadst the same,
> Yt were sufficient of it self alone
> Our great Elixir and our stone to frame.

C Of the preparation of the Materialls

Cap.2.

> Now hast thow stoar of matter to thie minde
> But nothing firm as yet to work uppon:
> Praepare therfore before thow doost begin
> In glas t'interr this rare and worthie stone
> Sometimes a porphire or a Serpentine,
> Wold steede the well to grind thie bodies smaall
> Calcined salts and romann vitrioll
> Do not refuse when spirits for them caall
> Sublime thow must and quicken often-times
> Make christall cakes and then recrudifie,
> Somme paines, some cost, somme dropps are like to faall
> Before thow hast a perfect Mercurie
> Besides all this thantinomian Horne
> For fierie Phoebus must praepared bee
> Whose bodie though the perfectest of them all
> Hath not attained, the higth of his degree

And Saturne heere must yeald to play his parte
Yf Phoebe do desire his Companie
Yea though he spend his lief and spirits all,
Yet may he not her iust request denie.
And though our work be still a worke of peace,
Yet Mars himself must breath his whottest fume
On Venus tender corps, els may wee not
With Hermes seale to close our egg presume
Unles wee flie to Iuppiter for ayde
Whoe helpes to make our heavenly earth to drie
Till it desire with moisture of his kinde
To glut & gorge itself abundantly.
And though that learned race of Hermes broode
Wth open mowth gainst Corofines doth crie
Yet strongest waters heere may doo the good
Yf thow doost know to use them wittely.

C Of the proportion of the Materialls to be used. Cap.3.

To have a matter royally prepard
Without proportion will the nought avayle,
Though all things els bee suted to thie will,
Yet heere unwarres thow maist comitt a foile
But how shall I this fearfull skill disclose,
When all the Fathers have wth one assent
Pronounste a curse on him that dares unfold
The secret oracles of their intent.
Yet hoping none but *trew adopted sons [*Hermes royall raice]
Shall to this high and subtile point aspire
Under the veile of our Philosophie
I will give somme content, to thie desire
Shall I with Ripley bid thee look to loame
Which Plaisterers doo dawbe uppon their waall?
Or shall I send thee to a sellie Nurse,
To gaze on that which shee her papp doth caall?

Shall I with Noaths flood thee terrifie
Leaste it surround the earth wthin thie glas.
Or rather bid thee feare those fatall flambes
Through which thie woorke and all the world must pas?
To weight and nombers I will thee reduce
A learned scale Ile put into that hand:
Whose weightes by figures of Arithmetique
Shall give thee light our course to understand.
Yf .3. to one, thenss that makes 4 to twelve
A nomber good commended ofte by many:
Yet one of Sol to 16 of an other
I hold the best not written of by any.
And when at length thow seekest to congeale
Then bodie .3. and spirit one is best:
Yf to dissolve, then of the spirit three
And bodie one, for that must bee the least.

C of the Philosophers Vessell.

Cap.4.

    Heere doth the glas of round or oval foorme
    Present itself proportions to receyve,
    Take heede therfore that want of iuste content
    Twixt glas and matter do the not deceyve.
    Thanncient wrighters all concurr in one
    That only one is heere sufficient:
    But vessells more I thee advise to take
    Least oyle and labour both shold be mispent.
    In foorme but one, in number more then two
    Els wilt thow want to multiplie thie base,
    A foode and ferment rich enough in kinde
    That with thie stone hath run an equall race.
    Somme court the earth, some water for thie glas
    And not the christall orbe which holdes the same
    Some for decocting of their Mercurie,

A threefold vessell to themselves doo frame
Bee sure thie glas wth glas bee sealed up
Trust not too lute that choisest witt can make,
Least subtile spirits eare thow canst beware
Their soddein flight through lute & loame do take
Beware thine Egg wherin thie chaos lies
Stands stedfast in a stronge unmoved place
Least Cannons roare, or motive violent
Do shake thie vessell & thie worck deface.
And whilst the wheele of our Philosophie
Doth turne itself, doo not thy glas unclose,
The wisest Men that ever wrought in mines.
Give Nature leave to binde and eake too loose
My last advice is so too place thie glas
That Phoebus may theron disperse his rayes,
A cheerefull sight it is for too behold
How every cooller in his season playes.

C Of the Philosophers fiers.

Cap.5.

Now listen well you learned vulcanistes
And marke what our Electrum doth require
Of owtward first, I will discourse at large
And then disclose our secrete inward fier.
No flamme so fierce but wee can use the same
No glowing Coales too whot when wee prepare
In limature of Mars, sometimes wee place
Sometimes in sande, *ground of all our Care [*hope]
And sifted ashes wee do not refuse
When gentle heates will fit and serve our turne
With Maries bath, or dew of Bath sometimes
Wee are content for feare our Azot turne
Amongst the rest the trewest in degree
The lampe I finde of all that I have sought

APPENDIX II

Yet cold I have, none other wold I use,
Then bath of Baath where Charnocks Mr wrought
In all thie fiers yet take this for a rule
That none sublime, although to sublimation
The rightest heate makes shortest accurtation.
Now soare aloft, and wright of highest heates
Bring Phoebus downe from his resplendent Spheere,
Amidst our Planetts philosophicall
In our whott heaven, let him the scepter beare
Yet Mercurie must kindle first the flambe
He is the match that setteth Sol on fire
With Venus help, whome Mars hath first inflambe
And lickte wth Lune taccomplish her desire.
The fierie Dragon heere doth first begin
The Lion greene pursueth on amaine,
And after him the dreadfull Basiliske
Then golden Lion doth furnish up the trayne.
One fier more which Authors all commend
Is that which dooth digest and is continuall:
Of gentle heate, shut up wthin thie glas
A subtile Spirite, but yet minerall.

C Of the Coollors appering in the woorcke. Cap. 6.

Imagine now our wheele begins to turne
And rowle abowt till all the plannetts seaven
Have wheeled themselves from Saturn unto Sol,
Without repose, within the Chimists heaven.
 Then choice and chaunge of Collors will appeere
First Saturn shews himself in sable hew
With bill and head of Crow as black as Jett
A perfect signe of putrifaction true.
Then Coollors meane twixt white and black break forth
Successively most glorious to bee seene:
The spotted Panther wth the Peacock tayle,

The rayneboe graced wth the Lion greene.
Thie base must often tymes bee circulate,
Before a perfect white thow shalt obtayne
And earth rotation in his circled coorse
Bringes all the Coollors round abowt againe.
But how to rubifie this fixed white
There lies a secret never yet disclosede
By pen or press to any wordly wight
      [*margin*: vide lapis philosph. no. 7.]
Save where a secret trust hath ben reposd
Yet if wee may beleeve Philosophie,
Digestion will in tyme bring foorth the same
O that I durst, whie shold I wish to dare,
Since no Man yet wold give it once a name.
Then let it pas, content the wth the white
A perfect Stone most pearle-like to the Eye
And rich enough to him that knoweth how
With kindly ferment yt to multiplie.
The cawse of Collors I cold heere define
And eake the Roote from whence they all do springe
But my desire is only to thine Eye
Advise of them in these few lines to bringe.
      [*crossed through*: finis H Platt. Miles.]

C of proiection .cap 7/

C Proiection is the last but not the least of all
This arte alone will Trie a Maisters skill
Whoe though he have a trew elixir made
Yet this unknowne he can not work his will.
The shortest wheele it is of all the rest,
And yet in greatest steede it will thee stande
For to invest our high commanding Kinge
With golden scepter in his royall hands.
Yet when thow hast a fluent medicine made

APPENDIX II

To flow like wax and of a *rubine hew, [*golden]
Then is it in thie choice which stone to make
Of all the six, for transmutation trew.
I teach not all yet none hath taught so much
Proiecting skill, to perfect up the stone
What heere doth want no wrighter dares supplie
But leaves the more unto thie self alone.
This secret point keepe of the Tirants hand
Whomme if it chance our *ruby to enioy, [*tincture]
Yet can he not before it be prepard
His neighbor Kings thrones at all annoy.
And if prepard, how can he multiplie
A blonted stone, and make it grow againe
For having lost his vertew generative
A finite nature doth it now conteyne.
Nay what if somme have made our stone aright,
Yet after cold not multiplie the same,
A fruitles worke and all for want of skill
A ferment fit, and kindly how to frame
Then how shall hee that fully knoweth not,
The base and grounde wheron the Artist wrought
Ferment this base till hee by wheeling oft,
Unto his highest pich the same have brought.

    Conclusion.
Yf notwthstanding all that I have writt,
There be no hope this golden fleece to win
Then have I miste, as other Archers have,
Els have I hit the Clowt if not the pin.

H. Platt Miles. 264 verses./

APPENDIX III

Transcription of the printed broadside on food and medicines for seamen, c. 1607

Certaine Philosophical Preparations of Foode and Beverage for Sea-men, in their long voyages: with some necessary, approved, and Hermeticall medicines and Antidotes, fit to be had in readinesse at sea, for prevention or cure of divers diseases.

And first for Foode, A cheape, fresh and lasting victuall, called by the name of Macaroni amongst the Italians, and not unlike (save onely in forme) to the Cus-cus in Barbary, may be upon reasonable warning provided in any quantity to serve either for change and variety of meat, or in the want of fresh victual. With this the Author furnished Sir Francis Drake and Sir John Hawkins, in their last voyage.

2. Any broth or Calase, that will stand cleare and liquid, and not gellie or grow thicke when it is cold, may also be preserved by this fire of Nature from all mouldinesse, sowrenesse, or corruption, to any resonable period of time that shalbe desired. A necessary secret for all sicke and weake persons at sea, when no fresh meate can be had, to strengthen or comfort them.

3. Now for Beverage: All the water, which to that purpose shall bee thought needefull to be carried to sea, will bee warranted to last sweete, good and without any intention to putrefaction, for 2, 3, or 4 yeeres together. This is performed by a Philosophical fire, being of a sympatheticall nature with all plants and Animals. In the space of one moneth, the Author wil prepare so many Tunnes therof, as shall be reasonably required at his hands.

4. By this meanes also both Wine, Perrie, Sider, Beere, Ale, and Vineger, may be safely kept at Sea, for any long voyage, without feare of growing dead, sowre or mustie.

5. And, as for Medicine, if any Nobleman, Gentleman, or Marchant, shall by his Physition be advised to cary any speciall distilled waters, decoctions, or iuyces of any plant or any other liquid vegetable or animall body whatsoever with him in any long voyage, this Author will so prepare the same onely by fortifying it with his owne fire of kinde, that he may be assured of the lasting and durabilitie therof, even at his owne pleasure.

6. Here I may not omit the preparation of the iuice of Limons with this fire: because it hath of late been found by that worthy Knight Sir James Lancaster to be an assured remedy in the scurby. And though their iuice will, by natural working and fermenting, in the end so spiritualize itselfe, as that it will keepe and last either simply of itself, or by the help of a sweete olive oyle supernatent yet this Author is not ignorant, that it hath lost much of his first manifest nature, which it had whilest it was conteined within his owne pulp and fruit (as is evident in the like example of wine, after it hath wrought long, which differeth exceedingly both in taste and nature from the grape out of which it was expressed) whereas being strengthened with this philosophicall fire, It retaineth still both the naturall taste, race, and verdure, that it had in the first expression: and so likewise of the Orange.

7. There is also a speciall powder for Agues, Quotidian, and Tertian: and sometimes it helpeth Quartans. Halfe a dramme is sufficient for a man: and quarter of a dramme for a child. It is taken, often at the second and seldome or never faileth at the third time. It is not offensive to the taste. It expelleth the disease, without any evacuation or weakening of the Patient.

8. A sweete Paste, for the head-ache: which commonly giveth ease, in one houres space, either upon the first or second taking, because it is specificall. the dose is the weight of 6d.

9. A safe, general & goode purging Powder, to be taken in white wine, working easily without any convulsion, or other offence to the stomacke. it is pleasant, and hath not any common or known purgative therin. It

weakeneth not the patient, neither doeth the body grow costive after it: which is usuall in most of the common purgatives. There have been so many trials made upon all sorts of complexions with this powder, as that it may well deserve the name of a generall purge: yet I can least commend it in Cholericke bodies. The dose is two drammes and an halfe at the time. This being taken in warme weather for three dayes together, in the Spring and Fall, will prevent both the Gowte and Dropsie, and most of those diseases that spring from rheumaticke causes: and if it cure them in eight or ten dayes, take it for advantage. It cureth the Pockes newly taken in five or sixs dayes: and in tenne or twelve dayes, at the most, it cureth a deepe rooted Pocke.

10. And if the plague, burning feaver, or small Pockes, or Measels happen to infect any of the Souldiers or Mariners, or others in the ship: then if, within sixe or eight houres after infection, a dose of my antidote powder (whereof eight grains are sufficient) be taken, it commonly preventeth the rage and violence of the Plague, by mastering the poyson, seldome suffering any sore to arise: and it disperseth and conquereth the matter of the small Pockes and Measels: whereby in a few houres it vanisheth, without making the Patient heart-sicke. And in the cure of any kind of poyson, no Vnicornes horne, no Bezoar Stone, no Terra Limmnea or Sigilleta, no mithridates &c. is able to match the same, though taken in a double proportion. It is an excellent remedie against swooning, or any sodaine passion of the heart.

11. There is also a medicine, which I will commend for the Sea (being a notable astringent powder) which stayeth any flux of blood in a short time, and often cureth the Piles and Emerhoides.

12. The essences of spices and floures (as of Cinnamon, Cloves, Mace, Nutmegs, Rosemary, Sage, &c.) being in the forme of powders, may with lesse danger be carried at sea, are more apt to be mixed and incorporated with Syrupes, iuleps or Conserves, are more pleasing to nature, and are more familiarly taken, and with better success then the chymicall oyles themselves, drawen by limbecke: their effects are answerable to the nature of the oyles.

Thus much I am bold to offer and publish for the benefit of seafaring men, who for the most part are destitute both of learned Physitions and skilfull Apothecaries: and therfore have more needs then others to cary their owne defensives and medicines about them. Which if it shall receive entertainemnt according to the worth therof and my iust expectation, I may happily be encouraged to prie a little further into Natures Cabinet, and to dispense some of her most secret Iewels, which she hath long time so carefully kept, onely for the use of her dearest children: otherwise finding no speedy or good acceptance of this my proffer (but rather crossed by malice or incredulity) I doe here free and enlarge my selfe from mine owne fetters: purposing to content my spirits, with my place and dignitie, and in likelyhood proove also more profitable in the ende, then if I had thenkelesly devoted my selfe to Bonum Publicum. In which course, happy men are sometimes rewarded with good words: but few or none, in these dayes, with any reall recompense.

Ut Deus per Naturam, sic Natura per ignem Philosophicum.

H.P. Miles.

APPENDIX IV

A manuscript advertisement for medical and other supplies for seamen

(BL, SL 1272, f. 30)

Waters, Oyles Electuaries & other compositions necessary for the Sea

A trosse procuring sleepe in burning feavers to be mixed wth some of the compositions following to be taken at severall times, the ounce	0–6–0
An Electuary to stay any violent vomit, the ounce	0–2–0
An Electuary in hot burning feavers, the ounce	0–3–4
A vulnerary Electuary healing any inward bleeding, by mixture or otherwise, as allso all woundes, from the Centre outwardly, the pounde	0–10–0
A sanguinary powder for the bloody flux or pissing blood, or other inward bleeding, the ounce	0–2–6
Aqua Bezoartica the pint 40s	2–0–0
Aqua Caelestis the pint	0–12–0
Cinnamon water the pinte	0–4–0
A Syrop for the Scurbie the pownd	0–5–4
An ointment for the Scurbie, the pound	0–10–0
A Lotion for the Scurbie to washe the gommes, the pinte	0–10–0
A purging powder for the Scurbie to be taken at severall times, the oz	0–10–0

APPENDIX IV

A syrop of Quinnces aromaticall, the pound	0–5–4
Water in the Scurbie to be mixed with the Syrop, the pinte	0–2–6
A pill for the bloody flux, to be taken at severall times, the ounce	0–10–0
An astringent [crossed through: powder] water for the flux, the pinte	0–2–0
Venis methridate the ounce	0–3–4
Cardans water the pinte	0–1–0
Rosa solis with pearle, the pinte	0–5–0
Rosa solis without pearle, the pinte	0–3–4
Aqua mirabilis the pinte	1–0–0
A syrop for a burning fever to be mixed wth ye electuary & cardans water ye pound	0–5–4
A syrop of grapes the pound	0–2–0
A water for the Cancer in the mowthe, the ounce	0–2–0
An ointment for burning with powder, the pound	0–6–0
An excellent balsamum for all greene woundes, the ounce	0–2–0
An excellent plaister for greene woundes, & old ulcers, the pound	0–16–0
A pill for a violent coughe & procuring sleepe, to be taken at severall times oz	0–10–0
Rose water the pinte	0–1–4
Rose viniger the pinte	0–1–0
Sweet sallet oile, the pinte	0–0–10
D Stevens water the pinte	0–5–0
Conserve of Berberries, the pound	0–3–0
Conserve of Roses, the pound	0–3–0
Greene ginger the pound	0–3–0
Syrop of violetts the pound	0–3–0
Severall purgatives by pilles, electuaries & cakes.	
Quinces and Cheries preserved the pound	0–2–0
Oile of vitrioll to make a cooling drinke, by mixing wth water the oz	0–1–6
Irishe aqua vita, or usqubath both to drinke alone in any extremity or mixt wth water to make a pleasant drinke, the pinte	0–1–0

A MANUSCRIPT ADVERTISEMENT

Certaine other necessaries for the Sea

Musket & Caliver bulletts extraordinary to pearce the plancks
    of any shipp that is thought to be musket free the pound   0–1–6
ffreshe & lasting victuall, serving instead of bread & meat
    whearof 4. ounces is a sufficient meale/I did furnish
    Sr ffr Drake therwth, ye pound   0–0–6
An ointment defending armoures & peeces from rusting.
    with this I did furnish Sr ffra Drake/ the pound   0–2–0
An ynk powder to be mixt with water, wine, vinegar &c.
    to make black ynk at your pleasure the ounce   0–0–6

List of Illustrations

Figure		Page
1	Title-page of *Divers new sorts of Soyle*, 1594	6
2	The title-page of *The Garden of Eden*	44
3	Pages from *Floraes Paradise* showing secrets taken from T.T.'s manuscript	44
4	Plat's pasta machine	114
5	The title-page of the 1628 edition of *Delights for Ladies*	152
6	Pages from *Delights for Ladies* (1628)	152
7	A wagon which can be taken apart and assembled quickly	291
8	The portable pump Plat claimed to have invented	291
9	Devices for expressing oils depicted in *The Jewell House of Art and Nature*	292
10	A copper press for moulding small objects as described in *The Jewell House*	292
11	A new and cheap lantern designed by Plat	307
12	Plat's coalballs, neatly piled up in a grate	307
13	An fine ear of barley that Plat grew in his garden using soap-ashes as fertilizer	308
14	Plat's bolting hutch	308
15	A glass desk Plat describes which enables outlines to be traced using light underneath the desk	322
16	ABC dice, Plat's idea for a way of teaching children the alphabet	322

Bibliography

MANUSCRIPTS

British Library
Add. Mss 72, 891
Sloane Ms SL 2170
Sloane Ms SL 2171
Sloane Ms SL 2172
Sloane Ms SL 2177
Sloane Ms SL 2189
Sloane Ms SL 2195
Sloane Ms SL 2197
Sloane Ms SL 2203
Sloane Ms SL 2209
Sloane Ms SL 2210
Sloane Ms SL 2212
Sloane Ms SL 2216
Sloane Ms SL 2223
Sloane Ms SL 2242
Sloane Ms SL 2243
Sloane Ms SL 2244
Sloane Ms SL 2245
Sloane Ms SL 2247
Sloane Ms SL 2249

National Archives
E134/8 Jas 1/East 2.
E210/10401 20 Eliz
PROB 11/112/114

Guildhall Library
MS 5485
Microfilm MF 324

London Metropolitan Archives
M92/III
A29341

PUBLISHED ARCHIVE SOURCES

Acts of the Privy Council, 1590–91, 1591, 1591–2, 1596–97.
Calendar of State Papers Domestic, 1595–7.
Calendar of State Papers, Ireland, 1599–1600.
Harleian Society Publications, vols., 93, 1913; 109/10, 1963.
Records of the Honorable Society of Lincoln's Inn, The Black Books vol I, 1897, vol II, 1898.
Richard Holdsworth, 'Directions for a Student in the Universitie', Emmanuel College, Cambridge, MS 1.2.27(1), ff.51–52. Reproduced as Appendix II of H.F. Fletcher, *The Intellectual Development of John Milton*, Ithaca, NY, 1961.
The Diary Of Samuel Pepys, ed. R. Latham & W. Matthews, 1970–83.
The Diaries of John Dee, ed. Edward Fenton, Charlbury 1998.
The Hartlib Papers [CD-ROM] Ann Arbor, 1995.

PRINTED WORKS

Unless otherwise indicated, the place of publication is London.

Abbot, L.W., *Law Reporting in England, 1485–1585*, 1973.
Adams, Simon, *Leicester and the Court*, Manchester 2002.
Allgood, Henry G.C., *History of Bethnall Green*, 1905.
Amherst, Alicia, *A History of gardening in England*, 1895.
Andrews, K.R., *Drake's Voyages*, 1967.
Anon., *Epulario, or the Italian Banquet*, 1598.
––, *The history and description of fossil fuel*, 1841.
––, *Alchemy in the English State Papers*, www.levity.com/alchemy/statpap.html.
Appleby, A.B., *Famine in Tudor and Stuart England*, 1978.
––, 'Nutrition and Disease: the case of London, 1550–1750', *Journal of Interdisciplinary History*, VI, I, 1975.
Archer, Ian W., 'Material Londoners?', in *Material London ca. 1600*, ed. Lena Cowen Orlin, Philadelphia, PA, 2000.
Arthington, Henry, *Provision for the poore, now in penurie*, 1597.

BIBLIOGRAPHY

Ash, Eric H., 'Queen v Northumberland, and the control of technical expertise', *History of Science*, xxxix, 2001.

Ashmole, Elias, *Theatrum Chemicum Brittanicum*, 1652.

Aubrey, John, *Aubrey's Brief lives*, ed. Oliver Lawson Dick, 1982.

Bacon, Francis, *Advancement of learning*, in *Works of Francis Bacon*, ed. Ellis, Spedding & Heath, Boston, 1860–64.

Barlow, W., trans., *Three Christian sermons made by Ludovike Lavatore, Minister of Zuricke in Helvetia, of Famine and Dearth of Victualls*, 1596.

Bier, A.L., & Roger Finlay, eds., *London, 1500–1700*, 1986.

Blith, Walter, *English Improver Improved*, 1655.

Bowers, R., & P.S. Smith, 'Sir John Harington, Hugh Plat, and Ulysses upon Ajax', *Notes & Queries*, vol. 54, no. 3, 2007.

Bradley, Richard, *General treatise of husbandry and gardening*, 1723.

––, *The Country Housewife and Lady's Director*, 1736, ed. Caroline Davidson, 1980.

Brickell, Christopher, ed., *Royal Horticultural Society Encyclopedia of Gardening*, 1992.

Bryden, D.J., 'Evidence from advertising for mathematical instrument making in London, 1556–1714', *Annals of Science*, vol. 49, 1992.

Bull, Stephen, 'Pearls from the dungheap: English saltpetre production 1590–1640', *Ordnance Society Journal*, vol. 2.

Calvert, Dr., ed., 'Richard Gardiner's 'Profitable Instructions', 1603' *Shropshire Arch. Nat. Hist. Soc.*, series II, vol. 4, 1892, pp. 241–2.

Campbell, Susan, 'Glasshouses and Frames', in *The Country House Kitchen Garden 1600–1950*, ed. C. Anne Wilson, Sroud, 1998.

––, *Charleston Kedding*, 1996.

Caraman, Philip, trans., *John Gerard*, 1951.

Cardano, Girolamo, *De subtilitate libri XX*, Lyons, 1554.

Chartres, John, & David Hey, *English Rural Society, 1500–1800*, Cambridge, 1990.

Clulee, Nicholas H., *John Dee's Natural Philosophy*, 1988.

Cole, Mary Hill, *The Portable Queen*, Amherst, MA, 1999.

Conner, Clifford D., *A people's History of Science*, New York, NY, 2005.

Cooper, Alix, 'Homes and Households', in *Cambridge History of Science*, vol. III, Cambridge, 2006.

Corbett, J.S., *Drake and the Tudor Navy*, 1898.
Cortese, Isabella, *I Secreti etc.*, Venice, 1574.
Coryate, Thomas, *Coryats Crudites*, 1611.
Dallas, E.S., *Kettner's Book of the Table*, 1877.
Debus, Allen G., 'Palissy, Plat, and English Agricultural Chemistry in the 16th and 17th Centuries', *Archives Internationales d'Histoire des Sciences*, 21, Rome, 1968.
––, *The English Paracelsians*, New York, NY, 1965.
Dickson, Donald R., 'Thomas Henshaw and Robert Paston's pursuit of the red elixir', *Notes & Records of the Royal Society of London*, vol. 51, I, 1977.
Donald, M.B., *Elizabethan Copper*, 1955.
Donno, Elizabeth Story, ed., *Sir John Harington's A New Discourse of a Stale Subject Called the Metamorphosis of Ajax*, 1962.
Driver, Christopher, ed., *John Evelyn Cook*, Totnes, 1997.
Eamon, William, *Science and the Secrets of Nature*, Princeton, 1994.
Earle, Peter, *The Making of the English Middle Class*, 1989.
Eden, R., *A very Necessarie and Profitable Booke Concerning Navigation*, 1579.
Elton, G.R., *England under the Tudors*, 1971.
Estienne, Charles, & Jean Liébault, *L'Agriculture et maison rustique*, Paris, 1586.
Estienne, Charles, *Maison rustique, or, The countrey farme ... translated ... by Richard Surflet ... Now newly reuiewed, corrected, and augmented, with diuers additions, out of the works of Serres his Agriculture, Vinet his Maison champestre ... and other authors. And the husbandrie of France, Italie, and Spaine, reconciled ... with ours here in England: by Gervase Markham, etc.*, 1616.
Evelyn, John, *A philosophical discourse of earth*, 1676.
––, *Sylva*, 1670.
––, *Terra*, a new edition, with notes by A. Hunter, 1787.
Falls, Cyril, *Elizabeth's Irish Wars*, 1996.
Fenton, Edward, ed., *The Diaries of John Dee*, Charlbury, 1998.
ffoulkes, Charles, *The Armourer and his Craft*, 1912.
Fisher, F.J., *London and the English Economy*, 1990.

BIBLIOGRAPHY

Fitzherbert, *Booke of Husbandry*, 1598.

Fleming, Laurence and Alan Gore, *The English Garden*, 1979.

Ford, Boris, ed., *Cambridge Cultural History of Britain*, Cambridge, 1992.

Fox, Harold G., *Monopolies and patents*, Toronto, 1947.

Frampton, John, trans., *Joyfull newes out of the Newe Founde Worlde*, Nicholas Monardes [1577], intr. Stephen Gaselee, 1925.

Fuller, John, *The Worthies of England*, 1811.

Fussell, G.E., *The Classical Tradition in Western European farming*, 1972

––, *Old English Farming Books*, 1978.

Fussell, G.E., & K.R. Fussell, eds., *Delightes for Ladies*, 1948.

Galinou, Mireille, ed., *London's Pride*, 1990.

Gapper, Claire, 'The London Plasterers' Company and decorative plasterwork in the 16th and early 17th centuries', *The Journal of the Building Limes Forum*, vol.9, 2002.

Garden History, vol. 27, I, 'Tudor Gardens', 1999.

Gardiner, Richard, *Profitable instructions for the manuring, sowing and planting of kitchen gardens. Very profitable for the common wealth and greatly for the helpe and comfort of poore people*, 1599.

Gerard, John, *Catalogus arborum, fruticum ac plantarum*, 1596.

––, *The herball or Generall historie of plantes*, 1597, 2nd ed., 1633.

Goodway, Keith, 'John Smith's Paradise and Theatre of Nature', *Garden History*, vol.24, 1, 1996.

Googe, Barnaby, *Foure Bookes of Husbandry*, 1577.

Goose, N., & L. Luu, *Immigrants in Tudor and Early Stuart London*, 2005.

Gotch, J. Alfred, *Early Renaissance Architecture in England*, 1914.

Graf, H.T., 'The Collegium Mauritanum in Hesse-Kassel and the Making of Calvinist Diplomacy', *Sixteenth Century Journal*, vol. XXVIII, no. 4, 1997.

Grassby, Richard, *Kinship and Capitalism*, Cambridge, 2001.

Graves, T.S., 'A note on the Swan Theatre', *Modern Philology*, vol. 9, no. 3, 1912.

Gray, Todd, 'Walled gardens and the Cultivation of Orchard Fruit in the South-West of England', in *The Country House Kitchen Garden 1600–1950*, ed. C. Anne Wilson, Stroud, 1998.

Griffiths, P. & M.S.R. Jenner, eds., *Londonopolis*, 2000.

Hadfield, Miles, *A History of British Gardening*, 1985.

Hakluyt, Richard, *Voyages* (8 vols.), 1962.

Hall, Bert S., *Weapons and Warfare in Renaissance Europe*, Baltimore, MD, 1997.

[Harington, Sir John], *Ulysses on Ajax. Written by Misodiaboles*, [1596] 1814.

Harkness, Deborah E., '"Strange" Ideas and "English" Knowledge', in *Merchants and Marvels: Commerce, Science, and Art in Early Modern Europe*, ed. Pamela H. Smith & Paula Findlen, 2002.

––, *The Jewel House: Elizabethan London and the Scientific Revolution*, 2007.

Harris, John, ed., *The Garden*, 1979.

Hartlib, Samuel, *Hartlib's Legacie*, 1655.

Harvey, John, *Early Nurserymen*, 1974.

Haynes, Alan, *Invisible power*, Stroud, 1992.

Heal, Felicity, & Clive Holmes, 'The economic patronage of William Cecil', in *Patronage, Culture and Power, The early Cecils*, ed. Pauline Croft, 2002.

Hennessy, G.L., *Novum repertorium ecclesiasticum parochiale Londinense*, 1898.

Henrey, Blanche, *British Horticultural and Botanical Literature*, 1975.

Hieatt, Constance B., *An Ordinance of Pottage*, 1988.

Hill, Thomas, *The gardener's labyrinth*, [1577], ed. Richard Mabey, Oxford, 1987.

Hoskins, W.G., 'Harvest fluctuations and English economic history, 1480–1619', *Agricultural History Review*, xvi, I, 1968.

Houghton, Walter E., 'The English Virtuoso in the Seventeenth Century', *Journal of the History of Ideas*, vol. 3, 1942.

Hoyle, Richard, 'Woad in the 1580s: alternative agriculture in England and Ireland', in *People, Landscape and Alternative Agriculture*, ed. Richard Hoyle, Exeter, 2004.

Hoyles, Martin, *Gardeners Delight*, 1994.

––, *Bread and Roses*, 1995.

Hulme, E.W., 'The history of the patent system under the prerogative and at common law', *Law Quarterly Review*, 12 (1896), 141–54; 16 (1900), 44–56.

BIBLIOGRAPHY

Hume, M.A.S., *The Year after the Armada*, 1896.

Hunter, A., *Georgical Essays*, York, 1803.

Hunter, Lynette, and S. Hutton, eds., *Women, Science & Medicine*, 1997.

Hunter, Michael, 'John Evelyn in the 1650s: A Virtuoso in Quest of a Role', in *John Evelyn's 'Elysium Britannicum' and European Gardening*, ed. Therese O'Malley & Joachim Wolschke-Bulmahn, Washington, DC, 1998.

Hyde Price, William, *The English Patents of Monopoly*, 1906.

Janacek, Bruce, 'Thomas Tymme and Natural Philosophy', *Sixteenth Century Journal*, vol. XXX, no. 4, 1999.

Jeffers, R. H., *The friends of John Gerard*, Falls Village, CT, 1967.

Jenner, Mark S. R., *Early Modern conceptions of 'cleanliness' and 'dirt' as reflected in the environmental regulation of London c.1530–c.1700'*, D.Phil. thesis, Oxford University, 1991.

Norden, John, *The Surveyor's Dialogue*, 1607.

Jonson, Ben, *The Alchemist*, ed. F.H. Mares, 1997.

Jourdain, Margaret, *English Interior Design, 1500–1830*, 1950.

Juel-Jensen, Bent, 'Some Uncollected Authors XIX, Sir Hugh Plat ?1552–?1611', *Book Collector*, 1959.

Keevil, John J., *Medicine and the Navy*, 1957.

Keller, Alex, 'Mathematical Technologies and the Growth of the Idea of Technical Progress in the Sixteenth Century', in *Science, Medicine and Society in the Renaissance*, ed. A. Debus, 1972.

Kerridge, Eric, *The Agricultural Revolution*, 1967

Kocher, Paul H., 'Paracelsan medicine in England', *Journal of History of Medicine*, Autumn 1947.

Lang, R.G., 'London's Aldermen in Business: 1600–1625', *Guildhall Miscellany*, vol. 3, 1970.

Laurence, John, *A new system of agriculture and gardening*, Dublin, 1727.

Lemire, Beverly, 'Consumerism in Preindustrial and Early Industrial England', *Journal of British Studies*, 27, 1988.

Leslie, M., & T. Raylor, eds., *Culture and Cultivation in Early Modern England*, 1992.

Levy, F.T., 'How information spread among the gentry, 1550–1640', *Journal of British Studies*, xxi, 1982.

Levy Peck, Linda, *Consuming Splendor*, 2005.
Loades, David, *The Tudor Navy*, 1992.
Lodge, Thomas, *A treatise of the plague*, 1603.
Love, Harold, *The Culture and Commerce of Texts*, Amherst, MA, 1993.
Macleod, Christine, *Inventing the Industrial Revolution*, Cambridge, 1988.
Markham, Gervase, *The English Housewife*, [1615], ed. M. R. Best, Montreal, 1986.
——, *The Whole Art of Husbandry*, 1631.
Mascall, Leonard, *A Booke of the Arte and Maner How to Plant and Graffe all Sorts of Trees*, 1572.
McDonald, Donald, *Agricultural Writers*, 1908.
Middleton, Thomas, *The Roaring Girl*, 1611.
Moran, Bruce T., 'The Alchemical World of the German Court', *Sudhoffs Archiv Zeitschrift Fur Wissenschaftgeschichte, Beiheft*, 29, Stuttgart, 1991.
Morley, Henry, ed., *Character Writings of the Seventeenth Century*, 1891.
Mortimer, John, *Whole Art of Husbandry*, 1716.
Mullet, Charles F., 'Hugh Plat: Elizabethan Virtuoso', *University of Missouri Studies*, vol. 21, 1946.
Newman, W. R., & A. Grafton, *Secrets of nature*, Cambridge, MA, 2001.
Nicholl, Charles, *The Chemical Theatre*, 1980.
Norton, Thomas, *The ordinal of Alchemy*, ed. John Reidy, Oxford, 1975.
Nutton, V., and R. Porter, eds., *The History of Medical Education in Britain*, Atlanta, GA, 1995.
O'Day, Rosemary, *Education and Society, 1500–1800*, 1982.
Oppenheim, M., *A History of the Administration of the Royal Navy*, 1896.
Oxford Dictionary of National Biography, Oxford University Press, 2004–2008; online edition, www.oxforddnb.com.
Paracelsus, *Selected Writings*, ed. Jolande Jacobi, New York, NY, 1951.
Pearce, Brian, 'Elizabethan Food Policy and the Armed Forces', *Economic History Review*, vol. 12, 1942.
Pelling, Margaret, 'Knowledge Common & Acquired', in *The History of Medical Education in Britain*, eds. V. Nutter and R. Porter, Atlanta, GA, 1995.

––, 'Thoroughly Resented? Older Women and the Medical Role in Early Modern London', in *Women, Science & Medicine*, eds. Lynette Hunter and S. Hutton, 1997.

––, *The Common Lot*, 1998.

––, *Medical Conflicts in Early Modern London*, 2003.

Pelling, Margaret, & Frances White, *Physicians and Irregular Medical Practitioners in London 1550–1640*, Online Database, http://www.british-history.ac.uk, 2004.

Pennell, Sara, 'Consumption and consumerism in Early Modern England', *Historical Journal*, vol. 42, 2, 1999.

Perrin, W.G., 'Boteler's Dialogues', *Navy Records Society*, 1929.

Perry, Charles, 'Couscous and its Cousins', *Staple Foods, Oxford Symposium on Food and Cookery 1989*, Totnes, 1990.

Phillips, Jeremy, 'The English patent as a reward for invention', *Journal of Legal History*, vol. 3, 1982.

Piper, David, 'The 1590 Lumley Inventory: Hilliard, Segar and the Earl of Essex, II', *Burlington Magazine*, vol. 99, no. 654, 1957.

Plat, Sir Hugh, *The Floures of Philosophie*, 1572, intro. Richard J. Panofsky, New York, NY, 1982.

––, *A Briefe Apologie of Certaine New Inventions*, 1593.

––, *The Jewell House of Art and Nature*, 1594.

––, *Diverse new sorts of Soyle not yet brought into any publique use*, 1594.

––, *A Discoverie of Certaine English Wants*, 1595.

––, *Sundrie new and Artificiall remedies against Famine. Written by H.P. Esq. upon thoccasion of this present Dearth*, 1596.

––, *The new and admirable Arte of setting of Corne*, 1600.

––, *Delightes for Ladies*, 1602, 1628; also Fussell, G.E., & K.R. Fussell, eds., *Delightes for Ladies*, 1948.

––, *A new, cheape and delicate Fire of Cole-balles*, 1603.

––, *Certaine philosophical preparations of foode and beverage for sea-men*, nd [c. 1607].

––, *Floraes Paradise*, 1608.

––, *The Garden of Eden*, 1653.

––, *The second part of the garden of Eden*, 1675.

Plattes, Gabriel, *A discovery of infinite treasure*, 1639.

Porritt, Edward, 'Five centuries of liquor legislation in England', *Political Science Quarterly*, vol. 10., no. 4, 1885.
Porta, G.B. della, *Magiae Naturalis*, Naples, 1558.
Porter, Roy, *London, a Social History*, 1994.
Prest, W. R., *The Inns of Court under Elizabeth I and the Early Stuarts, 1590–1640*, 1972.
Prezzolini, Giuseppe, *Spaghetti Dinner*, New York, NY, 1955.
Principe, L.M., & W.R. Newman, 'Some problems with the Historiography of Alchemy', in *Secrets of Nature*, ed. W.R. Newman & A. Grafton, Cambridge, MA, 2001.
Prior, Moody E., 'Bacon's man of science', in *Roots of scientific thought*, ed. P.P. Wiener and A. Noland, 1957.
Pumfrey, Stephen, and Frances Dawbarn, 'Science and patronage in England, 1570–1625', *History of Science*, vol. 42, 2004.
Raleigh, Sir Walter, *History of the World*, 1614.
Rhodes, Denis E., 'The Italian Banquet, 1598 and its origins,' *Italian Studies*, vol. 27, 1972.
Ripley, George, 'The Compound of Alchemie', in Elias Ashmole, *Theatrum Chemicum Botanicam*, 1652.
Roach, F.A., *Cultivated Fruits of Britain*, 1985.
Sacks, David Harris, 'The countervailing of benefits', in *Tudor Political Culture*, ed. Dale Hoak, Cambridge, 1995.
––, 'London's Dominion', in *Material London ca. 1600*, ed. Lena Cowen Orlin, Philadelphia, PA, 2000.
Santich, Barbara, *The Original Mediterranean Cuisine*, Totnes, 1995.
Saunders, Anne, ed., *The Royal Exchange*, 1997.
Scammell, G., 'The sinews of war : manning and provisioning English fighting ships c. 1550–1650', *Mariners' Mirror*, vol. 73, no. 4, 1987.
Scappi, Bartolomeo, *Opera Di M. Bartolomeo Scappi*, Venice, 1570.
Scarisbrick, Diane, *Tudor and Jacobean Jewellery*, 1995.
Schofield, John, *Medieval London Houses*, 1995.
Schuler, Robert M., *Alchemical Poetry 1575–1700*, New York, NY, 1995.
Scott, Reginald, *A Perfite platforme of a hoppe garden*, 1574.
Scott-Warren, J., *Sir John Harington and the book as gift*, 2001.

Serventi, Silvano, and Françoise Sabban, *Pasta: the story of a universal food*, New York, NY, 2002.

Shapin, Steven, 'The house of experiment', *Isis*, vol. 79, 1988.

––, '"A scholar and a gentleman": The problematic identity of the scientific practitioner in early modern England', *History of Science*, vol. 29, 1991.

Shapiro, James, *1599*, 2005.

Sharrock, Robert, *The History of the Propagation and Improvement of vegetables*, Oxford, 1660.

Sheail, John, *Rabbits and their History*, 1972.

Sherman, William H., *John Dee: the politics of reading and writing in the English Renaissance*, Amherst, MA, 1995.

Sinclair Rohde, Eleanour, *Old English Gardening Books*, 1924.

Singer, Charles, *The earliest chemical industry*, 1938.

Slack, Paul, *The Impact of Plague in Tudor and Stuart London*, Oxford, 1985.

Smith, John, *England's Improvement Reviv'd*, 1673.

Smith, Pamela H., *The Body of the Artisan*, Chicago, IL, 2004.

Somerville, Robert, *The Savoy*, 1960.

Speed, Adolphus, *Adam in Eden*, 1659.

Spurling, Hilary, *Elinor Fettiplace's Receipt Book*, 1986.

St Clare Byrne, M., *The Elizabethan Home*, 1949.

Starkey, David, et al., *The English Court from the Wars of the Roses to the Civil War*, 1987.

Stern, Tiffany, '"On each Wall and Corner Post"', in *English Literary Renaissance*, vol. 36, I, 2006.

Stow John, *A Survay of London* [1598], ed. Henry Morley, [1893].

Strong, Sir Roy, *The English Icon*, 1969.

Stuart, David, *The kitchen garden*, 1984.

Sugden, John, *Sir Francis Drake*, 1990.

Sykes, John, 'Product innovation in Early Modern London', *Past & Present*, no. 168, 2000.

Thick, Malcolm, *The Neat House Gardens, early market gardening around London*, Totnes, 1998.

––, 'Roots and other garden vegetables in the diet of Londoners, c. 1550–

1650, and some responses to harvest failures in the 1590's', *Staple Foods. Oxford Symposium on Food and Cookery 1989*, 1990.

Thirsk, Joan, ed., *Agrarian History of England and Wales*, vol. iv, Cambridge, 1967.

––, *Economic Policy and Projects*, 1978.

––, 'Making a fresh start', in *Culture and Cultivation in Early Modern England*, ed. M. Leslie & T. Raylor, 1992.

––, *Alternative Agriculture*, Oxford, 1997.

––, *Food in Early Modern England*, 2007.

Thomson, G.M., *Sir Francis Drake*, 1988.

Thornton, Jonathan, 'The History, technology, and conservation of architectural *papier mâché*', *Journal of the American Institute for Conservation*, vol. 32, no. 2, 1993.

Utterstrom, Gustaf, 'Climatic Fluctuations and Population Problems in Early Modern History', *The Scandanavian Economic History Review*, 3, 1955.

Vale, George F., *Old Bethnal Green*, 1934.

Varro, Marcus Terentius, *On Agriculture* (Loeb Classical Library), 1967.

Voss, Paul J., 'Books for sale: Advertising and patronage in late Elizabethan England', *Sixteenth Century Journal*, vol. 29, no. 3, 1998.

W.A., *A Booke of Cookrye*, 1591.

Wallis, Patrick, 'Apothecaries and medicines in Early Modern London', in *From Physick to Pharmacology*, ed. Louise Curth, 2006.

Webb, Henry J., *Elizabethan Military Science*, Madison, WI, 1965.

Webster, Charles, 'Alchemical and paracelsian medicine', in Charles Webster, ed., *Health, Medicine and Mortality in the Sixteenth Century*, Cambridge, 1979.

Wernham, R.B., ed., *The expedition of Sir John Norris and Sir Francis Drake to Spain and Portugal, 1589*, 1988.

White, Eileen, ed., *The English Cookery Book*, Totnes, 2004.

Williams, A.R., 'The Production of Saltpetre in the Middle Ages', *Ambix* vol. 22, pt. 2, 1975.

Williams, Alan, and Anthony de Reuck, *The Royal Armoury at Greenwich, 1515–1649*, 1943.

Williamson, James A., *Hawkins of Plymouth*, 1969.

Wilson, Arthur M., 'The Logwood Trade in the Seventeenth and Eighteenth Centuries', in *Essays in the History of Modern Europe*, ed. Donald C. McKay, New York, NY, 1936.

Wilson, C. Anne, ed., *The Country House Kitchen Garden 1600–1950*, Stroud, 1998.

Woodhouse, Elisabeth, 'Kenilworth, the earl of Leicester's pleasure grounds following Robert Laneham's letter', *Garden History*, vol. 27, I, 1999.

——, 'The spirit of the garden', *Garden History*, vol. 27, I, 1999.

Woodward, Donald, '"Swords into Ploughshares": Recycling in Pre-Industrial England', *Economic History Review*, 2nd Ser., vol. XXXVIII, 1985.

——, 'An Essay on manures', in *English Rural Society, 1500–1800*, ed. J. Chartres & D. Hey, Cambridge, 1990.

——, 'Straw, bracken and the Wicklow whale: the exploitation of natural resources in England since 1500', *Past & Present*, vol. 159, 1998.

Woolley, Benjamin, *The Queen's Conjuror*, 2001.

Worlidge, John, *Systema Agriculturae*, 1675.

Wrigley, E.A., 'A simple model of London's Importance', in E.A. Wrigley, *People cities and wealth*, 1987.

Zhang, J., et al., 'Greenhouse gases and other airborne pollution from household stoves in China', *Atmospheric Environment*, 34, 2000.

Zilsel, Edgar, 'The Origins of Gilbert's Scientific Method', in *The Roots of Scientific Thought*, ed. P. Weiner and A. Noland, New York, NY, 1957.

Index

advertising, 83, 135, 141, 207, 225, 233, 238, 286, 316, 319, 326, 340ff.
   in books and pamphlets 342–347
   Plat's use of, 384–385
   sophisticated copywriting, 384–385
agriculture, 8, 12, 34, 54, 80–114
   contemporary importance of, 80
   alternative, 103, 118, 161, 252, 253, 260, 329, 375, 377
Agrippa, Cornelius, 85, 185
Albany, Robert, 27
Albany, William, 27
alchemical medicines, for use at sea, 140
alchemical poetry, 182–198, 388ff.
alchemy, 7, 8, 9, 10, 26, 39
   as intellectual challenge, 190
   attempts to sell secrets of, 202–205, 245, 254
   critique of Quercetanus, 202
   discussions with John Dee, 179–181
   experiments in, 186–189
   gardening and, 52–54, 81, 136, 153, 169–205
   general interest in, 169–70
   historiography of, 171–172
   informants on, 182–185
   links with medicine, 198–205
   Paracelsian, 198–202
   Plat's initiation into, 178–179
   Plat's summation of, 1767, 191
   Plat's reading on, 184–186
   scepticism of, 172
   warning of false practitioners of, 173–176
Aldenham, Herts., 13
Alexis of Piedmont, 255
Alison, Roger, a cutler's boy, 222
Allet, Sir John, 33
Amherst, Amelia, 41
Andrews, —, Irishman, 266
Andrews, —, saltmaker, 51
animal husbandry, 101–105
apothecaries, 32, 39, 49, 62, 103, 105, 151, 156, 162, 184, 206, 207, 210–211, 213, 214, 215, 219, 224, 229, 235, 239, 245, 258, 261, 280, 326, 327, 339, 367, 400
Archer, Ian, 327
armour, 140, 142, 260, 261, 267, 285, 286, 288, 289, 297
   made from new materials, 301–302
   new way of cleaning, 302–304, 357, 318, 335, 341, 403
Arthington, Henry, 164
artichoke, 33, 47, 54, 59, 78, 92, 109, 159
   storing, 46, 160
artillery, 287, 298, 299, 303
artisans, 48, 243, 245, 247, 249, 250, 251, 252, 260, 261–262, 264, 265, 272, 275, 278, 318, 329, 332, 339, 340, 344, 352, 252, 354, 355, 369

INDEX

Arundell, Thomas, 1st Baron Arundell of Wardour, 142, 143, 287, 356
Ash, Eric, 177
Ashmole, Elias, 195, 197–198, 374
Auborn, Edward, 324
Aubrey, John, 183
Bacon, Sir Francis, 242, 246, 251, 252, 370, 378
Bacon, Roger, 185
Baeshe, Edward, 116, 117
Bakers' Company, 33
Banbury, Henry, 71
banquet, 64, 67, 153, 154, 157, 268, 310
banqueting house, 31, 64, 66
Barber Surgeons' Company, 207
Barn Elms (Barnes Elms), Surrey, 49, 58
Bartholomew Fair, London, 338
Barwick, Humphrey, 297
Basle, Switzerland, 199
Bateman (Batemen), Parson, of Newington, 166, 264
Beale, John, 376, 377
Beddington, Surrey, Carew Manor, 75
Beier, Lucinda McCray, 209
Belfeild, Mr, of St Albans, 236
bell-glass, 55
Bellingham, Charles, 41, 45, 371–375
Bethnal Green, Middx., Bishop's Hall, 30, 31, 62, 66
bills (posters), 212, 233, 339, 340, 341
bills (prescriptions), 211
Blackamore, Arthur, 348, 364
Blickling Hall, Norfolk, 334
Blith, Walter, 90
boat, portable, 304
Boate, Arnold, 376

Bodley, Josias, 287, 297
borrowed text, 150–154, 161–162
Bostocke, Lyonell, 89, 91, 95
Boteler, Nathaniel, 146
Bowes, Sir William, 183
Boyle, Robert, 243, 318
Bradbury, —, 141, 327, 338
Bradley, Richard, 104, 127, 128
Braithwaite, Richard, 18
Brewers' Company, 11, 12
brewing, 11, 16, 31, 32, 59, 91, 106, 147, 151, 163, 243–244, 262, 264, 306, 309, 315, 323, 324, 362
bridge, temporary, 289, 304
Briefe Apologie of Certaine New Inventions, 35, 315ff., 337
broadside, used as advertisement, 116, 135, 142, 207, 225, 237, 238, 241, 397
Bronsede, —, soap-boiler, 311
Brooke, Ralph, York Herald, 222, 237, 337
Broughton, Mr, 91
Bruges, Belgium, 91
Bruno, Giordano, 85
bullets, new and improved, 297–299
  chain-shot, 298
butter, 109, 119, 120, 125, 129, 130, 132, 133, 157–158, 314, 316, 352, 384
Caesar, Sir Julius, 135
Cambridge, St John's College, 15
camera obscura, 31
Campbell, Susan, 42, 62, 74
candying, 67, 148, 150, 263
card playing, Plat ambivalent on, 21
Cardano, Girolamo (Cardanus), 51, 76, 255, 262, 303
Carew, Sir Francis, 75

419

Carew, Richard, 71
carnations, 46, 47, 49, 54, 59, 60, 62, 63, 65–66, 67, 68
Catholics, suppression of, 24–26, 37
Cato, 77
Cavendish (Candish), Lady, 38
Cavendish, Sir Charles, 337
Cecil, William, 1st Baron Burghley, 35, 38, 48, 49, 56, 145, 280, 357
Cecil, William, 2nd Earl of Salisbury, 376
Certaine Philosophical Preparations of Foode and Beverage for Sea-men, 116ff., 142, 237, 238, 341, 397ff.
Chaloner (Challenor), Sir Thomas, 50, 69, 182, 256
Charnock, Thomas, 194, 195, 196
cheese, 27, 72, 119, 120, 129, 147, 157
    invented by Judith Plat, 158, 159
Chelsea Physic Garden, 62
Chetle, Nurse, 220
Child, Robert, 90, 112, 374, 376–378
Clapham, Surrey, 86
Clark, John, medical practitioner, 213, 217–219, 239, 280
Clifford, Lady Anne, 38, 321
Clowes, William, surgeon, 212, 217, 239, 327
coalballs, 31, 33, 35, 36, 37, 38, 96, 97, 167, 285, 286, 307, 312–314, 319, 320, 321, 329, 343, 344, 354, 361, 362, 365
cockney, Plat regarded himself as, 14
Coggaine, —, woadman, 269
College of Physicians, 207, 212, 213, 215, 218
Colman, Mrs, 220
Columella, 55, 77, 81, 88

Combe Royal, Kingsbridge, Devon, 58
comfits, 105, 150, 151, 153, 154, 155, 162, 252, 268, 310, 385
common-placing, in Plat's time, 381–382
    Plat's method, 383–387
compost, 49, 54, 85
    secret, invented by Plat, 97–98, 167, 286, 346, 347, 375
congelative water, 83–84
Conradus, Master, 14
Constable, Sir Henry, 106
cookery, 8, 9, 72, 115, 121, 123, 252, 270, 310, 368
    Plat's knowledge of not great, 147, 149, 150, 153, 155, 161,
Cooper, —, goldsmith, 266
Cooper, Alix, 279
Cornet, Charles, 340
Cornwallis, Sir Thomas, 47
Cortese, Isabella, 295
Coryate, Thomas, 204
cosmetics, 8, 9, 149, 252, 256, 286, 368
Cotford, —, goldsmith, 258
Court, 9, 24, 32, 33
    social activities of, 34
    Plat's activities in, 34–38, 46, 50, 71, 115, 149, 150, 161, 166, 183, 197, 213, 219, 220, 231, 234, 235, 256, 257–260, 306, 321, 328, 344, 354–356, 368
Court officials, source of technical information, 258–259
couscous, 124–125, 142, 305
Coventry, Warwickshire, 335
Cowcroft, Middlesex, 323
cow-keeping, 31, 111
crab-apples, 48
Crosley, —, apothecary, 156

INDEX

Crosse, Captain, 145
cryptography, 9
cucumber, 55, 56, 121
Culpepper, Sir Cheney, 375
Cumberland, Countess of, 38, 321
Curtis, Robert, 214
customs administration, Richard Young's improving, 24–26, 164, 329, 358
Dacus (Dakeham, Dakeson), James, of Blandford, Dorset, 46, 51, 257
dancing, 19, 21, 33
Darcey, Edward, Groom of the Privy Chamber, 222, 231
Darcy, John, 3rd Baron Darcy, 50, 58
Davenport, Henry, 335, 357
Davies, Sir John, 197
Dawbarn, Frances, 355
Dawson, —, printer, 262
dearth, 35, 97, 112, 120, 121, 141
  of the 1590s, Plat's public response to, 162–168, 309, 320
Debus, Alan, 82, 200
Dee, John, 26, 85, 178–183, 185, 187, 188, 197
Dekker, Thomas, 214
Delightes for Ladies, 29, 39, 72, 115, 130, 147ff., 186, 210, 368, 376
della Porta, Giambattista (Giovanni), 51, 93, 167, 185, 295
Denny, Sir Edward, 51
Devereux, Robert, 2nd Earl of Essex, 34, 342
Dickson, Donald R., 374
Digges, Thomas, 182
Discoverie of Certaine English Wants, 29, 304, 344
Diverse new sortes of Soyle not yet brought into any publique use, 81ff.
Doile, Dr —, 337
doves, 105
Drake, Sir Francis, 34, 118, 119, 125, 134, 135, 141, 145, 256, 257, 303, 341, 364
drunkenness, 19–20
  palliatives for, 22
Dudley, Anne, Countess of Warwick, 17
Dudley, Robert, Earl of Leicester, 17
Duncombe, —, 89, 91
dung, 36, 37, 38, 54, 55, 56, 60, 61, 65, 81, 82, 83, 86, 88, 89, 93, 94, 95, 97, 99, 100, 108, 111, 189, 282, 293, 313
dunghill, cover for, 96–97
dwarf trees, 46, 49, 50, 57, 58, 60, 62, 63, 64, 65, 67, 69, 72, 74, 75, 76
dyeing, 147, 151, 252, 253, 260, 264, 268–277, 279, 281, 282, 386
East Greenwich, Kent, 323
Eden, Richard, 248
education, 11, 13, 14
  Cambridge syllabus, 15, 35
  legal, in Plat's time, 17–20, 28
  of gentry, 23, 169, 245, 327, 329, 330
Egerton, Sir Thomas, 36
eggshells, etching on, 8
elixir, 53, 171, 198, 373, 390, 395
Elizabeth I, Queen, 197
Elkinton, Thomas, 240, 348, 349, 352
English, —, a German, 51, 57
English, —, the travayler, 263
English Garden, The, 42
Epulario, or the Italian Banquet, 121
Erith, Kent, 86, 300
Estienne, Charles, 55, 81
Evelyn, John, 41, 85, 112, 127, 243, 374, 377

exhibition, of inventions, 31, 35, 36, 280, 321, 337, 344
exotic plants, 46, 47, 49, 65, 76, 78, 85, 109
experiments, 8, 29, 31, 32, 43, 45, 54, 56, 81, 83, 89, 115, 141, 172, 175, 179, 182, 184, 186, 197, 220, 225, 250–253, 268, 270, 273–284, 302, 314, 367–369, 376–377, 384
    farming, 95–96
    plan of future, 98
Faber, Martin, 179, 184
fabric, rainproof, 316–318
fairs, plans to sell at, 338
Falls, Cyril, 118
farm, ideal, plans for, 107–112
farming, see agriculture
fashionable dress, 20, 28
Faulkner, Thomas, 372
fencing, 19, 340
fertilizers, 8, 54, 60, 65, 76, 80, 81, 83
    regional variations of, 89–93, 97, 98, 287, 305, 310, 329, 354, 365
Fioravanti, Leonardo (Firouanta), 255
fire, Greek, 295–296
'fire of nature', use as preservative of perishable liquids, 135–139
fish-farming, 105
Fitzherbert, John, 70, 81, 87
Fleetwood, William, Recorder of London, 220
Fleming, Laurence, 42
Fletcher, Richard, Bishop of Worcester, 50, 236
Floraes Paradise, 29, 38, 39, 41ff., 81, 84, 151, 207, 226, 237, 254, 343–346, 352, 371
Floures of Philosophie, 15, 16, 17, 45

Flower, —, gardener, Bethnal Green, 46
flush-lavatory, 36–38
food
    advertised by Plat, 141–142
    decay at sea, 117
    for Irish military expeditions, 118–119
    for Low Countries military expeditions, 118
    lobbying for contracts for, 142–146
    meat preservation, 130–133
    military, 115ff.
    Plat's writings on, 147–168
    preservation of, 8, 128–129
    problems with military victualling, 116–120
    sold to Drake and Hawkins, 144–145
    summary of Plat's work on, 115–116
    water preservation, 133–136
Fortescue, Sir John, 143, 327, 356
fortifications, temporary, 299–300
Fowle, —, Gentleman-Keeper of the Queen's House, 50
Foyster, Elizabeth, 12
'front matter', 342–343
    of books, 7
fruit-trees, 31, 41, 47, 52, 58, 59
    contemporary interest in 70–74, 109
    dwarf, 57, 74–76
Fussell, George, 81
gambling, 19–20
Gapper, Claire, 333
garden, indoor, 31, 50, 52, 62, 63, 65
Garden of Eden (*Floraes Paradise*), 45ff., 371, 375, 376
garden ornaments, 67–69

INDEX

gardeners, 32, 41–79, 100, 106, 164, 312

gardening, 8, 30–31, 34, 41–79, 100, 102, 109, 151, 164, 192, 247, 252–253, 260, 286, 312, 343, 367, 371, 375–377, 379, 380

gardens, 16, 20, 30–32, 34, 41–79, 89, 92, 100, 102, 105, 106, 109, 111, 163–164, 192, 210, 234, 281, 305, 308, 371

Gardiner, Richard, 52, 164

Garret, —, miller, 50

Garret (Garrett), James, 49, 219, 258, 261

Gascoigne (Gascoine), Thomas, 106, 157, 178, 183, 230, 258, 314

gentleman, gentlemen, 9, 10, 11, 12, 15, 18, 19, 21, 22, 23, 28, 32, 33, 34, 35, 38, 45, 46, 48, 49

    London, 31–33, 71, 104, 107, 350

    offer agricultural advice, 80, 96–98, 100, 104, 106, 107, 109, 112, 115, 124, 129, 150, 161, 164, 166, 169, 173, 179, 183, 185, 192, 202, 205, 206, 207, 209, 214, 217, 221, 228, 242, 243, 245, 246

    provide garden advice, 50ff., 70, 71, 77

    sources of technical information, 256–258

    veracity of information from, 257–259, 261, 262, 264, 268, 275, 278–280, 287, 289, 299, 314, 320, 321, 323–327, 333, 337, 338, 345–348, 350, 352, 354, 355, 367, 369, 372, 381, 383, 398

Gerard, John, 48, 49, 50, 54, 66, 71, 73, 74, 78, 209, 234

Gerard, John, S.J., 24

Gesner, Conrad, 198, 206, 216, 316

gift, New Year's, 33

Gilbert, Adrian, 183

Gilbert, William, 251, 251

glass, 21, 54–59, 76, 94, 123, 137, 182, 183, 187–189, 194, 203, 244, 252, 261, 262, 277, 281, 303, 310, 331–332, 358, 373, 384, 387

glasshouse, 50, 59, 156n

Godfrey, —, dyer, 270, 272, 273

godliness (Puritanism), 13, 14, 15, 17, 20, 22, 23, 39–40, 320, 381

Goodman, Thomas, gardener, 47, 49

Goodrus (Gooderons, Gowthrowse), William, surgeon, 49, 135, 343

Googe, Barnaby, 51, 77, 94, 343

Gore, —, sister-in-law of Sir Hugh Plat, 230

Gore, Alan, 42

grafting, 43, 47, 48, 70

    fruit-trees, 72–74, 253, 383

Granvelle (Grannile), Antoine Perrenot de, Cardinal, 220

grapes, 46, 47, 50, 57, 59, 76–77, 402

Gregorie (Gregory), Arthur, 257, 258, 259, 282, 296

Grene, Mr, refiner, 220

Grindal, Edmund, Archbishop of Canterbury, 166

Grocers' Company, 207

Guisborough, Yorks., 182

gunpowder, 125, 252, 286, 288, 290, 295, 297, 306

guns, 288

    improvement of, 290, 297–299, 301, 302

Hackney, Middlesex, 46, 47, 50

Hadfield, Miles, 42

Hampton Court Palace, 69, 332

Hanbury, William, 312

Harding, Richard, 361
Harington, Sir John, 25, 26, 36, 37, 38, 97, 260, 354
Harkness, Deborah, 355, 366ff.
Harriot, Thomas, 85, 182, 256, 356n
Harsley, Goodwife, Bethnal Green, 230
Harsley, Mr, of Bethnal Green, 89
Harsnett, Samuel, 172
Hartlib, Samuel, 112, 373–377
Hartmann, Johannes, 204
harvest failure, in England in the 1590s, 9, 80, 98, 116, 141, 162–168, 328
Hastings, Lady, 38
Hawkins, Sir John, 125, 141, 145, 341
Hawkins, Richard, 141, 146
health, Londoners' attitude towards, 208–211
see also medicine
heat
 used in gardening, 55–62
 waste, use of, 306–309
hedges, 16, 52, 67
Hendrickes, —, goldsmith, 263
Henry VIII, King, 70
Henshaw (Hinshaw), Thomas, 373–375
Herbert, Mary, Countess of Pembroke, 183
Heresbach, Conrad, 51, 81
Hesse-Kassel, Moritz, Landgrave of, 168, 178, 184, 202–205, 240, 241, 345, 349–353
Hill, Mrs, shoemaker's wife in Bread Street, 229
Hill, Thomas, 55, 69, 106, 312, 343
Hill, William, Auditor of the Exchequer, 50, 74, 157, 182, 220, 222, 230, 257, 258, 282, 303, 314, 317

Hobby (Hobbey), Sir Edward, 105
Hoby, Margaret, 210
Hodges, Nathaniel, 371
Hodges, Rhoda, 372
Hodges, Thomas, Dean of Hereford, 371–373, 375
Hodges, Thomas, jr., 201, 371–373, 375
Honrick, Gerard, 293
Hooker, John, 71
hops, 92, 106, 109, 110, 123, 163, 168, 306
hot-bed, 47, 50, 51, 55, 56, 57
hot-house, 59–62
hot walls, 50, 58, 59
Hounce, —, a Dutch conjuror, 21
Howard, Anne, née Dacre, Countess of Arundel, 210
Howard, Thomas, 21st Earl of Arundel, 231
Hoxton, Middlesex, 335
Hoyles, Martin, 42
Hu[t]chins, —, dyer, 270
Hunter, Alexander, 312
Hunter, Lynette, 210
Hutten, Ulrich von (Sir Ulrick Hutton), 220, 229
imperial expansion, England's, 115
incendiaries, 295–297
indoor plants, 31, 50, 52, 62–63, 65, 156
information-seeking, Plat's methods of, 259–278
Inns of Court, life at, in Plat's time, 17–19, 22–23, 245, 324, 355
inventions, 9, 29, 31, 35, 36, 37, 38, 97, 101, 115, 126, 142, 146, 157, 243, 247, 260
 attempts to sell to the Crown, 353–357, 368
 exhibition of, 279–280

INDEX

military, 287
  promotion of, 319–321, 325, 328–331, 335, 342, 343–346, 351
  types of, 285–286
Ireland, military food for, 126
Irish whiskey, plan to sell, 339
irregular medical practitioners, 212ff.
Italians, in London, 32, 122–123
Italy, 50
  source of crops and information, 51, 102, 123, 124, 232, 329, 331, 384
Jacob, Master, see Verzilini, Giacomo
Janacek, Bruce, 170
Jarfield, Richard, preacher, 235
Jarrett, see Gerard, John
Jeames, —, 'a duch gardener', 45
Jefferies, —, 'the gardener at Sheene', 47, 106
Jefferyes, Mr, 'of Mawlin, Suffolk', 364
jelly, meat, as victual, 126–129, 144
Jenks, Harry, servant, 235
jewel house, plans for a, 336–338
Jewell House of Art and Nature, 20–22, 29, 34, 43, 80ff., 115, 122, 124, 148, 153–155, 157, 158, 167, 201, 224, 243, 244, 248, 250, 252–254, 264, 309, 311, 316, 320, 342–344, 361, 368, 376
jewels, 28, 39, 329
Jonson, Ben, 170
Jordan, Dr [Edward Jorden], 101, 217
Josselin, Ralph, 208, 209
Jung, Carl, 172
Kelley, Edward, 181
Kelway, Mr, trumpeter to the Queen, 220, 260
Kemmish (Kemish), —, apothecary, 156, 184, 206, 219, 239, 261, 327

Kensington, Middlesex, St Mary Abbots, 375
Kentish Town, Middlesex, 323
Ker, Mathew, 206
Kerridge, Eric, 87, 91
Kimberley, Mr, 88
King, Clement, gardener to Lord Burghley, 49
Kirwin, —, gardener, Newington Butts, 48
Lancaster, Sir James, 139
Laude, —, goldsmith, 266
Laurence, John, 113
law, Plat's lost books on, 324, 326
  Plat's possible practice in, 28–29, 35, 173, 326, 359, 360, 383
law courts, 29, 32–33, 381
law suit, by Crown against Judith Plat, 28, 361
Lawson, William, 69
lemon juice, 116, 138
  to prevent scurvy, 139
Lewes, Dr, Judge of the Admiralty, 213
Libavius, Andreas, 204
library, Plat's, 370ff.
Liébault, Jean, 55, 81
Liège, Belgium, 62, 312, 313
liquorice, 51, 105–106, 109, 111, 127, 129, 138
Lisle, Lady, 210
Little Moreton Hall, Cheshire, 334
Lockie, Rowland, painter, 261–262
Lodge, Thomas, 20, 135, 229–230, 233, 340
London
  Allhallows, Bread Street, 27
  Birchin Lane, 323

Candlewick Street, 23
Cornhill, 229, 236, 324
Garlick Hill, 30, 324
Gutter Lane, 235
Holborn, John Gerard's garden, 48, 74, 78
Lincoln's Inn, 17–20, 22
Old Swan brewery, 11
Royal Exchange, 335–336
St Anthony's grammar school, 14
St James Garlickhythe, 13, 14, 323-324
St Mary Abchurch, 235, 236
Thames Street, 323
London's Pride, 42
Lull, Raymond, 246
Lupton, Thomas, 255, 303
macaroni, 116, 119–124
    machine for making, 122, 140–144, 167, 326, 335, 337, 341, 364, 384, 397
McDonald, Donald, 41, 42
magic, 23, 85, 93, 167, 171, 173, 190, 197, 199, 221–222, 224, 229, 255
Maier, Michael, 371
manuscripts, Plat's, 8, 9, 10, 24, 26, 28, 29, 31, 32, 43, 45, 52, 56, 60, 70, 80, 81, 94, 98, 100, 115, 124, 126, 142, 150–153, 155, 157, 160, 161, 162, 178, 180, 185, 197, 201
    later ownership of, 371ff.
    Plat's notes on trades, 252ff., 280, 282, 286, 287, 299, 311, 319, 324, 335, 336, 341, 360, 366, 368, 369
marketing, Plat tries to sell goods directly to the public, 330–338
    see also moneymaking
Markham, Gervase, 126, 161, 210, 254
marl, 83, 85, 87–88, 97, 167, 250, 353

marriages, Plat's, 16, 23
    second, 27, 28, 29, 94
Marshall, William, 100
Martyn, Thomas, Professor of Botany, 372
Mascall, Leonard, 51, 70, 71, 105
Mayerne, Sir Theodore de, 371
medical books, 211
medical practice, Plat's, 216ff.
medicine, 6, 8, 9, 26, 32, 39, 105, 115ff.
    for use at sea 140–141, 216ff.
    household, 209–210
    how Plat practised, 228–231
    in London of Plat's time 207–226
    links with alchemy, 206–242
    Plat's fees for dispensing, 236–237
    'spagiricall', 226
Meggs, —, soap-boiler, 311
Mendoza, Bernardino de, 34, 166, 257
Metamorphosis of Ajax, 25, 36, 37
Milan, Italy, 233
military secrets, 356–357
Mirror of Alchemy, 185
Mizauls, Antoine (Mizaldus), 255
Monardes, Nicholas, 167
moneymaking, 323–365
monopolies, see patents
Moran, Bruce T., 204
Mortimer, John, 113
Mortlake, Surrey, 179
Mosanus (Mosan), Godfrey, 178, 179, 184, 187, 217
Mosanus, James, 204, 213
mouldings, 147, 154, 157, 252, 253, 264–268, 281, 286, 292, 318–320
    plans to sell, 330–334
mowing, grain crops, 101

INDEX

mummia, 224
Munster, Sebastian, 333
muskmelon, 50, 56, 109, 261
'muskmelon' gentleman, 50, 366–367
Nantwich, Cheshire, 86
Napier, John ('Nepper the Skott'), 51, 86, 157
Naples, Italy, 123, 124
Nathias, —, alchemist, 183, 184
nature, 51, 53, 54, 64, 82, 85, 111, 142, 183, 185, 190–193, 198, 239
    depiction of, 202, 248–249, 400
    fire of, 135ff., 189, 198, 254, 397
naval expeditions, increasing size of, 118, 139, 141, 144, 145, 160, 281, 299, 341, 364
New and admirable Arte of Setting of Corne, 90ff., 342
New, cheape and delicate Fire of Cole-balles, 201–202, 343
Newman, W.R., 172
Nicholas of Cusa, 250
Nicholl, Charles, 171, 196
Nicholson, —, Hackney, 47
Noel, Baptist, Viscount Campden, 375
Nonsuch Palace, Surrey, 69
Norden, John, 89
Norris, —, cook, confectioner, 263, 268
North, Sir Thomas, 364
Northborough, John, Bishop of Bristol, 183
Norton, Thomas, 177, 178, 185, 196
note-taking, 29–30, 252–255, 275–278
    Plat's method of, 383ff.
nurserymen, 47–48, 65, 71, 261
oatmeal, 90
    for military victualling, 125–126, 130, 143–144
occult, 171, 185, 204, 372
O'Day, Rosemary, 18
Old Swan brewery, 11
Oughtred, William, 374
out-of-season garden produce, 47, 109
Overbury, Sir Thomas, 19, 20, 34
Palissy, Bernard, 51, 82, 83, 84, 85, 87, 250, 250
Pallisser, Sir Michael, 231
papier mâché, 302, 318–320, 330–334
Paracelsus (Phillippus Aureolus Theophrastus Bombastus von Hohenheim), 81, 85, 166, 169, 185, 199, 200, 201, 247, 250
parsnips, 105, 109, 129
    cakes of dried parsnips, 59n, 160
Parsons, —, apothecary, 103, 104, 156
Partridge, Edward, 263
pasta, as military victual, 119–124
    see also macaroni
Paston, Robert, 375
patents, 36, 357–363
    for 'Ox Shin-bones', 129
    for saltpetre, 360–361
    Plat's plans for new types of, 361–363
patterns, for mouldings, 333–334
Peacham, Henry, 334
Pelling, Margaret, 30, 208, 209, 211, 213, 214, 216
Percy, Henry, 9th Earl of Northumberland, 182
Petty, William, 325
Phillips, —, alchemist, 183
philosophers' stone, 170, 171, 180, 183, 192-198, 204, 285

philosophical garden, 52-3
philosophy of science, 242-284, 367-270
    Plat not greatly interested in, 245
    Bacon's 369-370
physician, Plat as, 8
    see also medicine
Pichfork, —, apothecary, 235
Piemontese, Alessio (Alexis of Piedmont), 255
Pill, Mr, of Exeter, sometime a draper of Bristol, 128
plague, 231–236
Plat, Alice, 13
Plat, Captain Anthony, 134, 135, 145, 364
Plat, Henry, 13
Plat, John (son of Sir Hugh Plat), 256
Plat, Judith, née Albany (second wife), 27–29, 39, 323
Plat, Margaret, née Young (first wife), 23, 27
Plat, Richard, 11–14, 39, 323
Plat's agricultural publications, later appreciation of, 112–113
Plattes, Gabriel, 112
playgoing, Plat indulged in, 20
playwright, 20
pleasure gardens
    Plat's ideal, 67–69
    public, Plat frequented, 20
plums, 48, 70, 151
poetry, Plat's early works on, 15–17
Pointer, Richard, 71
Pointer, Vincent, nurseryman, 48, 49
Pope, Mr, of Queenborough Castle, 183
Popham, Sir John, Lord Chief Justice, 217
portable still, for use at sea 135

portrait, of Plat's father, 12
potato, 78
pots, garden, 51, 58, 62, 65, 66, 75, 83, 155
potting (plants), 65, 66
poultry, 106, 110, 376
practice-books, Plat's, 221ff., 227
prescriptions, 211
preserving, food, 32, 45, 116, 119, 125, 130, 135–139, 147–148, 150, 155–156, 208, 244, 253, 255, 263, 281
Principe, L.M., 172
professional cooks, 156–157
progress, scientific, Plat's ideas of, 247–249
property, investment in, 11
Protestant refugees, 45, 260–261
pruning, fruit-trees, 70–73
publications, of Plat, 29
Pumfrey, Stephen, 355
pump
    garden, 50, 69
    portable, 260, 291, 311, 344, 360
purgation, 225
'qre', Plat's notes replete with, 275–278
Quercetanus (Joseph Duchesne), 185, 187, 202, 204
quince, 46, 50, 58, 72, 127, 140, 150, 155, 234, 402
rabbits, 111, 102–105, 109, 110, 264
Raleigh, Sir Walter, 34, 142, 143, 182, 190, 198, 203, 351, 356
Ramism, 16
recipes, 150ff.
reed mats, 57
reputation, Plat's
    in his lifetime, 320–321, 342, 367
    posthumous, 41, 45, 81, 375–377

INDEX

Rich, Richard, of Lee, 49, 157, 219–220, 257
Rich, William, 213
Ripley, George, 53, 185, 194, 195, 196
Robenson, Mr, 87
Robinson, —, cousin to Sir Hugh Plat, 337
Rohde, Eleanour Sinclair, 42
Romero, Señor, of Naples, 50, 76, 123, 124, 129, 135, 157, 220, 234, 235, 258, 266, 288, 297, 300, 301, 303, 313
root crops, 46
Rotheram, Mr, of Bedfordshire, 88, 91
Rowland, servant to Sir Hugh Plat, 235
Royal Exchange, 329, 335–336, 338, 365
Russell, —, distiller, 184, 219
Russell, Goodwife, 236
Sackville, Thomas, first Baron Buckhurst, 142, 143, 146, 356
St Albans, Herts., 96, 236, 323, 324
St Pancras, Middlesex, 323
salad, invented by Judith Plat, 27
salt, common, 51
    as food preservative, 117ff., 149, 160, 188, 220, 305n, 359
    used to improve soil fertility, 85ff., 102, 105
salt, essential, vegetative, theory of, 82–83, 92, 85, 100, 175, 189, 200, 202, 223, 250, 277, 296
saltpetre, 50, 94, 147, 252, 255, 260, 263–264, 278–279, 290–295, 305, 306, 309, 344, 346, 347, 358–362, 364
sand bags, 285, 299–300, 347
Sawer, Edmund, Auditor Hill's man, 222
Sawte, —, physician, 50

Scammell, Geoffrey, 118
Scappi, Bartolomeo, 122, 155
Scarlet, Thomasina, 212
scholars, unsuited to scientific enquiry, 245–247
scholastic philosophy, 190, 192, 245–247
schools and schoolboys, in London, 14
scientific research, in London, in Plat's time, 242ff., 366–367
Scott, Reginald, 106, 172
Second part of the garden of Eden, 41ff., 371, 376
secret weapon, 304
Secreta de pampinei, 254
secrets
    attempt to sell to Landgrave of Hesse, 349–353
    attempts to sell to the Crown, 353–357
    books of, Plat's use of, 255
    evidence of success in selling, 347–348
    obtaining money for, 346,ff.
    Plat's attitude towards, 244–245
    Plat's overall success selling, 363–365
seeds, garden, 78–79
Segar, Sir Francis, 183–184, 202, 204, 205, 240, 253, 258, 350, 352, 353
Segar, William, 184, 258, 337
Seneca, 77
setting, seeds in rows, 98–101
Shapin, Steven, 245, 258, 280
Sharrock, Robert, 376
Shattenden, —, the woad-man, 221, 269
Sheen, Surrey, 69
Sheffield, Edmund, 3rd Baron Sheffield, 231

shoe-blacking, 314–315
shoe leather, plan to sell via Dutch cobblers, 339–340
shop, Plat's plans for, 335–336
Shrewsbury, Salop, 164
Simpson, Parson, 50
Skinner, Stephen, 375
Skinner, Sir Vincent, 231
Skipwith, Raff, 218
'skirter', Plat an example of 30
Slack, Paul, 231
Sloane, Sir Hans, 371, 372
Smith, —, gardener, 47
Smith, John, 111, 112
smoke screen, 296–297
soap-ashes, waste, use of, 57, 90, 91–92, 94, 113, 286, 305, 308, 310–331, 339
soil, fertility, 52–54, 82ff.
sources, Plat's acknowledgement of, 41–43
Southwark, Surrey, 323
spade cultivation of fields, 99–101
Speed, Adolphus, 90, 104, 110, 112
Spencer, 'old', gardener at Merchant Taylors' Hall, 48
Spenser, 'the Travayler and coster', 129
steam-heating, Plat's method, 9
see also heat
steep, for seeds, 92–96
Stern, Tiffany, 340
Stourbridge Fair, Cambridge, 338
stove, for hot-house, 59–62
Stow, John, 14, 17, 18, 27
straw mats, 57
Strupp, Joachim, 168
Stuart, Lady Arabella, 38
sugar-peas, 46, 51, 257

sugar plate, Plat's way of making, 8, 148, 150, 154, 268
sun, heat from, 58
Sundrie new and artificiall Remedies against Famine, 116ff., 147ff.
Switzer, Stephen, 41
Sykes, John, 334
T.T., 43, 52, 151–153, 161, 253
Taverner, John, 52, 99
tax administration, public hostility to, 26
Taylor, —, surveyor, 50
technology, 7, 9
food, 147–148, 177, 242–255, 279, 284, 285ff.
Tey, John, a servant, 258
theatre, 20
see also playgoing
Thirsk, Joan, 88, 103, 148
Thorp, Robert, vintner, 229
Throgmorton, —, Sir Thomas Cornwallis' gardener, 47
Thurland, Thomas, 153
Tilton, —, dyer, 272
Tottenham, Middlesex, 323
trades
history of, 162, 243, 249, 251ff.
investigated by Plat, 260–261
tradesmen, London, 32, 155
transmutation, 170–171, 173, 182, 184, 376
trees, forest, 41
fruit, 31, 41, 47, 52, 57–59, 70–74, 109
see also dwarf trees
Trigge, William, 212
Trithemius, Johannes, 199
Trout, Richard, 219

INDEX

Turner, William, 52
Tusser, Thomas, 81, 210, 343
Twickenham, Middlesex, 47, 48
Tymme, Thomas, 153, 170
Ulysses on Ajax, 37
unslaked lime, 56, 57
Valetius, Franciscus, 82, 85
van Helmont, J.B., 374
Varro, 77, 81
Vaughan, Thomas, 374
Venner, Richard, dramatist, 20, 315
Vere, Sir Francis, 38, 321
Verney, Lady Margaret, 209
Verzilini, Giacomo ('Master Jacob of the glasshouse'), 50, 58
vessel, wooden, to boil in, 203, 286, 315–316, 329, 344–345, 365
victualling, naval, 34, 115ff., 160–161, 385
vines, 31, 45, 50, 54, 58–59, 63, 70, 74, 76–77, 106, 111
Virgil, 81, 94
virtuoso, 242–243, 284
Vives, Juan Luis, 250
Waad, Armigal, 183, 184
wager, offer of, 35
Walker, —, female cousin of Sir Hugh Plat, 230
Walker, Obadiah, 374
wall fruit, 46, 50, 58
Walsingham, Sir Francis, 49, 58
Warde, —, secretary to the Queen's Council, 257
Warner, —, nurseryman, 71
warren, suburban, 102–105
waste, 90–91
  ideas for using profitably, 304ff.
Watson, —, surgeon, 50, 89, 258

wealth, 11–12, 16, 23, 39, 320
  evidence of Plat's, 322–325
Webb, Mr, 89
Webb, Henry J., 288
Webber, Mr, 'one of Her Ma[jes]ties Privie Kitchin', 156–157, 259, 261
Webster, Charles, 190
Webster, John, 246
Wecker, Johann Jacob (Wickerus), 255
Wentworth, Thomas, 1st Baron Wentworth, 47
Wheeler, John, 327
Whitney, Geoffrey, 333
wholesaling, Plat's plans for, 339
Widowes Treasure, 151
Wilkins, John, Bishop of Chester, 371
Wilkinson, —, peterman, 264, 293
Williams, Mr, schoolmaster, Highgate, 217
Williamson, James, 118
wills
  of Henry Plat, 13
  of Sir Hugh Plat, 361
  of Richard Plat, 13
Wilton, Wiltshire, 183
wine, winemaking, 8, 31, 54, 77, 109, 135–138, 154, 155, 157, 160, 167, 252–256, 260, 262, 264, 275, 276, 281, 286, 287n, 326, 330
  made by Plat, 34, 38, 254, 321, 335, 338, 345, 346, 347, 386–387, 398
Winter, Mr, 101
woad, 32, 54, 80, 89, 90, 81, 92, 102, 109, 110, 220, 259, 268–270, 305
Wolfe, Reynold, 334
Woodhead, Abraham, 374
Woodward, Mr, of Warwickshire, 129

Woodward, Donald, 89
Wootton, John, 222
workshop, Plat's, 279–281
Worlidge, John, 113, 377
Wortley, Mr, of Yorkshire, 89
Wright, Katherin, Mr Barnes the mercer's wife, 237
Young, Richard, 24, 25, 26, 27, 36, 37, 178, 179, 234, 235, 256, 356
Zouche, Edward la, 11th Baron Zouche, 50